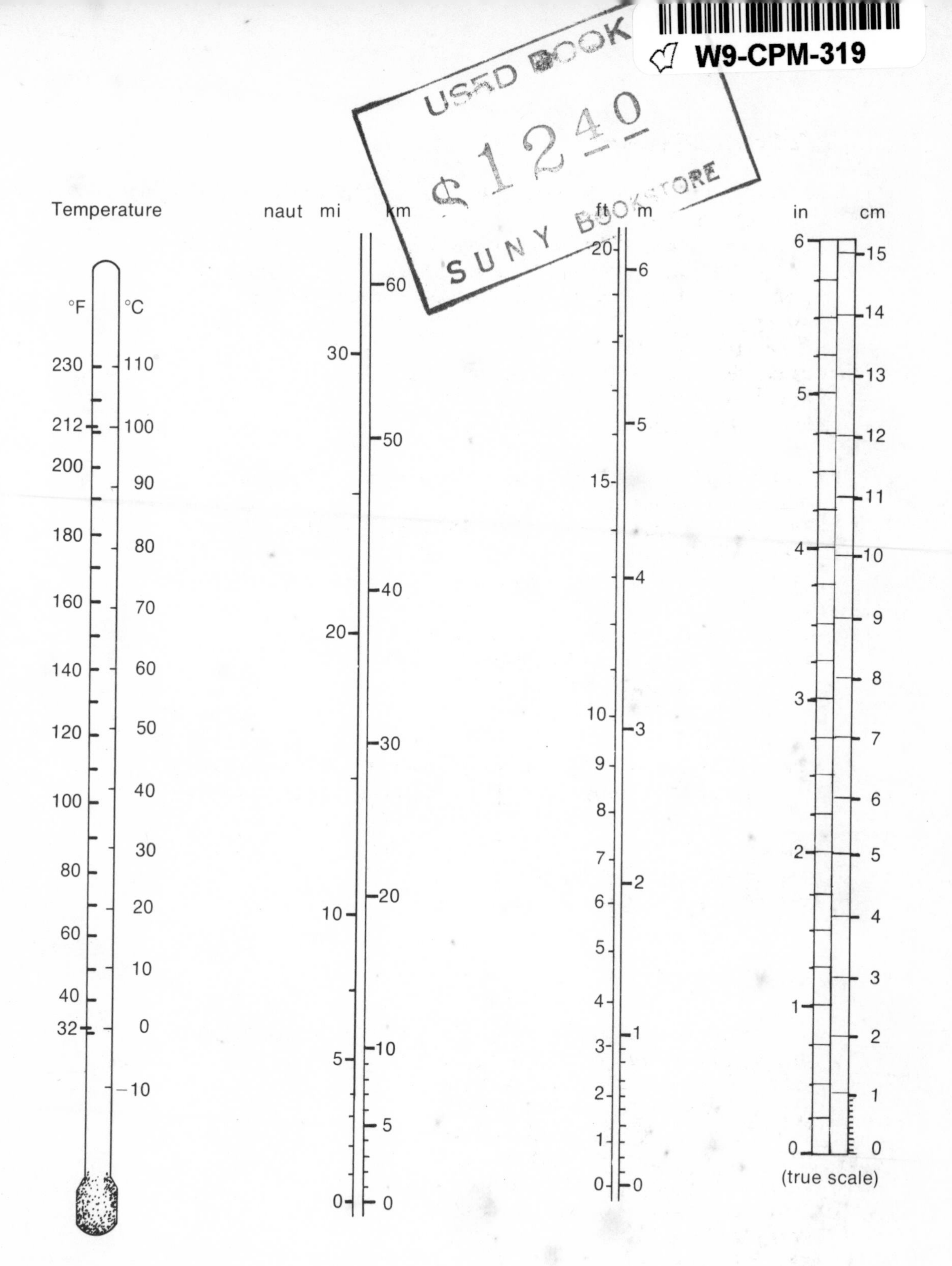

Kathy McCormac

ELEMENTS OF OCEANOGRAPHY

J. Michael McCormick
Montclair State College

John V. Thiruvathukal
Montclair State College

W. B. SAUNDERS COMPANY Philadelphia • London • Toronto

W. B. Saunders Company: West Washington Square
Philadelphia, PA 19105

1 St. Anne's Road
Eastbourne, East Sussex BN21 3UN, England

1 Goldthorne Avenue
Toronto, Ontario M8Z 5T9, Canada

Library of Congress Cataloging in Publication Data

McCormick, Jon Michael, 1941–

Elements of oceanography.

Bibliography: p.

Includes index.

1. Oceanography. I. Thiruvathukal, John V., joint author. II. Title.

GC16.M32 551.4′6 75–10388

ISBN 0–7216–5900–4

Front cover illustration: Japanese Prints, Hokusai, Katsushika: The Great Wave at Kanagawa.

Courtesy of the Metropolitan Museum of Art, the Howard Mansfield Collection, Rogers Fund, 1936.

Elements of Oceanography

ISBN 0-7216-5900-4

Last digit is the print number: 9 8 7 6 5 4 3

Preface

The oceans, although they hold vast resources of food, minerals, petroleum, and energy, are under ever increasing environmental stress. The prospects for gaining further benefit from the seas depend on human activities and political decisions. These are processes in which we all take part. Therefore, it is of tremendous importance for all of us to be aware of the nature and potentials of the oceans and of their frailty. In that sense, this book is aimed at the college student as citizen, voter, and future doctor, lawyer, teacher, artist, politician, and so on.

This text is designed for a beginning oceanography course and does not require any previous scientific background. The metric system of measurements is emphasized throughout the text, and appropriate conversion tables are given at the end of the book. Technical terms are defined where appropriate as well as in the Glossary. We have endeavored to describe the essentials of the nature of the oceans and their potential for wise exploitation as well as the dangers that lie in their misuse. We hope that this has been done in a way that gives relevance to the various principles of oceanography.

As no text is written in a vacuum, we are indebted to many of our colleagues in the scientific community. Many of the photographs were graciously provided by researchers in diverse fields of oceanography and marine biology. We are especially indebted to:

Lamont-Doherty Geological Observatory
Allan Bé
National Marine Fisheries Service
Harold "Wes" Pratt
Oregon State University
David Stein
The Laboratory, Plymouth, England
Douglas P. Wilson
Scripps Institution of Oceanography (Deep Sea Drilling Project)
M. N. A. Peterson
University of Rhode Island
Paul Hargraves
James Kennett
John Sieburth
Woods Hole Oceanographic Institution
Susumu Honjo

In addition, Vicky Briscoe of the Woods Hole Oceanographic Institution and Anne Nixon of the Lamont-Doherty Geological Observatory were especially helpful in providing us with institutional material for many of the illustrations. We would also like to acknowledge the help of Dean Bumpus and Gilbert Rowe of the Woods Hole Oceanographic Institution and Oswald Roels of Columbia University, who provided important information.

Portions or all of the manuscript were read and sometimes reread and criticized by Larry Leyman of Fullerton College, James McCauley of Oregon State University, Thomas Worsley of the University of Washington, James Marlowe of Miami-Dade Community College, Richard Mariscal of Florida State University, and R. Gordon Pirie of the University of Wisconsin–Milwaukee. Their help is gratefully appreciated. We, of course, must take full responsibility for any errors that remain. Our students are also appreciated for their willingness to serve as critics of some of the early manuscript. In addition, we would appreciate receiving comments and suggestions from the readers of this text.

All the way through this project the staff of W. B. Saunders Co. has been encouraging and helpful. Many of the fine drawings were done by Grant Lashbrook and John Hackmaster. Special thanks are due Richard Lampert, editor, and Amy Shapiro, manuscript editor, who have shown incredible patience, endurance, and good sense. We also extend a special note of gratitude to our wives, Beverly and Mary, whose encouragement and help kept us going.

Contents

Elements of Oceanography

The *Atlantis*, the first ship of the Woods Hole Oceanographic Institution. (Courtesy of WHOI.)

CHAPTER 1

History of Oceanography

The oceans undeniably have been there since the dawn of the human species and for eons before that. Ever since people started wondering about the world around them, the oceans must have exerted a great fascination. Preliterate people must have observed the daily and monthly cycles of tides, and they certainly obtained a good bit of their food from the oceans nearby. Probably some of the more daring fishermen occasionally ventured out of sight of land, becoming, in a sense, the first explorers.

We can appreciate how strongly the oceans influenced primitive people by considering the central role of the sea in Homer's *Odyssey.* Odysseus faces storms, fierce currents, shipwreck, and drowning throughout the story, which is considered to be one of the central myths of the Middle Eastern "cradle of civilization."

Even today, the power of myths concerning the mysterious sea persists. Books such as *Moby Dick* and *Jaws* tell about terrible creatures from the depths of the ocean. Even in horror movies, the most incredible monsters always seem to rise from the ocean.

Although oceans have been utilized by people for thousands of years, it was not until the eighteenth century that a systematic study of the oceans was begun. Progress in oceanography has accelerated since then, but a great deal remains to be learned. Much of the earliest interest in the oceans was oriented toward navigation. In the nineteenth century, interest diversified to include economic problems such as the transatlantic cable crossings and fisheries, as well as purely scientific curiosity. The twentieth century saw a new era of international cooperation through efforts such as the International Council for the Exploration of the Sea, the International Geophysical Year, and the International Indian Ocean Expedition.

Today the study of the oceans is a discipline in which all modern sciences come together. This multiscience involves the study of physics, chemistry, geology, and biology. No oceanographer is an expert in all of these branches. However, many oceanographical problems are interdisciplinary in nature and require cooperation among scientists of varied backgrounds.

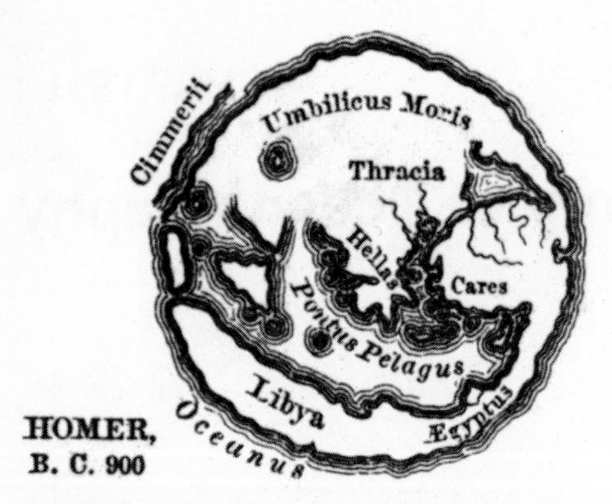

FIGURE 1.1 Homer's conception of the world (about 850 BC). (Courtesy of Bettmann Archive.)

EARLY EXPLORATION

The first explorations of the oceans were aimed mainly at searching for new trade routes and riches and describing the shapes of the seas. By 1000 BC, the Mediterranean Sea was pretty well known, at least near shore and in the vicinity of islands and shallows. Many of the early sailors had a fear of leaving sight of land. Homer (about 850 BC) envisaged the world as comprising the lands around the Mediterranean, with an all encircling ocean beyond, which had not been explored (Fig. 1.1).

The Phoenicians were the best navigators in the Mediterranean about 1000 BC. They could navigate by the stars, at night. They are believed to have sailed to Spain (founding the city of Cadiz) and possibly even as far north as the coast of Cornwall in England, the location of ancient tin mines.

At about 600 BC, King Necho of Egypt reportedly sent a crew of Phoenicians down the Red Sea and around Africa, returning by way of the Straits of Gibraltar about three years later. Accounts of this journey were not taken seriously or were not known to the early scholars. Thus Ptolemy (about 150 AD) assumed there to be unexplored land linking Southeast Asia and Africa (Fig. 1.2), across

FIGURE 1.2 Ptolemy's conception of the world (about 150 AD). (Courtesy of Bettmann Archive.)

Table 1.1 MAJOR EARLY EVENTS IN SEA EXPLORATION

1000	Leif Ericson crosses the Atlantic Ocean to Canada
1492	Christopher Columbus crosses the Atlantic to West Indies
1500	Pedro Alvares Cabral discovers Brazil
1513	Juan Ponce de Leon describes the Florida Current
1515	Peter Martyr discusses the origins of the Gulf Stream
1519–1522	Ferdinand Magellan circumnavigates the earth and attempts deep-sea soundings in the Pacific but does not reach bottom
1569	Gerardus Mercator develops a projection of the world on which north-south and east-west directions are represented by straight lines. This type of projection is still used for navigational charts.

what is known today as the southern part of the Indian Ocean.

Ptolemy's conception of the world persisted through the Dark Ages until 1497, when the Portuguese explorer Vasco da Gama reached India by sailing around Africa and through the Indian Ocean. Other important early events in charting the seas are summarized in Table 1.1. These events provided the backbone for the scientific study of the seas.

THE BEGINNING OF SCIENTIFIC STUDY OF THE SEAS

In 1770, Postmaster General Benjamin Franklin, in order to improve mail service between England and the colonies, authorized the production of a chart of the Gulf Stream, in the Atlantic Ocean. This enabled navigators to choose the fastest route by taking advantage of favorable currents. Franklin's chart of the Gulf Stream was all that was available to navigators until the nineteenth century. He believed that ocean currents were caused mainly by wind and that some currents are deflected by obstacles such as continents.

Lieutenant Matthew Fontaine Maury, a navigator on board the USS *Falmouth,* had been frustrated in his attempt to find detailed wind and current charts. When in

1842 he was made Superintendent of the Depot of Charts and Instruments (the Depot was eventually divided into the Naval Hydrographic Office and the Naval Observatory), he decided to compile the types of charts he had sought earlier. Maury collected data on winds and currents from the logs of naval ships and plotted the information on a chart of the Atlantic Ocean. What he saw was a generally clockwise flow of water around a relatively calm Sargasso Sea. In 1855, he published *Physical Geography of the Sea,* considered to be the first major oceanography text.

One of the first expeditions to investigate the biology of the oceans in addition to collecting navigational data was that of the *Beagle* (1831–1836) with Charles Darwin aboard. The *Beagle* sailed around South America to the Galapagos Islands off Ecuador. During the voyage, Darwin collected enough data to last him a lifetime. His conclusions about natural selection and evolution gained him a place in the history of biology, and his theory of the origin of coral reefs is still widely accepted, with few modifications.

In 1840 Edward Forbes, a professor at the University of Edinburgh, reported that his dredging investigations in deep water indicated that the abundance of life decreased with depth, until an "azoic zone" was reached at about 600 meters. This view is, of course, erroneous. In fact, Sir John Ross (in 1818) had already collected animals from nearly 2,000 meters. However, Forbes was a well-known marine biologist, and his views received widespread attention and encouraged others to attempt even deeper dredging.

During 1839 to 1843, James Ross (a nephew of John Ross), on the ships *Erebus* and *Terror,* sailed the Antarctic Ocean and charted its southern limits. This was the first expedition to do more than a few deep-sea soundings. He also collected animals from the sea floor and found life at all depths.

It might appear that such deep dredging was of little practical importance to the governments that sponsored these expeditions. However, toward the end of the nineteenth century, there was a sudden interest in the successful laying of transatlantic telegraph cables. More information was needed regarding depths, deep currents, bottom composition, and especially animals that might bore into and destroy the protective covering of the cable.

In 1847, Louis Agassiz, a zoologist, accompanied the US Coast Survey on its work in Massachusetts Bay. This was the first of a series of expeditions in which he was to participate with the Coast Survey. He collected sediments and animals from the sea floor and later studied the nature of coral reefs. His work gained prestige for himself and the Coast Survey and probably did much to encourage government spending on oceanographical research.

FIGURE 1.3 The HMS *Challenger,* which conducted oceanographic work around the world in the nineteenth century. (Courtesy of Bettmann Archive.)

Professor Wyville Thompson of the University of Belfast, Ireland, led expeditions to the sea near Britain in 1868 and 1869. These expeditions produced much evidence of a great variety of marine life off the coast, and temperatures in the depths as cold as 0° Celsius (°C) were recorded. The reports that followed, combined with a British fear of being outdone by the Americans, led to the greatest deep-sea expedition of all time, the *Challenger* expedition (1872–1876), the first major scientific journey organized solely for the study of the oceans. The HMS *Challenger* (Fig. 1.3), led by Thompson, traveled almost 70,000 nautical miles along a winding course around the world. The results, bearing on physical, chemical, geological, and biological oceanography, took 23 years to publish and filled 50 heavy volumes. In spite of equipment that was primitive by today's standards, the crew was able to bring back samples of mud and animals from depths of more than 8,000 meters. Some of the more striking accomplishments of the expedition include the now recognized fact that the relative composition of seawater is constant from one place to another in the open sea. In addition, knowledge of the drifting and bottom-dwelling life was expanded tremendously, and almost 5,000 new species were identified.

In the late nineteenth century, it was known that driftwood and the remains of an American ship crushed by ice had drifted from the Siberian Sea through the Arctic Ocean to the east coast of Greenland. This led the Norwegian explorer Fridtjof Nansen to plan an expedition to study the circulation of Arctic waters. Nansen believed that a ship could be constructed that would withstand the crushing forces of the ice. In spite of the considerable skepticism of other experts, he persuaded the Norwegian government to help finance the trip. His ship, the *Fram* (Fig. 1.4), was a 38-meter, double-ended, three-masted schooner with a reinforced hull four feet thick.

In 1893, Nansen sailed northeast until the *Fram* was frozen into the ice north of Siberia. Eventually the *Fram* drifted northwest to 84° N latitude, about 600 kilometers (km) from the North Pole. At this point, Nansen left the ship with Frederick Johansen and tried to reach the North Pole by dog sled. Meanwhile the *Fram,* with the rest of the crew, drifted more or less southwest toward Spitzbergen (north of Norway). Nansen and Johansen had to turn back before reaching the Pole and were luckily picked up by an English expedition near Franz Josef Land. The ship had

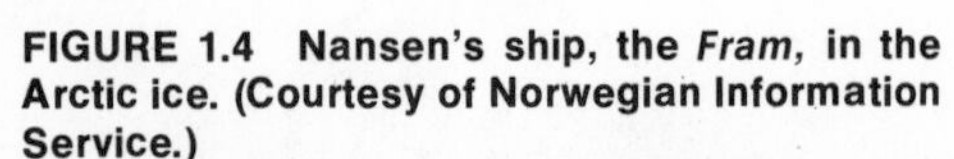
FIGURE 1.4 Nansen's ship, the *Fram,* in the Arctic ice. (Courtesy of Norwegian Information Service.)

drifted less than two km a day over depths greater than 3,000 meters. The entire voyage lasted about three years. Very careful records were kept of the drift, wind, and depth. Nansen observed that the ship tended to drift not quite with the wind but somewhat to the right of the wind. In 1902, V. W. Ekman reported that this phenomenon was due to the effect of the earth's rotation (See Appendix C, Coriolis force) on all objects moving over its surface. This knowledge has been of great value in the study of ocean currents.

In Table 1.2 are listed some additional milestones in the history of oceanography.

THE DEVELOPMENT OF FISHERIES RESEARCH

Although people have obtained food from the oceans for thousands of years, it was not until the 1860s that governments began to consider financing of expeditions to study the erratic catches of fish in European waters. Some experts believed that the problem was due to "overfishing." Others thought that varying currents or other characteristics of the ocean waters were responsible for the poor catches observed in certain years. Research was

Table 1.2 EARLY MILESTONES IN OCEANOGRAPHY

1802	Nathaniel Bowditch publishes his *New American Practical Navigator.* This book has been revised many times and remains a standard work on navigation to this day.
1818	John Ross dredges animals from nearly 2,000 meters while exploring Baffin Bay for a passage to the Pacific
1835	Gaspard Gustave de Coriolis publishes a classic paper on the effects of the earth's rotation on fluids in motion
1843	Alexander Dallas Bache becomes superintendent of the US Coast Survey (later called the US Coast and Geodetic Survey*) and expanded work on physical oceanography and the nature of sediments and bottom-dwelling life
1865	Johann George Forchhammer analyzes samples of seawater from a variety of locations and finds that the ratios between the major salt constituents are nearly constant from sample to sample

*Now incorporated into the National Oceanic and Atmospheric Administration (NOAA) of the Department of Commerce.

aimed at determining natural causes for the appearance or disappearance of the fish.

A Norwegian zoologist, G. O. Sars, believed that the fish, especially the young ones, were carried by currents. He studied the biology and migrations of cod and herring. Sars discovered that cod eggs float and drift with the currents and that the young fish feed on plankton (drifting organisms, which for the most part are very small). Plankton are also the food for other young and many adult fish, such as herring. He considered that the presence of water that was good for the plankton at a time when plankton were needed by the newly hatching fish might result in an especially successful brood of fish. These fish would mature and grow during a period of a few years and could eventually result in large commercial catches.

After a number of projects failed to solve the problem, and realizing that the fish migrate through the waters of many nations, an effort was made to coordinate the research work of all the northern European fishing nations. In 1899, King Oscar II of Sweden called an international conference to discuss the situation. Representatives of Great Britain, Germany, Denmark, Norway, and Sweden attended. By 1902, Finland, Holland, and Russia had joined the others to form the International Council for the Exploration of the Sea. The Council coordinated efforts in physical and biological oceanography as they applied to the fishery problem. The Council is still active today but on a much reduced scale.

In the late nineteenth century Johannes Peterson, a Danish zoologist, designed a successful method of tagging flatfish, whereby two buttons are attached to the fish with a silver wire that runs through the back. He applied the method of tagging and recapture to observe the migration of fish. Similar techniques are still used today.

The Norwegian biologist Johan Hjort brought statistical techniques to the study of fish production in the early twentieth century. He determined the changing age distribution of various populations of fish over a period of years. He was able to recognize certain broods of fish from one year to the next as they grew and matured. These data indicated that a particularly good year for fishing might be due to conditions present in the breeding grounds years before. Also, early detection of a large new brood could be used to predict future fishing success. Since young fish eat plankton, he reasoned, conditions favorable to plankton growth should be required for a sizable brood of fish.

Although the work of the Council was ambitious, the fisheries problem was never completely solved. During this period of intensified research, however, much basic information was acquired about currents, water characteristics, and the ecology of plankton and fish.

MODERN OCEANOGRAPHY

The first major oceanographical expedition to be conducted in a truly systematic way was that of the *Meteor,* a German research vessel. The *Meteor* surveyed the physical, geological, chemical, and biological aspects of the Atlantic Ocean from 1925 to 1927 in a series of transects, sampling the water at predetermined depths. The most up-to-date technology available at the time, including echo-sounders, was used. The vertical profiles of temperature, salinity, and oxygen that were obtained from this expedition for the Atlantic Ocean are still widely quoted.

Various governments have continued to engage in oceanographical expeditions and research. However, in recent years there has been a shift in these activities toward private and public research institutions and universities, frequently with considerable governmental financial support.

Much of the progress in oceanography in the United States has been greatly enhanced by the work at a

FIGURE 1.5 The Woods Hole Oceanographic Institution. Note the modern research vessels in the harbor. (Courtesy of WHOI.)

number of institutions formed specifically for that purpose. Scripps Institution of Oceanography (now part of the University of California) in La Jolla, California, was established in 1905. The Woods Hole Oceanographic Institution was founded in 1930 at Cape Cod, Massachusetts (Fig. 1.5). Both of these institutions have engaged in worldwide voyages of scientific discovery and have become first-rate academic institutions as well, educating oceanographers for the future. Today, more than a dozen universities conduct large-scale research and education in oceanography (see Appendix A, Major Oceanographic Institutions).

In 1966 the United States Congress passed the Sea Grant College and Programs Act, which outlined the government's commitment to long-term nonmilitary funding of marine science education and research. Colleges and private institutions are eligible for support from the Sea Grant Program. This program is similar in many ways to the Land Grant College Program initiated in the last century. Both are committed to education and public service regarding our natural resources.

Oceanography today is truly international. Many nations have again realized that international cooperation results in a greater economy of finances, avoids duplication of effort, and may speed the solution of problems. The International Council for the Exploration of the Sea (discussed earlier) was one of the first such efforts. More recently, the International Geophysical Year (1957–1958) and the International Indian Ocean Expedition (organized under UNESCO, 1959–1965) involved many nations in pursuit of knowledge on the high seas.

As with so many areas of human knowledge, the oceans and oceanographical information have also been used for selfish purposes by various nations. As we learn more about fisheries biology, mining of the sea floor, use of military submarines, and even the possibility of using the sea floor as a site for weapons; setting ground rules for the use of the oceans has become more critical and at the same time more difficult. These modern problems are so extensive that an entire chapter (Chap. 12) is devoted to them.

SUGGESTED READINGS

Bailey, Herbert S., Jr. 1953 (May). The Voyage of the *Challenger*. *Scientific American*.
Deacon, Margaret. 1971. *Scientists and the Sea, 1650–1900, a Study of Marine Science*. London, Academic Press.
Eiseley, Loren C. 1956 (February). Charles Darwin. *Scientific American*.
Idyll, C. P. (ed.). 1969. *The Science of the Sea, a History of Oceanography*. New York, Crowell.
Schlee, Susan. 1973. *The Edge of an Unfamiliar World, a History of Oceanography*. New York, E. P. Dutton.

The piston corer, a device used to sample sea-floor sediments. (Courtesy of Woods Hole Oceanographic Institution.)

CHAPTER 2

Geology of the Ocean Basins

As the need for exploration and exploitation of natural resources grows, the knowledge of what lies beneath the oceans (70 per cent of the earth's surface) becomes increasingly important. The survival of the human race may be at stake. Unfortunately, the major nonrenewable natural resources—petroleum and minerals—are found in only limited quantities near the earth's surface. We have nearly exhausted most of these surface supplies and are just beginning to turn our attention to the ocean bottom in search of new resources.

Many problems in oceanography, such as the origin of seawater and the origin and nature of ocean basins cannot be studied without an understanding of what the interior of the earth is like, which in turn requires some knowledge of the origin of the earth itself. Except for the rocks brought up from a few deep wells on land and in shallow water areas, there is no direct information about the nature of the earth's interior. Some of these wells have penetrated to depths of about 10 km (about 30,000 ft), but this is an insignificant value compared to the average radius of the earth (6,370 km). The deepest holes drilled in deep water have penetrated only about one kilometer below the seafloor. Even volcanoes cannot provide information about the earth's deep interior, because they are believed to originate within about 100 km of the earth's surface, and because the material that they extrude has been contaminated by the upper layers.

Present ideas about the origin and nature of the ocean basins are the result of a revolutionary theory in the earth sciences, the ***continental drift theory***, which was revived in the 1960s, based primarily on data collected about the ocean basins themselves.

ORIGIN OF THE EARTH

There are many theories about the origin of the earth. The ***dust cloud hypothesis*** is currently popular among astronomers. This theory states that about 4½ billion years ago, the earth (as well as other members of the solar system) developed from a diffuse cloud of dust and gas. This so-called dust cloud was made up mostly of hydrogen but also contained at least all of the elements found on earth and perhaps others no longer present. Individual members of the solar system, including the planets and their satellites, were formed by contractions taking place within local mass concentrations in the dust cloud. But most of the materials condensed at the main center of the cloud and eventually became the sun. The sun's temperature and density increased owing to gravitational contraction and both eventually became high enough to initiate the nuclear reactions by which the sun now radiates energy into space. Similarly, the earth's interior became very hot, although not sufficiently hot or dense to start nuclear reactions, and it melted. The heavier metallic elements, with iron probably dominating the mixture, settled toward the center of the earth, leaving the lighter materials above it. The result is a layered earth. Solar radiation increased the surface temperatures of planets in proportion to their distance from the sun. This heating, coupled with the radiation pressure from ***solar wind*** (a continuous stream of charged particles emanating from the sun), caused the earth to lose large quantities of the lighter elements and its original dust cloud cover.

The earth, then, was completely molten at some time during its early history, and it has been cooling ever since. However, only a portion of the heat lost by the earth today is the original heat; the other portion is believed to be derived from the disintegration of radioactive materials concentrated near the upper part of the earth. Although the earth is believed to be about 4.5 billion years old, the oldest known rock is only about 4 billion years old. The absence of older rocks may be due to the molten early history of the earth, or to their subsequent destruction or alteration by geological processes.

As the interior of the earth cooled, various minerals crystallized and aggregated to form the different rocks of the crust. Cooling at the surface helped to retain the atmosphere and eventually caused the condensation of the water now filling the ocean basins. The origin of the earth's atmosphere and of seawater are discussed in the next chapter.

INTERNAL STRUCTURE OF THE EARTH

Physical Nature of the Earth's Interior

Figure 2.1 illustrates a cross section of the earth, showing its four major layers: the ***crust,*** the ***mantle,*** the ***outer core,*** and the ***inner core.*** The crust varies in thickness from about 6 to more than 70 km. Below the

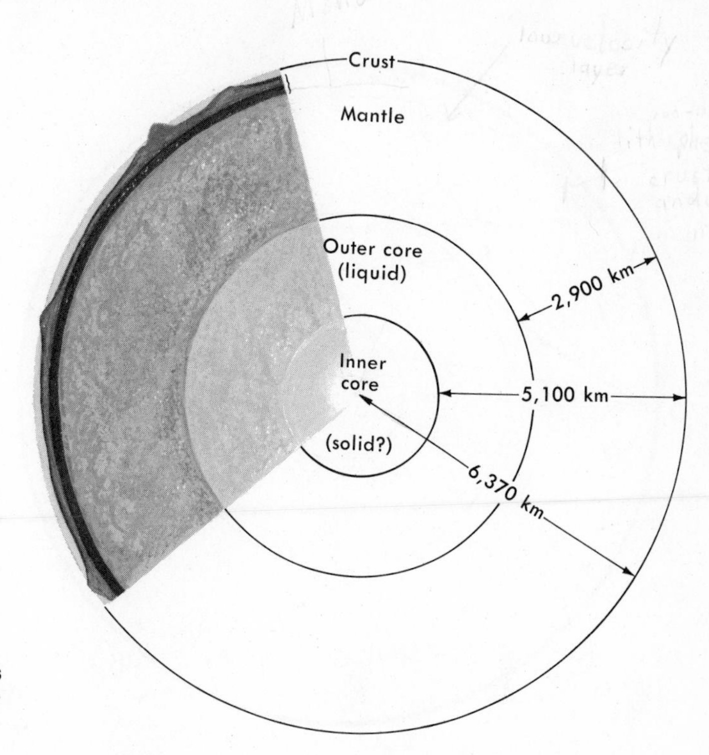

FIGURE 2.1 Cross-section of the earth, showing its major regions as interpreted from the study of earthquake waves.

crust and extending to a depth of 2,900 km is the mantle. The outer core extends from a depth of 2,900 to 5,100 km and resembles a liquid, whereas the inner core, extending from 5,100 to 6,370 km (the center of the earth), most resembles a solid.

Theories about the physical nature of the earth's interior are based on studies of earthquakes and nuclear explosions. During an earthquake (or nuclear blast), numerous shock waves or ***seismic waves*** are produced at the source. The three main types of seismic waves are ***primary*** or ***P*** waves, ***secondary*** or ***S*** waves, and ***surface*** waves. The P and S waves travel within the earth, whereas the surface waves, as their name implies, travel along its surface. We will discuss the behavior of P and S waves to illustrate the nature of the earth's interior.

About one million earthquakes are produced within the earth each year, and hundreds of seismographs located all over the world record the seismic waves produced by these earthquakes. P waves travel faster than S waves and hence are recorded first by seismographs. In addition, P waves can travel through solids and liquids, whereas S waves can be transmitted only through solid media. The physical difference between the two types of wave is that in the case of P waves, the particles of the medium vibrate back and forth in the same direction that the wave travels, while in the case of the S wave, the particles vibrate at right angles to the direction of the wave.

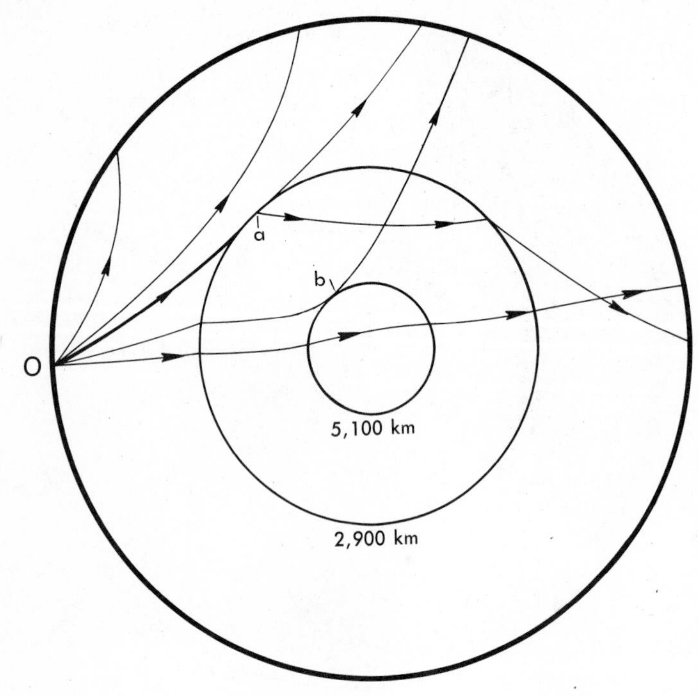

FIGURE 2.2 Paths of some P waves within the earth. O is the source (earthquake) from which these waves are being radiated. The P wave at a depth of 2,900 km (point a) is sharply deflected downward by an abrupt decrease in the wave speed. The sharp deflection upward at a depth of 5,100 km (point b) is caused by an abrupt increase in the wave speed at this depth. In each zone, continuous increase in wave speed with depth causes the waves to be refracted in curved paths.

The P wave velocity within any medium is dependent on the density, rigidity and incompressibility of the medium, whereas the S wave velocity depends only on density and rigidity. The speeds of both P and S waves increase as the values of these factors are increased. It is because liquids have no rigidity that they are unable to transmit S waves.

Figure 2.2 shows the paths of the shortest travel times of some typical P waves within the earth. If the earth's interior were homogeneous and the P wave speed remained constant within it, all seismic waves would travel in straight lines rather than in the curved paths shown in the figure. The curved paths are produced by ***refraction***

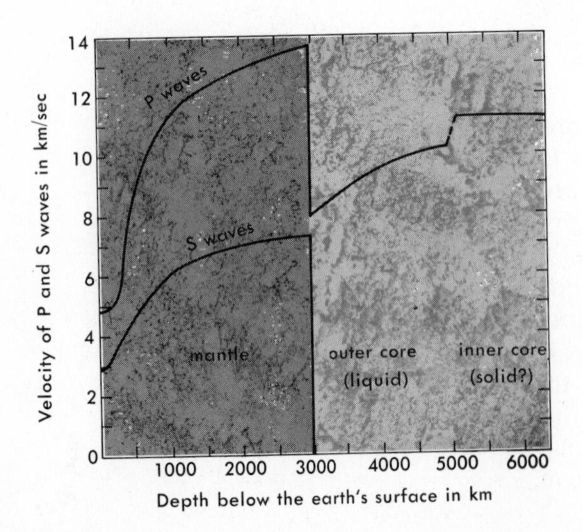

FIGURE 2.3 Variation of P- and S-wave velocities within the earth. P-wave velocity decreases abruptly at a depth of 2,900 km and increases abruptly at a depth of 5,100 km. No S waves are detected in the core (below 2,900 km). (Modified after Garland, 1971.)

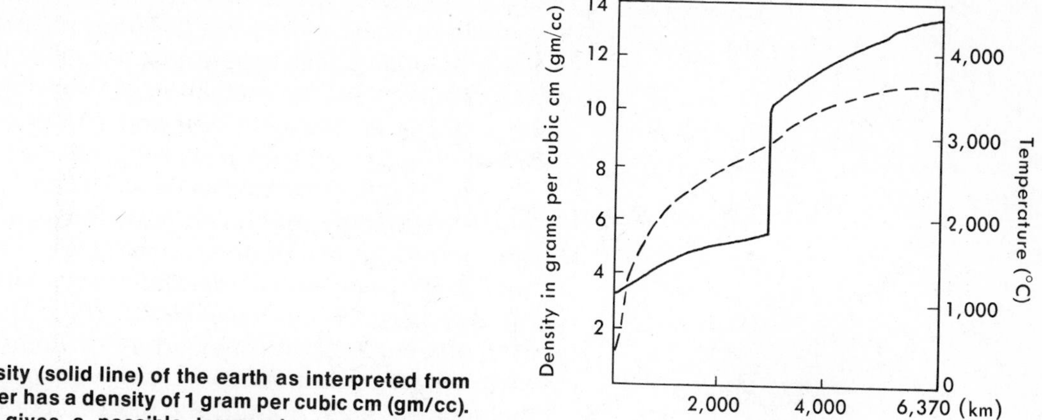

FIGURE 2.4 Density (solid line) of the earth as interpreted from seismic data. Water has a density of 1 gram per cubic cm (gm/cc). The dashed line gives a possible temperature curve for the earth's interior. (Modified after Garland, 1971.)

(bending) of the seismic waves, which is caused by a continuous increase in the wave speed within each zone (see Appendix B for a discussion of wave refraction). The sharp downward deflection of the wave at a depth of 2,900 km (point a) is apparently caused by an abrupt decrease in the wave speed at this depth. The wave is sharply deflected upward at a depth of 5,100 km (point b), apparently by an abrupt increase in the wave speed at this depth. The paths of S waves are similar to those of the P waves, except that no S waves have been observed to travel below a depth of 2,900 km. This would lead one to the conclusion that the region below this depth (the core) behaves as a liquid. However, seismologists interpret the abrupt increase in the P wave speed at the depth of 5,100 km as indicating that the inner core is probably solid. They postulate that the enormous pressure at the earth's center forces it into the solid state, in spite of the fact that it is the hottest zone. Figure 2.3 shows the variations of P and S wave speeds within the earth obtained from earthquake studies. Figure 2.4 is a density model of the earth as interpreted from seismic data.

Chemical Nature of the Earth's Interior

Seismic studies of the earth are indirect and provide information only about the physical nature of the earth's interior. We may never be certain about its exact chemical make-up. However, clues to its chemical composition come from an entirely different field, the study of meteorites. The two common types of meteorites falling on the earth are stony and metallic meteorites. Stony meteorites are silicates of iron and magnesium, and they resemble certain igneous rocks near the earth's surface. The density and seismic properties of these meteorites are comparable to those of the mantle under appropriate conditions, and geologists believe that it is likely that they are in fact very similar. The metallic meteorites, mixtures of approximately 90 per cent iron and 10 per cent nickel, resemble the core in their physical properties.

The origin of meteorites is still a controversial problem in astronomy. But if the meteorites give us an accurate picture of the earth's interior, we can say that the mantle is composed of iron-magnesium silicates and that the core is a mixture of nickel and iron. Other geophysical data also support the meteorite-like composition of the earth's interior. For instance, some geophysicists believe that an electrically conducting metallic core would help to explain the origin of the earth's magnetic field. If one assumes an initially molten earth, it may be easier to visualize the settling of heavier metallic materials toward the center of the earth. Of course, all of these ideas, no matter how elegant or scientifically compelling, are speculative. Many other substances could also account for the observed seismic speeds and the resulting densities of the mantle and the core.

Crust of the Earth

The crust of the earth is the incredibly thin skin covering the mantle. Proportionately, it is far thinner than the skin of an apple. It is, however, the most complex and variable part of the earth. The existence of the crust as a separate entity was first recognized in 1909 by Andrija Mohorovičić, a Yugoslav seismologist. He observed that the speeds of P and S waves increase abruptly when going from the crust to the mantle. For example, P waves travel at a speed of about 7.5 km/sec in the lowest part of the crust, but just below it, in the upper mantle, the speed is about 8 km/sec. The "boundary" between the crust and the upper mantle where the seismic speed changes is called the ***Mohorovičić discontinuity,*** commonly known as the ***Moho.***

Crustal thickness varies considerably. In oceanic areas, it is usually between 6 and 15 km, averaging about 10 km. This includes the ocean depth, which averages about 3.8 km. The crust in continental areas averages about 35 km in low-lying areas but may be greater than 70 km in mountainous regions. However, the crust is unusually thin in some parts of the world, such as in portions of northwestern United States, where it is less than 20 km thick. Figure 2.5 shows a typical cross-section of the crust across continents and oceans. Seismic studies indicate two distinct layers within the crust under continents, whereas oceanic crust consists of a single major layer. The upper

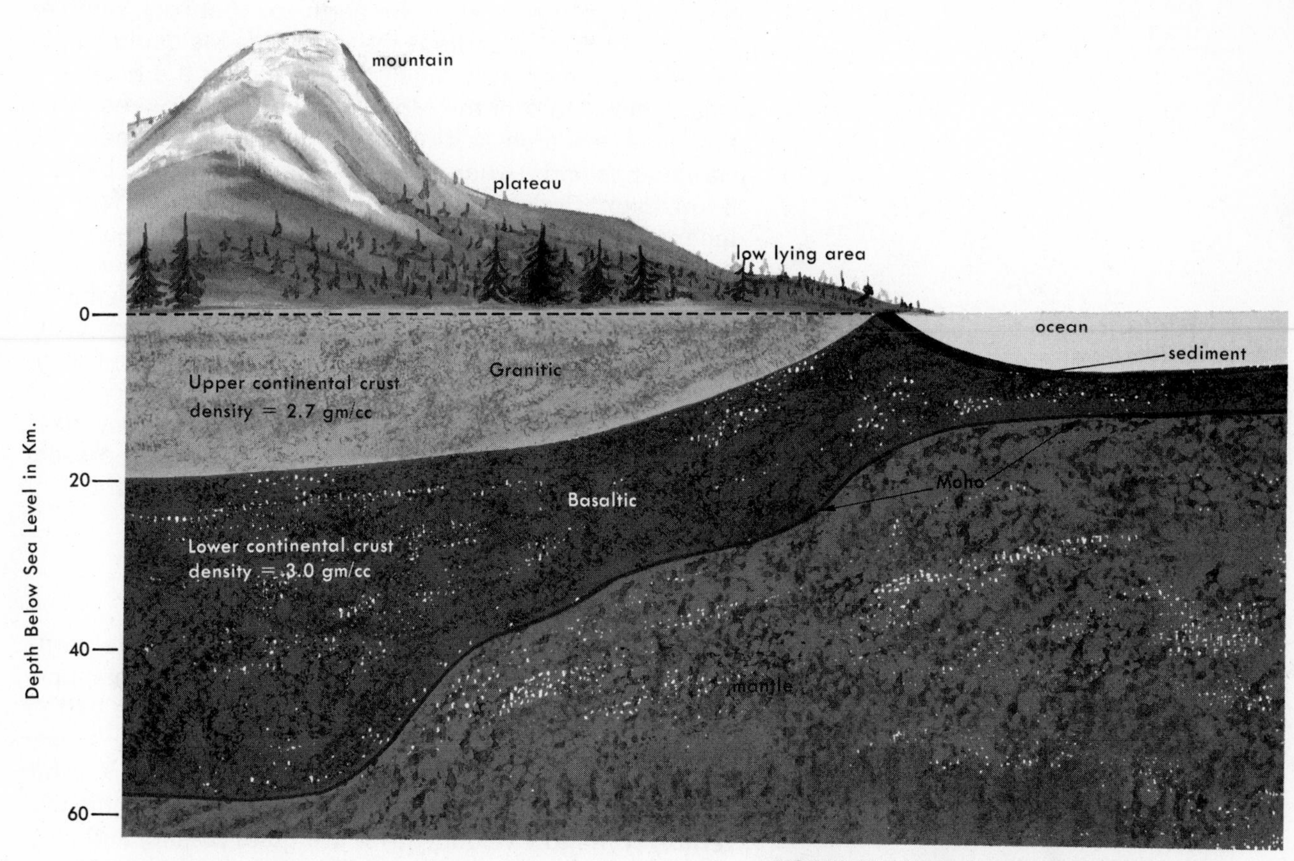

FIGURE 2.5 Typical cross-section of the crust of the earth. The continental crust has two layers, whereas the oceanic crust consists of one major layer. In continental areas, the crustal thickness increases with increasing elevation. The Moho is almost a reflection of the topography. The upper continental crust is granitelike, and the lower continental crust as well as the oceanic crust is basaltlike.

continental crust (about half the thickness of the crust) has an average density of about 2.7 grams per cubic centimeter (gm/cc), or 2.7 times that of water, whereas the lower continental crust, which is believed to be similar to the oceanic crust, has an average density of about 3.0 gm/cc.

The chemical make-up of the crust is poorly known. No drill hole has penetrated the whole crust, either in continental or oceanic areas. Deep-sea drillings indicate the presence of basalt (a dark-colored volcanic rock) below the oceanic sediments. However, these drill holes have penetrated only about a kilometer of rock. It is generally believed that the upper continental crust is composed of granitelike rocks (granite is a light-colored rock formed within the earth by the cooling of molten materials) and that the lower continental crust as well as the oceanic crust comprise basaltlike rocks. Granite and basalt are common near the surface of the earth, and in addition, they have the densities and seismic velocities observed in the crust.

The Mohole Project

In 1959, the United States government undertook an expensive study, known as the ***Mohole Project,*** to drill a hole through the crust of the earth, so that rock samples from the crust and perhaps the upper mantle could be obtained. As the crust is much thinner under the oceans, it was decided to drill the Mohole in an oceanic area. After extensive research, a drilling site was selected near the Hawaiian Islands, where the crust was believed to be only 6 km thick. Considerable effort was made to perfect the technology needed to drill such a hole in the ocean. The problem included not only drilling at the high temperatures and pressures that exist deep within the earth but also the stabilization of the drilling platform. Although some progress was made, and two small test holes were drilled, large cost overruns caused the project to be abandoned in 1965. The Soviet Union has announced that they have started drilling several Mohole-like holes. At the present writing their fate is unknown.

The Low Velocity Layer

In addition to the major structures within the earth, another section, the ***low velocity layer,*** exists in the upper mantle between about 100 and 300 km below the earth's surface. This universal layer is characterized by slower seismic speeds than the layers above and below it. What causes such lowering of seismic speeds? Perhaps the materials of the low velocity layer are quite plastic, and even partially molten. The entire earth above the low velocity layer is called the ***lithosphere.*** It should not be confused with the crust of the earth. The lithosphere includes the crust and the uppermost mantle and has a thickness of about 100 km in oceanic areas and perhaps a much greater thickness under continental areas. The low velocity layer is extremely important in explaining both the origin of volcanoes and the drifting of continents, as will be discussed later in this chapter.

Isostasy

Land areas of the earth are constantly undergoing erosion. At the present rate of erosion, most of the continental areas of today would be completely eroded down to sea level in a few tens of millions of years. How, then, could have continents survived so long? Furthermore, geophysical studies such as measurement of the gravitational field of the earth indicate that mountainous areas do not generally represent greater accumulations of mass as compared to the mass under lowland areas. Similarly, ocean basins do not represent mass deficits. Mountains and oceans represent only "apparent" excesses and deficits of mass, respectively, near the surface of the earth; the major portions of the earth are in mass equilibrium. This means that the mass under a square meter of the earth's surface is the same everywhere. This concept of mass balance within the earth is known as ***isostasy.***

Seismic studies indicate that the lithosphere in mountainous areas, in addition to having a thicker crust, generally extends deeper into the low velocity layer. The lithosphere may be thought of as "floating" on the "plastic" low velocity layer, as icebergs float in the ocean. Just as a large iceberg extends deeper into the water and rises higher into the air than a small iceberg, a mountainous lithosphere extends deeper into the low velocity layer than does the lithosphere in lowland or oceanic areas. When a mountain range undergoes erosion and the lithosphere becomes thinner, not only is the mountain top lowered but also the bottom of the lithosphere rises. This is analogous to what happens to a melting iceberg (Fig. 2.6). A similar mechanism of mass balance maintains isostasy in an area which is receiving large quantities of sediments.

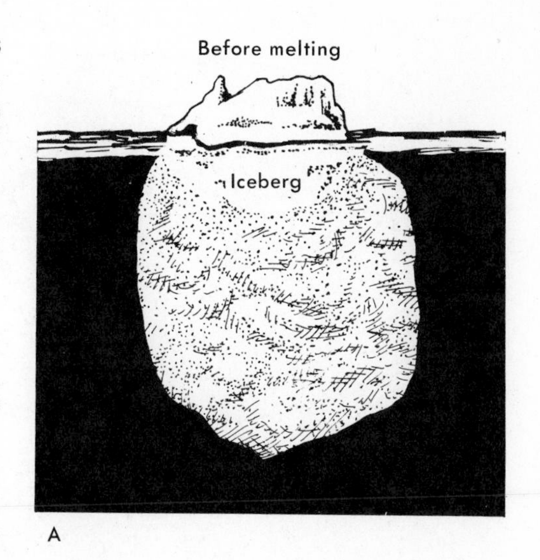

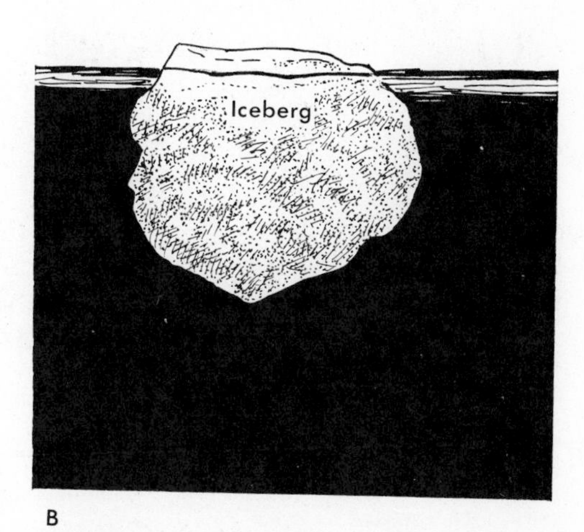

FIGURE 2.6 The melting iceberg analogy of isostasy. When a mountain range undergoes erosion, the bottom of the lithosphere rises.

CONTINENTAL DRIFT, SEA-FLOOR SPREADING, AND PLATE TECTONICS

Continental Drift

Until the early 1960s, most geological concepts were explained on the basis of the permanency of continents and ocean basins. That is, it was believed that continents and oceans remained in place throughout geological time, except for some continental areas which were repeatedly submerged under shallow seas and for the growth of certain continents along their edges. Some scientists, however, held the view that all the continents were once joined together as a supercontinent called ***Pangaea*** (Fig. 2.7), which broke apart, and that these continental fragments have been drifting over the earth ever since. ***Laurasia*** and ***Gondwanaland*** were names given, respectively, to the northern and southern hemispheric portions of Pangaea.

The greatest proponent of continental drift was a German scientist, Alfred Wegener, who in 1912 revived the concept and provided considerable evidence in support of it. Two years earlier, an American scientist, Frederick B. Taylor, had advanced a similar idea but was not as successful as Wegener in publicizing his theory. However,

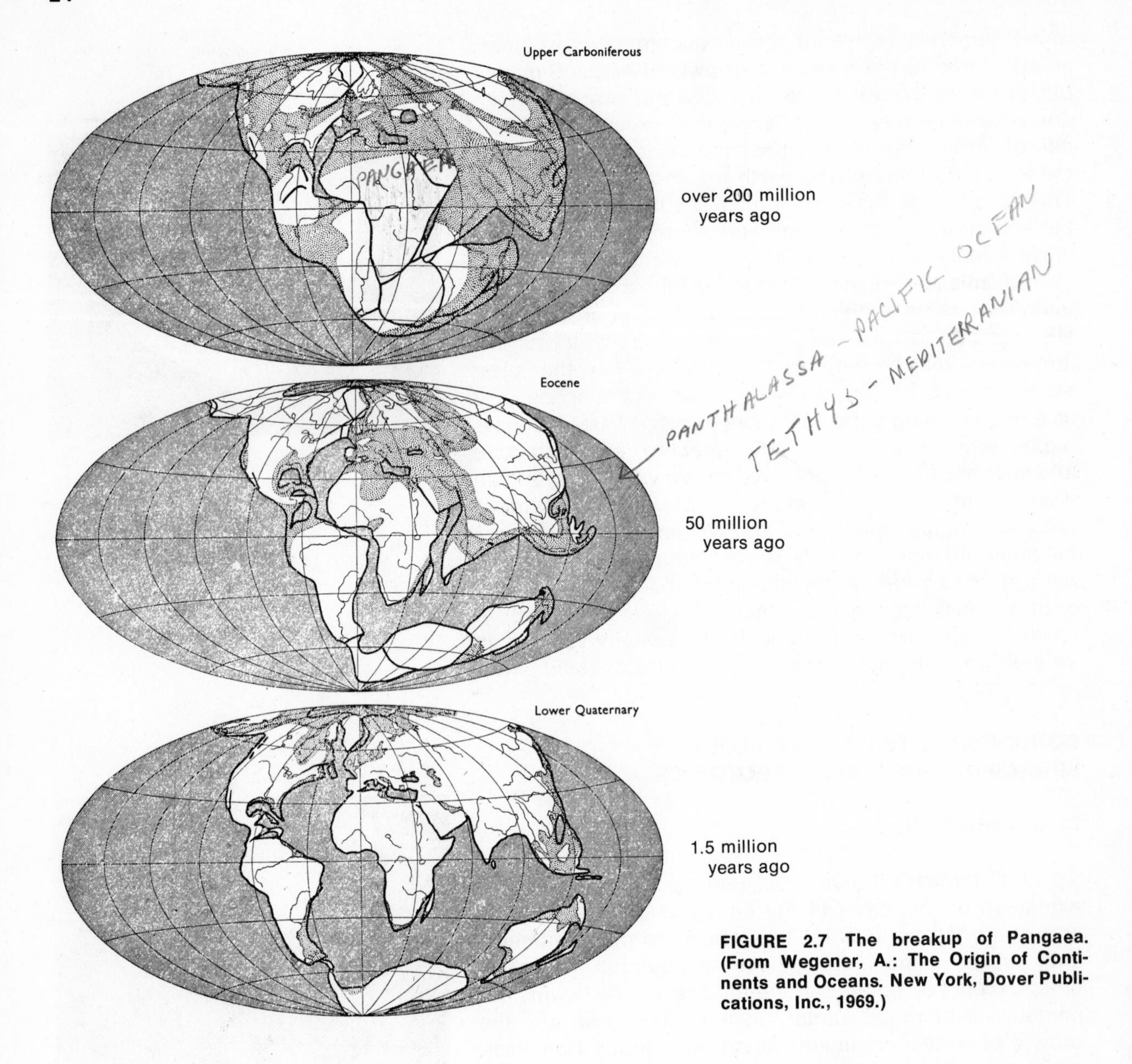

FIGURE 2.7 The breakup of Pangaea. (From Wegener, A.: The Origin of Continents and Oceans. New York, Dover Publications, Inc., 1969.)

ideas remotely resembling "continental drift" had been proposed as far back as 1620 by Francis Bacon. According to Wegener, the breakup of the supercontinent started about 200 million years (m.y.) ago. His most notable piece of evidence was the fact that continents seemed to fit together like pieces of a jigsaw puzzle (Fig. 2.8). Additional supporting arguments offered by Wegener and others for continental drift included the following: (1) geological similarities between continents (similar rock assemblages and mountain structures of the same age); (2) similar fossil assemblages of over 200 m.y. in many continents which were supposedly joined together; (3) the presence of coal in Antarctica and ancient coral reefs in high latitudes (both impossible to explain unless these regions were once located in low latitudes or warm climatic regions); and (4) evidence of glaciation more than 200 m.y. ago (Fig. 2.9) in areas now located near the equator. In addition, continental drift was able to explain the formation of folded mountain ranges such as the Himalayas by the collision of moving land masses. Although much of this evidence seems reasonable, geologists were unwilling to accept continental drift, mainly because of the lack of a known mechanism capable not only of breaking up the supercontinent but also of making the continents move large distances. Thus many of the arguments in favor of continental drift were considered coincidental relationships, or were explained alternatively.

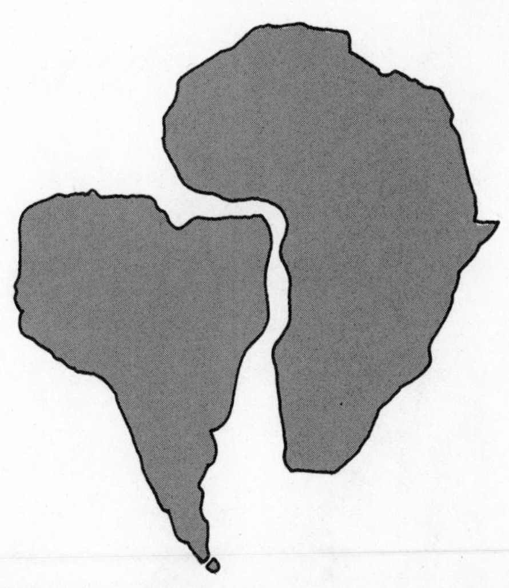

FIGURE 2.8 Fitting of South America and Africa. Most continents may be fitted in this way, providing the most notable evidence for continental drift.

Sea-Floor Spreading

In 1961 and 1962, two American geologists, Robert S. Dietz and Harry H. Hess, independently proposed a new and revolutionary hypothesis called ***sea-floor spreading,*** which explained many of the features of the ocean floor as well as the distribution of continents. According to this theory, pairs of rising convection currents, in addition to forming ***mid-ocean ridges*** also spread laterally away from these ridges. ***Oceanic trenches*** are produced, where the convection currents descend. (Mid-ocean ridges and oceanic trenches will be discussed in detail later in this chapter.) Figure 2.10 illustrates sea-floor spreading in a

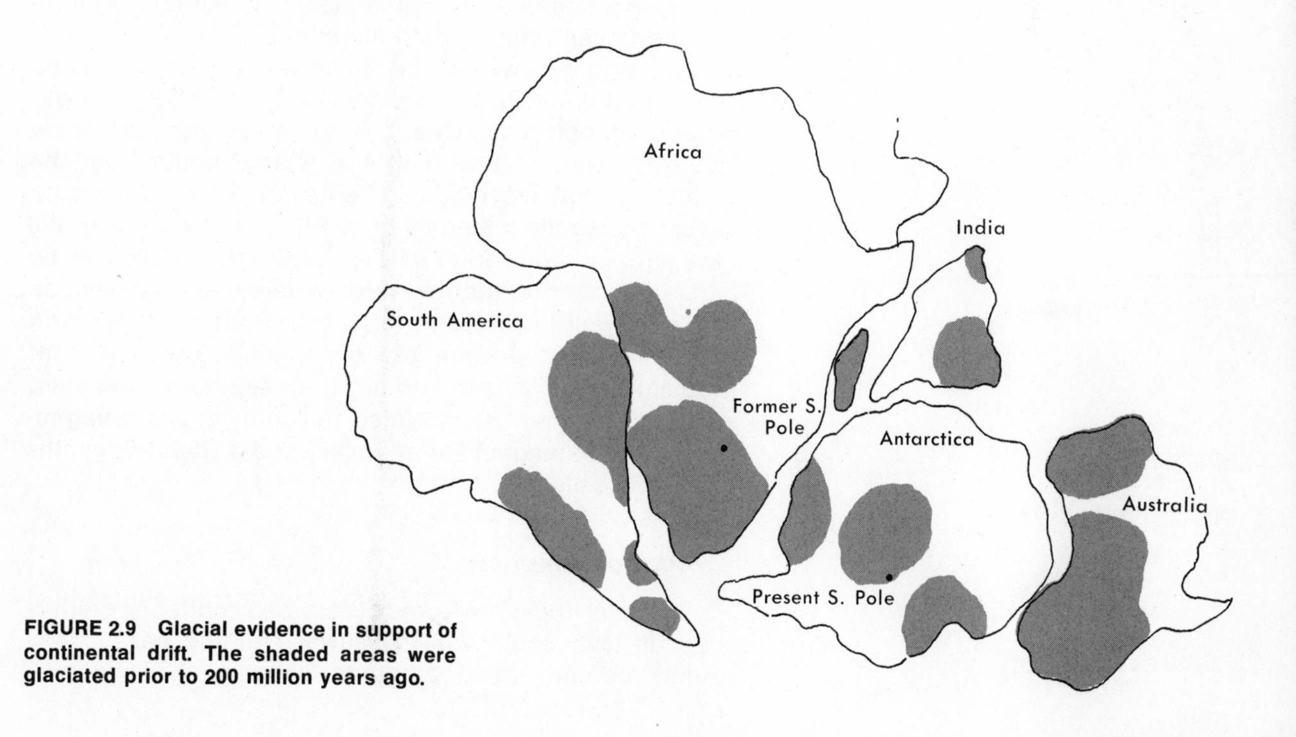

FIGURE 2.9 Glacial evidence in support of continental drift. The shaded areas were glaciated prior to 200 million years ago.

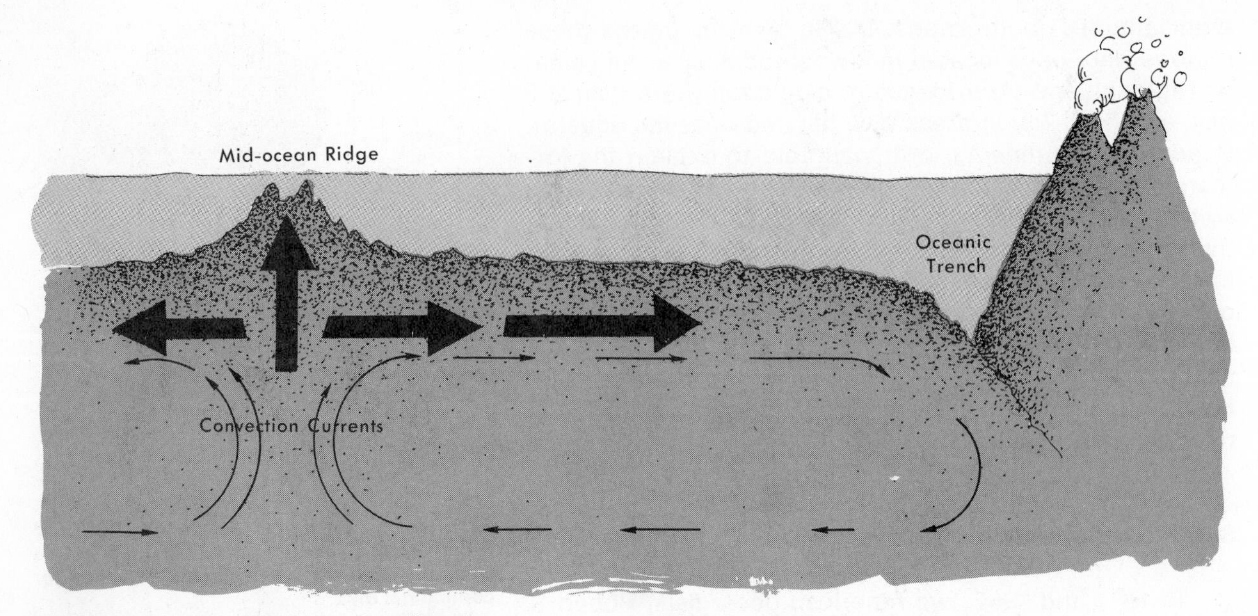

FIGURE 2.10 The sea-floor spreading hypothesis. Mid-ocean ridges are formed where convection currents rise, and trenches are formed where they descend into the mantle. New sea floor is created at the ridge and is continuously moved away as convection continues.

hypothetical ocean. The rising convection currents bring hot molten mantle materials to the mid-ocean ridge. Some of this material might reach the ridge surface in the form of volcanoes and lava flows, and the rest might solidify before reaching the sea floor. If the convection process is assumed to be continuous, the center of the ridge must always make room for the materials rising from below. Thus, new ocean floor is created near the center of the ridge. Previously formed ocean floor will move away slowly, at a rate of a few centimeters per year (cm/yr) from the mid-ocean ridge in both directions.

The origin as well as the dimensions of the postulated convection currents in the mantle are still unsolved problems in geophysics. One has to speculate that some localized source of heat in the mantle could heat the rocks and that the resulting reduction in density would cause the mantle materials to rise toward the surface. We have already seen earlier in this chapter that rocks may be partially molten in the low velocity layer, and convection may very well originate in this zone. Perhaps convection currents extend deeper into the mantle. The important problem, however, is to find proof for sea-floor spreading, and it comes from two sources, the study of ***paleomagnetism,*** ancient magnetism of rocks, and ***paleontology,*** the study of fossils.

PALEOMAGNETISM

Igneous rocks are produced when molten materials such as lava cool near the surface of the earth. They usually contain small amounts of magnetic minerals, which can be thought of as tiny compass needles that line up with the earth's magnetic field when cooled to about 600° C or lower. The resulting rock has a weak magnetic orientation, which can be measured using sensitive instruments called magnetometers. The orientation is fairly stable under the temperature conditions existing near the surface of the earth and will not change appreciably even if the earth's magnetic field changes drastically. Sedimentary rocks also possess magnetism. Some of the constituent particles of sediment are magnetic, and when settling to the sea floor, they align themselves with the earth's magnetic field, eventually resulting in magnetized sedimentary rocks.

By the early 1960s, paleomagnetic studies had revealed that the earth's magnetic field has reversed itself many times during geological history. That is, the present north and south magnetic poles of the earth have switched places repeatedly. These 180° reversals are evident from the exact opposite magnetizations of rocks. Figure 2.11 shows the worldwide reversals of the earth's magnetic field for the past 80 m.y. Evidence indicates that reversals have taken place throughout most of the earth's history, although the frequency of reversals has increased during the past 50 m.y. to about five reversals per million years, almost double the previous rate. A glance at Figure 2.11 indicates that the reversals are not regular, but the cause of these random changes is not known. Even the origin of the magnetic field itself is not well understood. It is believed to be caused by convective motions within the earth's core, and hence the reversals of the magnetic field should also be related to variations of these motions.[1]

The reversals of the magnetic field provide the best evidence for sea-floor spreading, as suggested by British scientists F. J. Vine and D. H. Matthews in 1963. If the sea floor has been spreading, the materials reaching the mid-ocean ridges, when cooled, should be magnetized by the then-existing earth's magnetic field. According to the sea-floor spreading concept, ocean floor is continuously created at the ridges and moves away in opposite directions, and if the magnetic field reverses, the new rocks formed at the ridges should have a reversed magnetization compared with the older rocks at a distance from the ridge. This should result in ocean-floor rocks that are

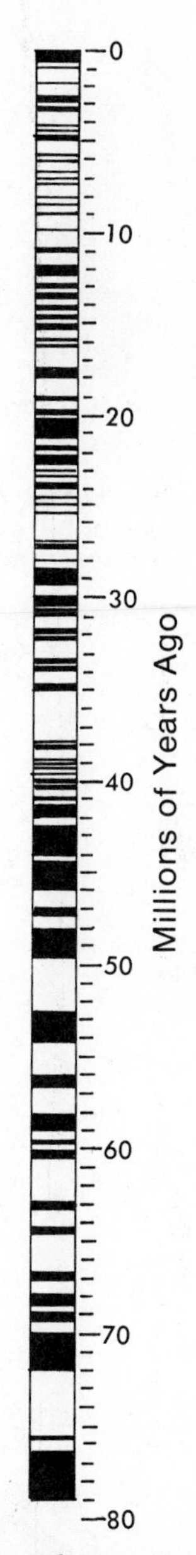

FIGURE 2.11 Times of magnetic reversals for the past 80 million years. The shaded regions indicate times when the magnetic field was normal (that is, oriented as it is today); and the unshaded areas correspond to times when the magnetic field was reversed. (After Heirtzler et al., 1968.)

[1]It is generally believed that magnetic reversals require about 10,000 to 20,000 years to complete. This transition time may have significance for life on earth, according to some scientists. The earth's magnetic field is responsible for the ***Van Allen radiation belts,*** which protect the earth from intense radiation from outer space. If during reversals, the earth's magnetic field is very weak or absent, the increased radiation reaching the earth may cause mutations of genes. Some scientists claim that a relationship exists between the times of reversals and the extinction or appearance of certain species. More research, however, is required to substantiate this claim as there is some dispute as to the amount of radioactivity reaching the earth due to the loss of the magnetic field, as well as to the validity of the correlation mentioned here.

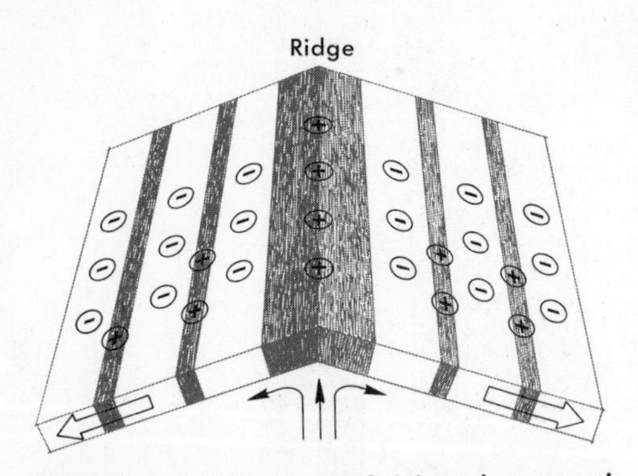

FIGURE 2.12 The normal (+) and reversed (−), magnetizations of the sea floor. Note the symmetry of the magnetizations with respect to the ridge.

magnetized in alternating normal and reversed directions, located symmetrically with respect to the mid-ocean ridges, as shown in the theoretical scheme of Figure 2.12.

The actual magnetism of ocean-floor rocks can be deduced from measurements made by magnetometers flown by airplanes or towed behind ships. Figure 2.13 shows a typical magnetic field pattern produced by the rocks of the ocean floor over a portion of the Mid-Atlantic Ridge, and similar results are obtained from magnetic measurements made over the other mid-ocean ridge systems. Figure 2.14 shows the results of magnetic studies in the northeast Pacific. These magnetic measurements can be used to determine the rate of spreading of the ocean floor in the past (Fig. 2.15). To do this, we need to know the times of reversals and the distance between corresponding rocks and mid-ocean ridges. The rates obtained range from 1 to more than 8 cm/yr, with an average value of about 2 cm/yr.

Paleontology

As marine organisms have evolved through time, and species has replaced species, a fossil record was created in marine sediments which provides a way of determining the age of sediments. By extension, it is also possible to use fossils to date the age of the volcanic rocks below them. As almost all volcanic rocks of the sea floor are thought to have been produced on the crests of mid-ocean ridges, the oldest sediments above them are only slightly younger. If fossils are used to date the basal sediments above the volcanics, one has a very good estimate

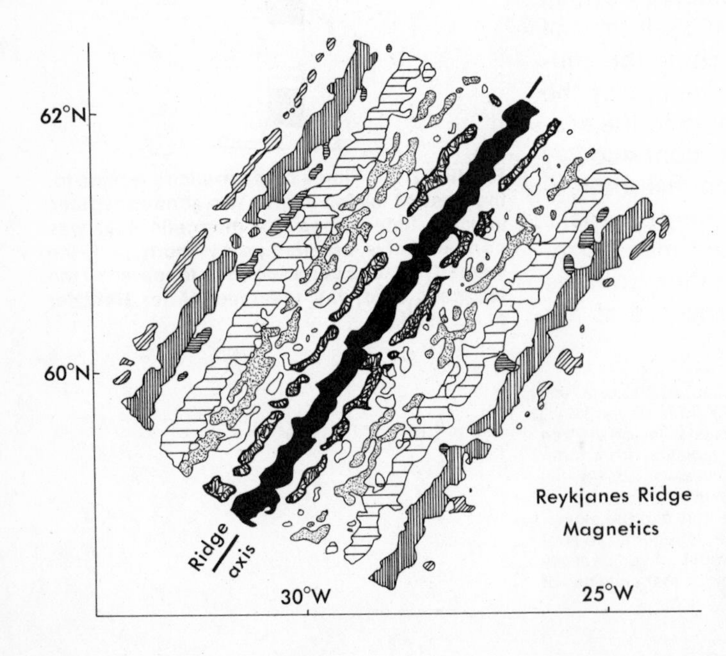

FIGURE 2.13 The magnetic field produced by the ocean-floor rocks over the Reykjanes Ridge near Iceland. Note the similarity to Figure 2.12. The outlined areas represent predominantly normal magnetization. The caps of the rocks increase away from the ridge. (After Heirtzler et al., 1966, and Vine, 1968.)

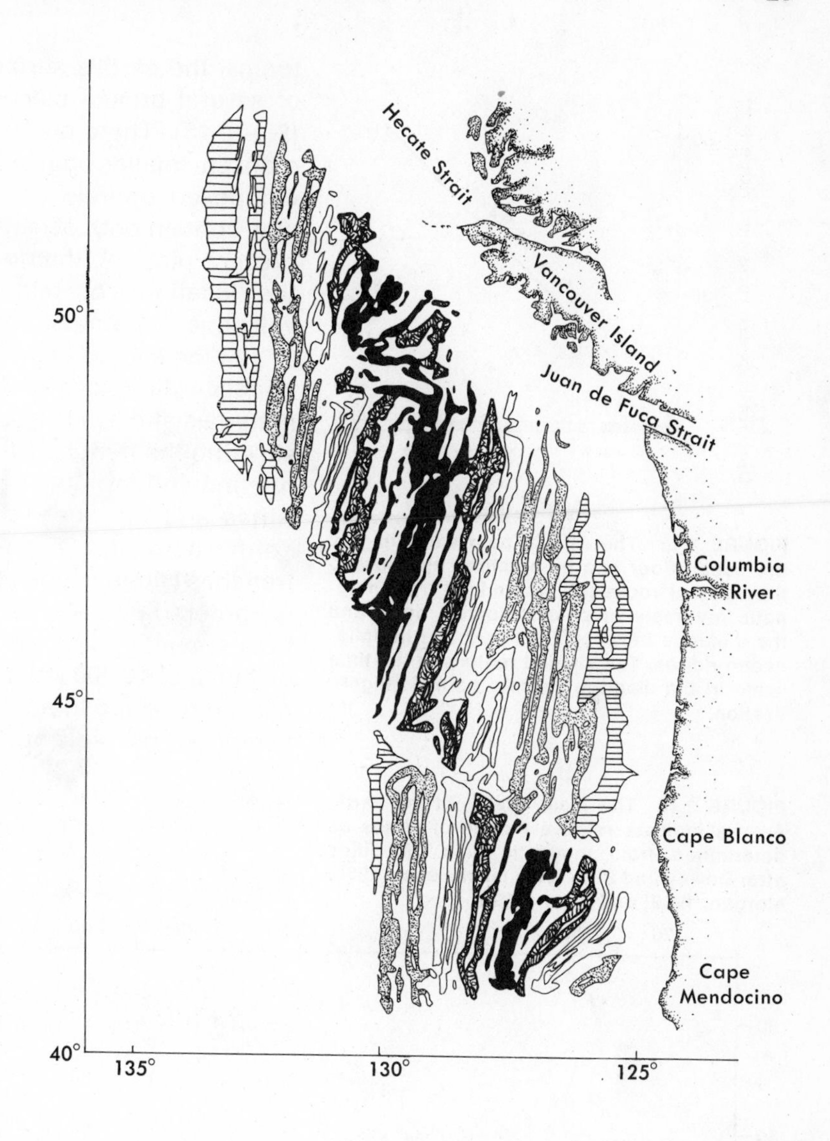

FIGURE 2.14 Magnetic field produced by the rocks on the floor of the northeast Pacific. Note the symmetry with respect to the ridges (the darkest shaded stripes). The ages of the rocks increase away from the ridges. The larger cross-hatched areas are about 8 to 10 million years old. (After Raff and Mason, 1961, and Vine, 1968.)

of the age of the volcanics. Comparison of results from this technique and magnetic technique show remarkably close agreement (Fig. 2.15), lending further credence to the theory of sea-floor spreading.

Plate Tectonics

In the late 1960s, the sea-floor spreading theory was modified to take into account all major geological structures of the earth. The new concept is known variously as ***new global tectonics*** or ***plate tectonics.*** The term "tectonics" refers to the large-scale structures of the earth as well as to their origin. The concepts of sea-floor spreading are included in plate tectonics, and most of the earlier discussions are also valid here. According to plate tec-

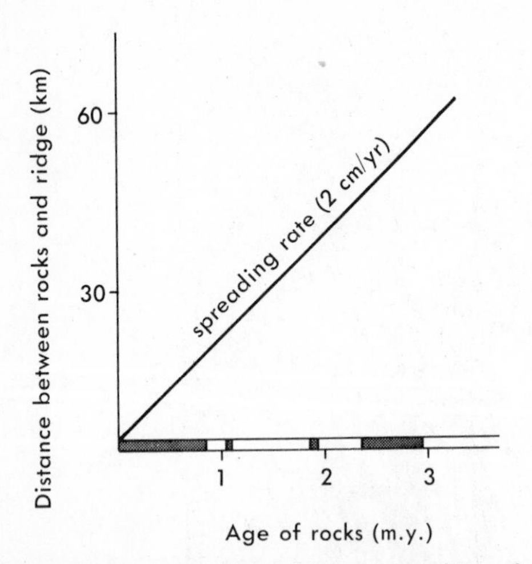

FIGURE 2.15 The rate of past spreading of the ocean floor can be obtained by knowing the ages of rocks (as determined from magnetic reversals or paleontological data) and the distance between the rocks and the mid-ocean ridges. The shaded regions of the time scale in the diagram indicate normal magnetization.

tonics, the earth's surface is pictured as being composed of several broken pieces of the lithosphere called ***plates*** (Fig. 2.16). There are about seven large plates and about a dozen smaller ones. Some, such as the Pacific plate, are composed entirely of oceanic lithosphere, while others may contain both oceanic and continental lithosphere. The present plate boundaries are determined on the basis of the locations of mid-ocean ridges, oceanic trenches, volcanoes, and most importantly, earthquakes.

At the mid-ocean ridges, the plates are moving away in opposite directions while continuously creating new plate materials at these ridges. The leading edges of many plates sink into the trenches. Some, such as the African plate, are not marked by any trenches. This plate includes both Africa and the ocean floor surrounding it; together they "drift" as a unit. The Atlantic Ocean has only two small trenches within it, whereas almost the entire Pacific Ocean is bordered by trenches. Assuming an average rate for seafloor spreading of 4 cm/yr and a maximum ridge-to-trench distance of 10,000 km, the entire floor of the Pacific Ocean can be renewed in about 250 m.y. On the basis of this assumption, no present Pacific ocean-floor rock should be

FIGURE 2.16 The major plates of the earth. Arrows indicate relative motion of plates as determined from magnetic data. (Modified after Dewey and Bird, 1970; Isacks et al., 1968; Morgan, 1968; and Vine, 1969.)

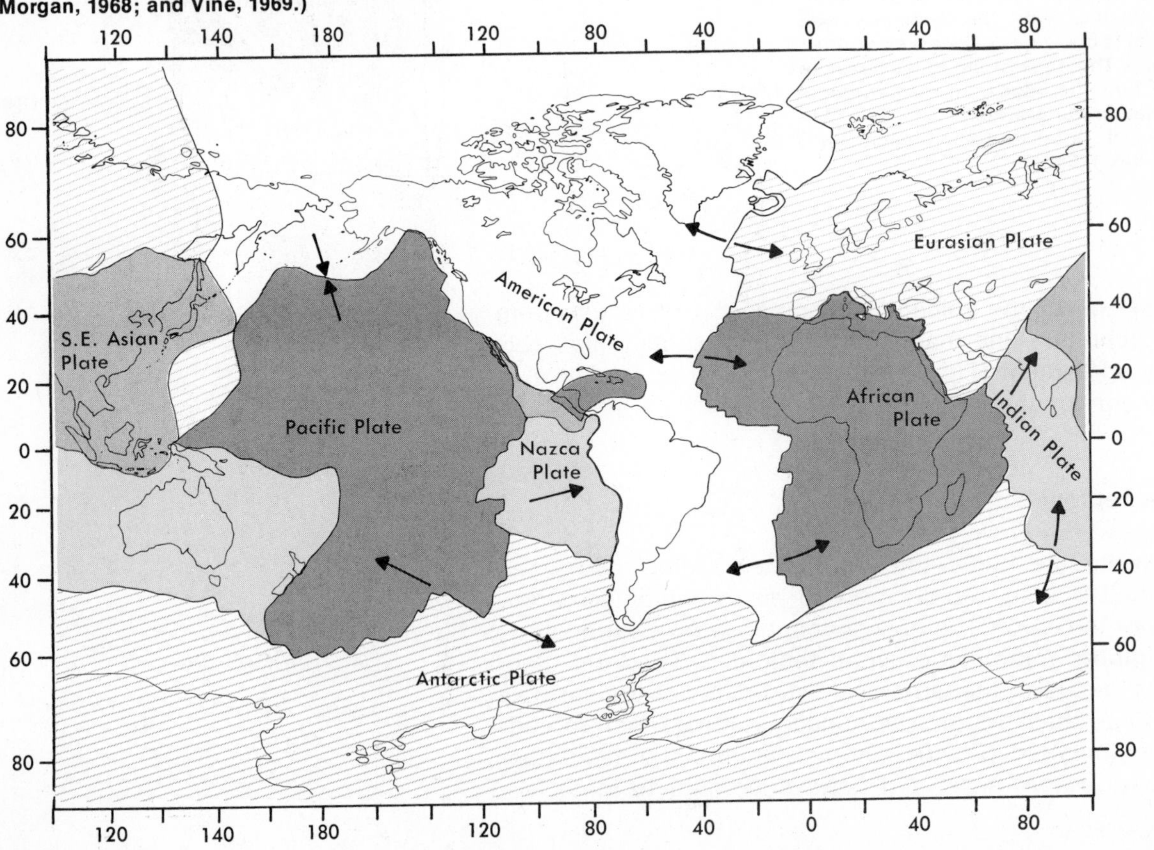

FIGURE 2.17 Mid-ocean ridges are offset by fractures called transform faults. Between points a and b, rocks on both sides of the fracture are moving in opposite directions, while in other parts of the fracture zone the movements are in the same direction.

older than 250 m.y. In fact, no rocks older than 200 m.y. have been found in any ocean.

According to present ideas, the initial break up of Pangaea and subsequent plate motions started about 200 m.y. ago and have been continuing since that time. The separation of Greenland from Europe, for example, occurred only about 60 m.y. ago. The Red Sea, the Gulf of Aden, the Gulf of California, and the East African Rift Valley (not yet a sea) were formed within the last 20 m.y. Of these areas, only the Gulf of Aden and the Gulf of California have ridges in them at present.

The mid-ocean ridges, and hence the plates, are intersected by many fractures (Fig. 2.16) called ***transform faults*** (Fig. 2.17). The rocks on both sides of the fault (between a and b) are moving in opposite directions as a result of the general plate movements away from the ridges. This differential motion is responsible for earthquakes in the fault zone. The San Andreas fault of California is considered to be a transform fault, and the shallow but great earthquakes occurring along it are the result of differential motion.

Implications of Sea-Floor Spreading and Plate Tectonics

Sea-floor spreading and plate tectonics have far-reaching significance not only in oceanography but also in geology and biology. The origin of the ocean basins, the age of the ocean floor, the origin of the mid-ocean ridges and oceanic trenches, the distribution of deep-sea sediments, and many other features of the ocean floor can be explained in terms of plate tectonics, especially since

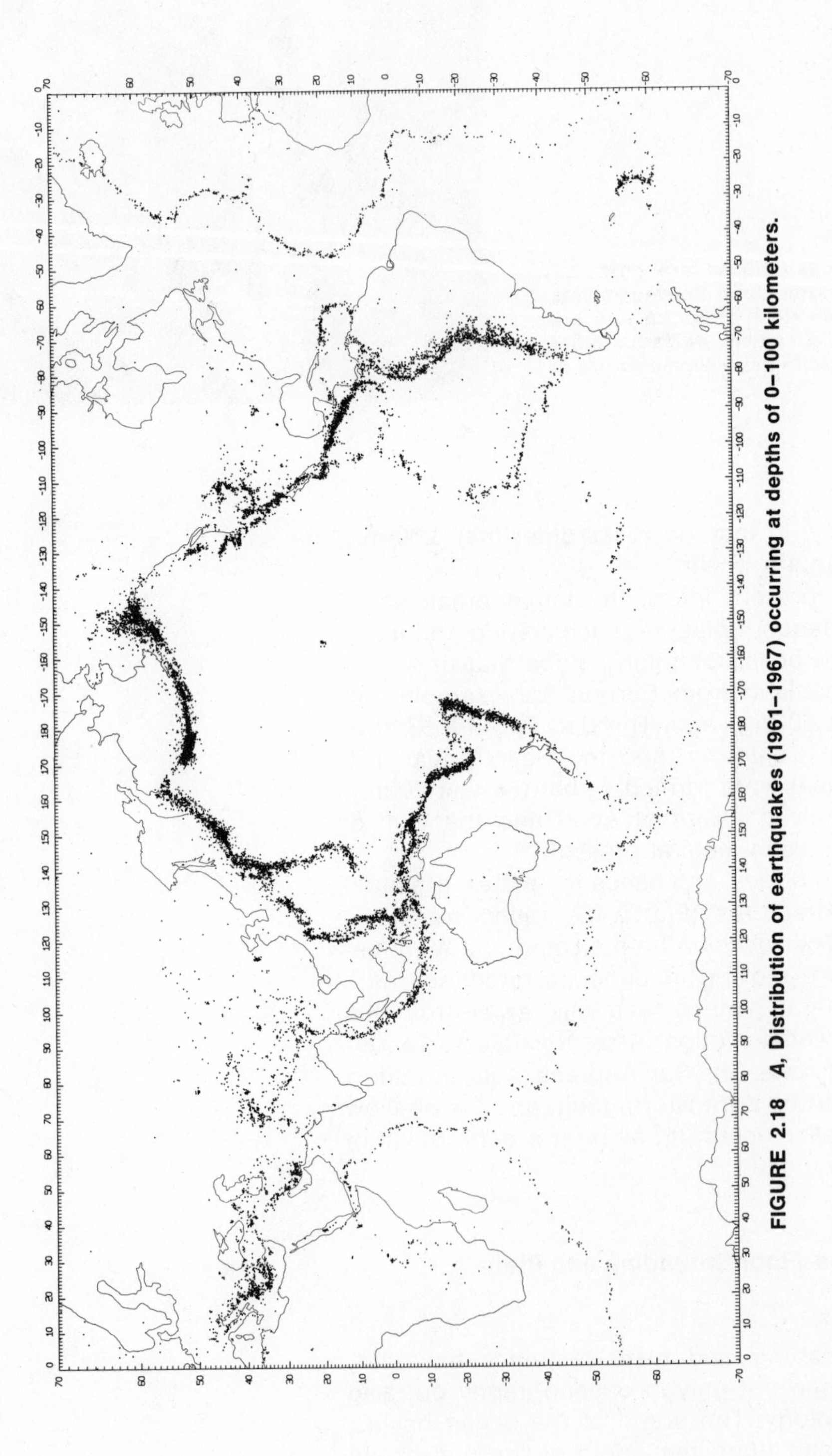

FIGURE 2.18 A, Distribution of earthquakes (1961–1967) occurring at depths of 0–100 kilometers.

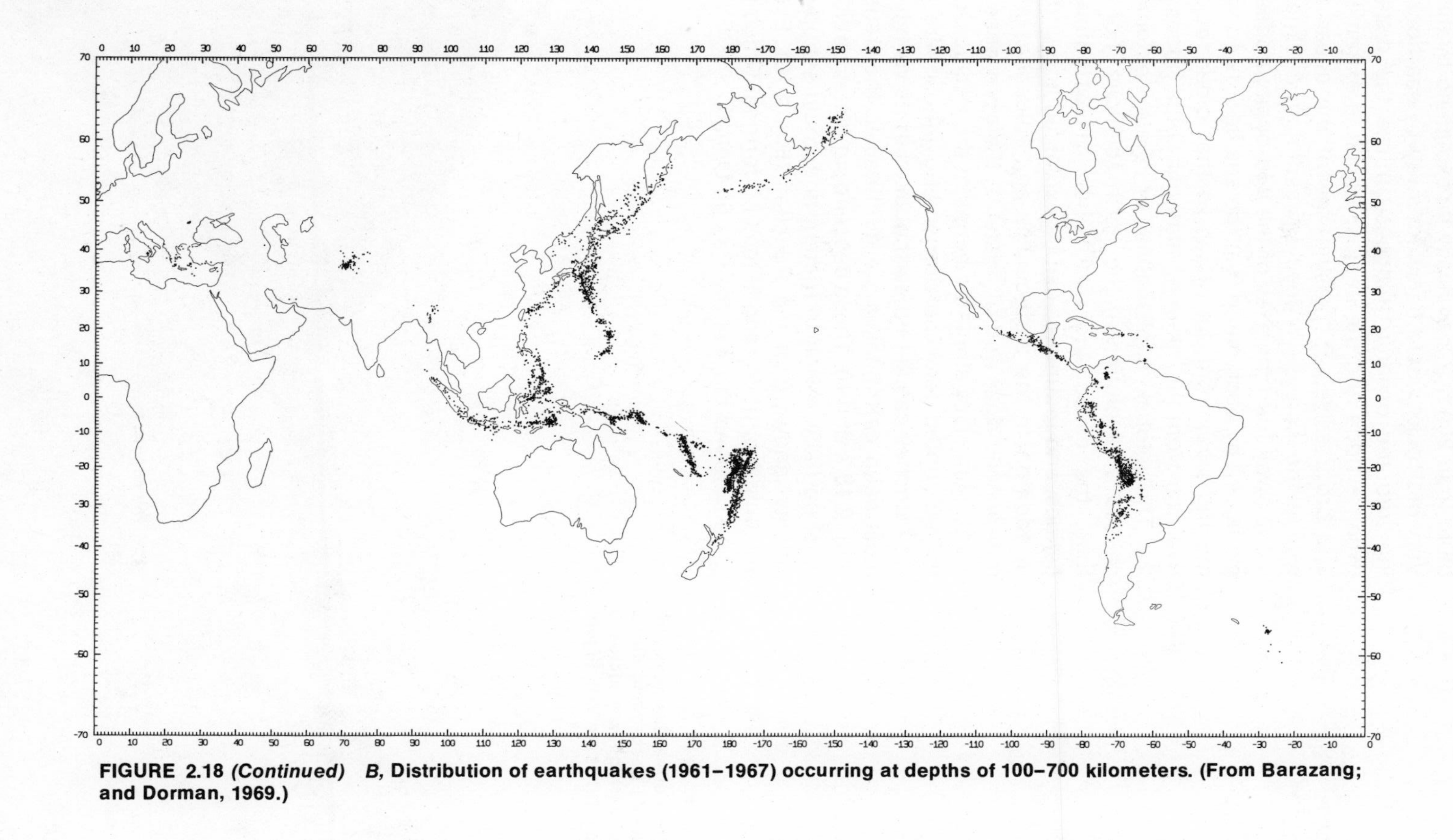

FIGURE 2.18 (*Continued*) ***B,*** **Distribution of earthquakes (1961–1967) occurring at depths of 100–700 kilometers. (From Barazang; and Dorman, 1969.)**

no other satisfactory explanations exist. For example, the fact that there are apparently no rocks on the ocean floor older than 200 m.y. can easily be explained by this theory. Very old rocks would be swept away by sea-floor spreading into the oceanic trenches, unless the distance between a ridge and the nearest trench is extremely great. The progressively increasing age of the ocean floor, as well as the increase in deep-sea sediment thickness as we move away from the axes of the mid-ocean ridges, can be explained by postulating that the sea floor is continuously moving away from the ridges so that older rock is relatively far from the "young" material at the ridges.

The global implications of plate tectonics have caused a revolution in the earth sciences. For the first time, the cause and distribution of earthquakes, volcanoes, and mountains can be explained and understood. In addition to the earthquakes and volcanoes along the mid-ocean ridges and rift systems, their presence in other areas, especially along the edges of the Pacific Ocean can be accounted for. The descending lithosphere beneath the trenches can set up earthquakes in its vicinity. In fact, most deep earthquakes are confined to the trench areas (Fig. 2.18 *a* and *b*). These descending plates could melt, at least partially, owing to friction and could set up volcanic activity landward of the trenches (Fig. 2.19). Eventually, volcanic mountain ranges may be formed, which has happened around most of the Pacific Ocean.

FIGURE 2.19 Production of earthquakes and volcanic activity by plates descending into a trench. The depths of earthquakes increase landward of the trench. Compare with Figure 2.18.

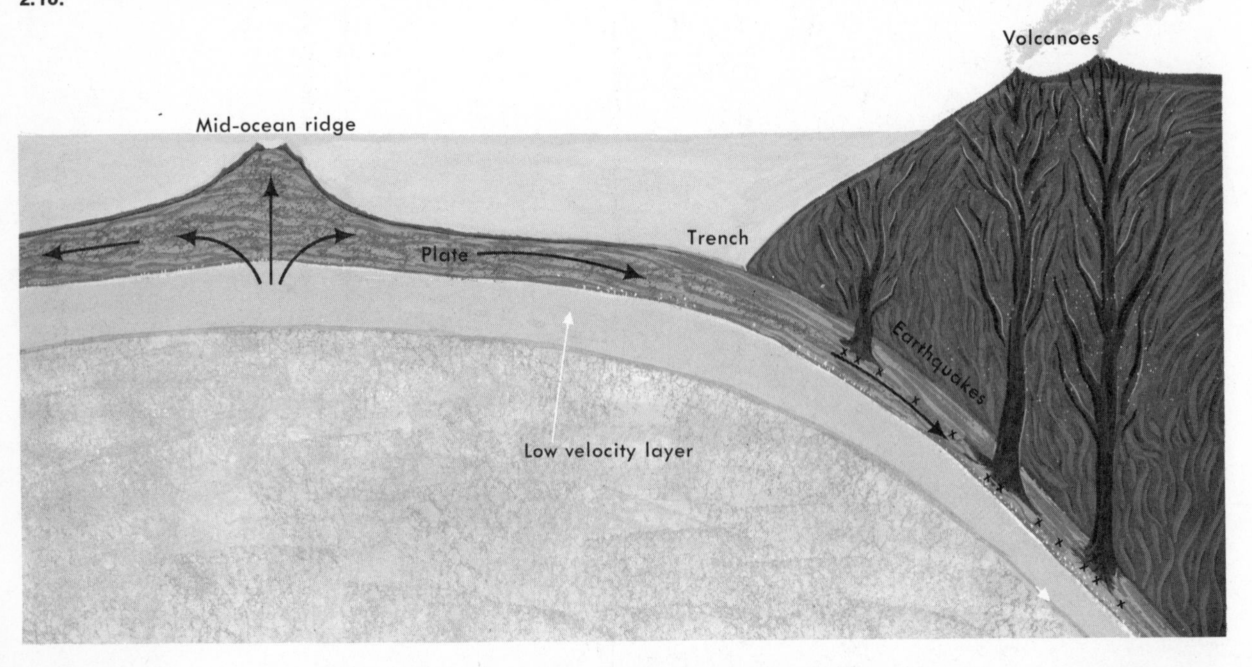

The largest mountains, such as the Alps and the Himalayas, are not volcanic in nature. They are composed mostly of folded (crumpled) sedimentary rock, possibly formed when two plates collided against each other. The Himalayas, for example, are believed to have been produced when India broke off and moved away from Africa (Gondwanaland) and collided with Asia (see Fig. 2.7). We may use the same explanation for folded mountains that are older than 200 m.y., the time at which the present episode of continental drift started; that is, there may have been other episodes of continental drift and sea-floor spreading in the past, resulting in the formation of many very old mountain ranges such as the Appalachians and Urals.

Collisions, as well as the breaking of plates, are of great significance in evolutionary biology. Such events create new environments and destroy pre-existing environments. Populations may become split between two continents that were once joined, and this might result in the development of new species. If a land mass migrates into a different climatic region, some species might become extinct or undergo dramatic adaptations. And, quite clearly, the joining of two previously separated land masses can create interactions among species which had never before been in contact with each other.

Origin of Ocean Basins

The preceding discussion of plate tectonics and sea-floor spreading can be used to explain, for example, how the Atlantic Ocean was formed after the American plate broke away from the European and African plates. The Atlantic Ocean will continue to grow in size if these plates continue to move. Similarly, we may visualize the Red Sea, the Gulf of Aden, and other such areas as the sites of future large ocean basins if present plate motions continue. The East African Rift may represent the site of a future ocean (see also Figure 2.34, p. 51). Other oceans and seas, such as the Mediterranean Sea, may decrease in size, and some may even cease to exist, if opposing plates collide.

Many questions still remain to be answered. How and when did the Pacific Ocean originate? An ancestral Pacific must have surrounded Pangaea. As seen earlier, the rocks of the Pacific Ocean are not older than 200 m.y., a situation which can be explained by the rejuvenation of the Pacific floor by sea-floor spreading and the descent of older rocks into its trenches.

The first ocean basins were formed after the earth's surface cooled and continents began to form. A primitive crust was probably formed by a differentiation process;

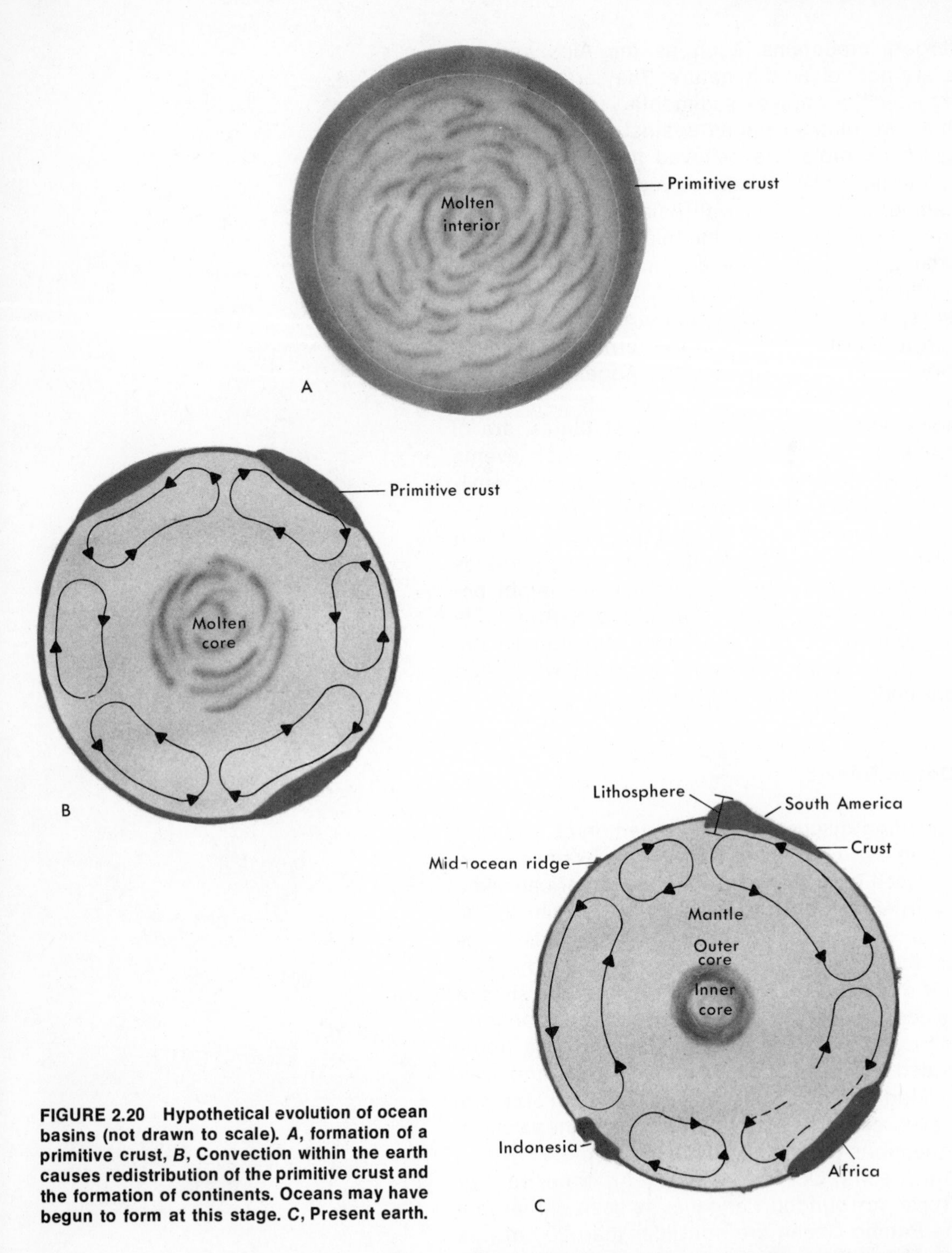

FIGURE 2.20 Hypothetical evolution of ocean basins (not drawn to scale). *A*, formation of a primitive crust, *B*, Convection within the earth causes redistribution of the primitive crust and the formation of continents. Oceans may have begun to form at this stage. *C*, Present earth.

that is, the separation of lighter rocks from the heavier materials below them. Convection currents, which have been active since the molten beginning of the earth, may have caused this primitive crust to move in a manner similar to the process of plate tectonics (Fig. 2.20). As the earth continued to cool, water vapor condensed and accumulated in the newly formed depressions of the crust.

These were perhaps the beginnings of "seas." Most of the water was derived from within the earth, and the seas grew into oceans along with the subsequent evolution of oceanic crust and the growth of continents. The process of plate tectonics has modified, and continues to modify, the shape and distribution of the ocean basins.

STUDYING THE OCEAN FLOOR

The topography of the vast ocean floor is perhaps not as complex as that of the earth's surface, but it is also not

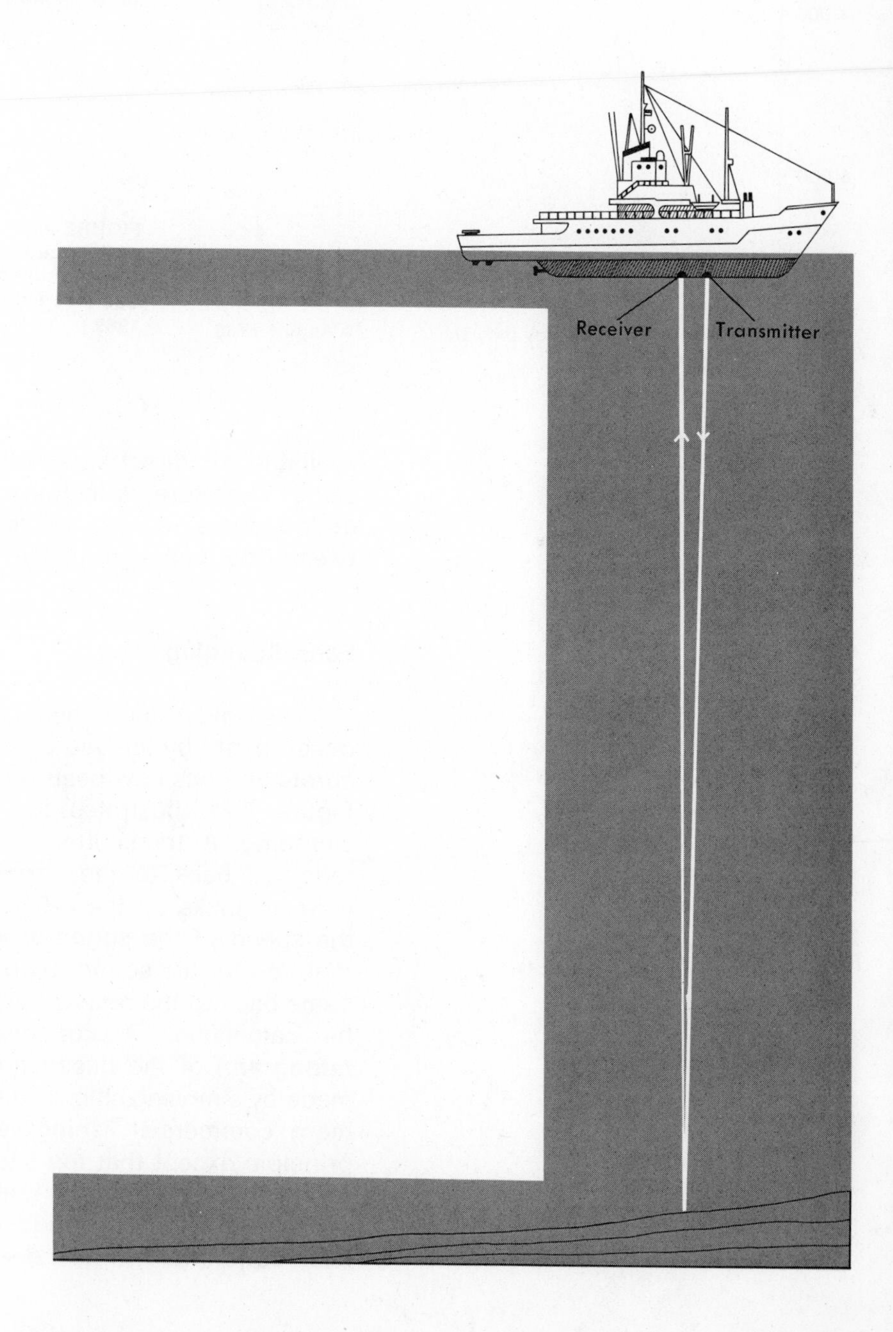

FIGURE 2.21 The principle of echo sounding. A transmitter sends a sound wave, which is reflected back to the surface by the ocean bottom and is picked up by a receiver. By knowing the total time involved and the speed of sound in the ocean (1,500 m/sec), water depth can be determined.

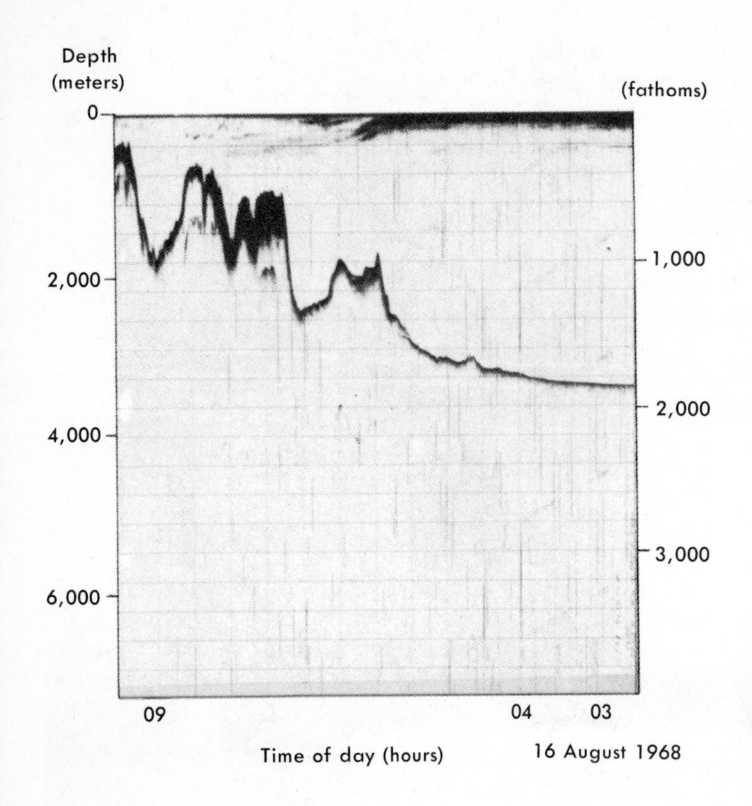

FIGURE 2.22 An echogram, a continuous record of the ocean bottom produced by a moving ship. Note the deep scattering layers near the surface. See also Figure 9.29. (From Lowrie and Eskowitz, 1969.)

available for direct observation except upon rare occasions. Therefore, with the exception of several hundred drilled cores, almost all of our information about the ocean floor comes to us by indirect means.

Echo Sounding

The old time-consuming system of determining ocean depth by the use of a weighted wire is not very accurate and has now been replaced by electronic methods. Figure 2.21 illustrates the principle involved in ***echo sounding.*** A transmitter sends a sound wave which is reflected back to the surface by the ocean bottom. A receiver picks up the reflected sound wave. By knowing the speed of the sound wave in the water and the time required for the sound to go down to the ocean floor and come back to the receiver at the surface, ocean depth can be calculated. A continuous record (***echogram*** or ***fathogram***) of the ocean floor profile (Fig. 2.22) can be made by a moving ship. The so-called fish-finder on board many commercial fishing vessels operates on the same principle except that the sound waves are bounced back from fish as well as from the bottom. In fact, some fish are so uniform in their schooling habits that they produce a false bottom known as a ***deep scattering layer.***

Continuous Seismic Profiling

Another method of obtaining not only the ocean bottom profile but also information about the sediments and rocks below the ocean floor is called ***continuous seismic profiling*** (or ***sparker profiling***). The principle involved is similar to that of echo sounding, but a stronger energy source is used. Part of the sound energy is reflected back by the ocean bottom, but the other part enters the bottom and is eventually reflected back to the sea surface by various rock or sediment layers (Fig. 2.23). As the ship

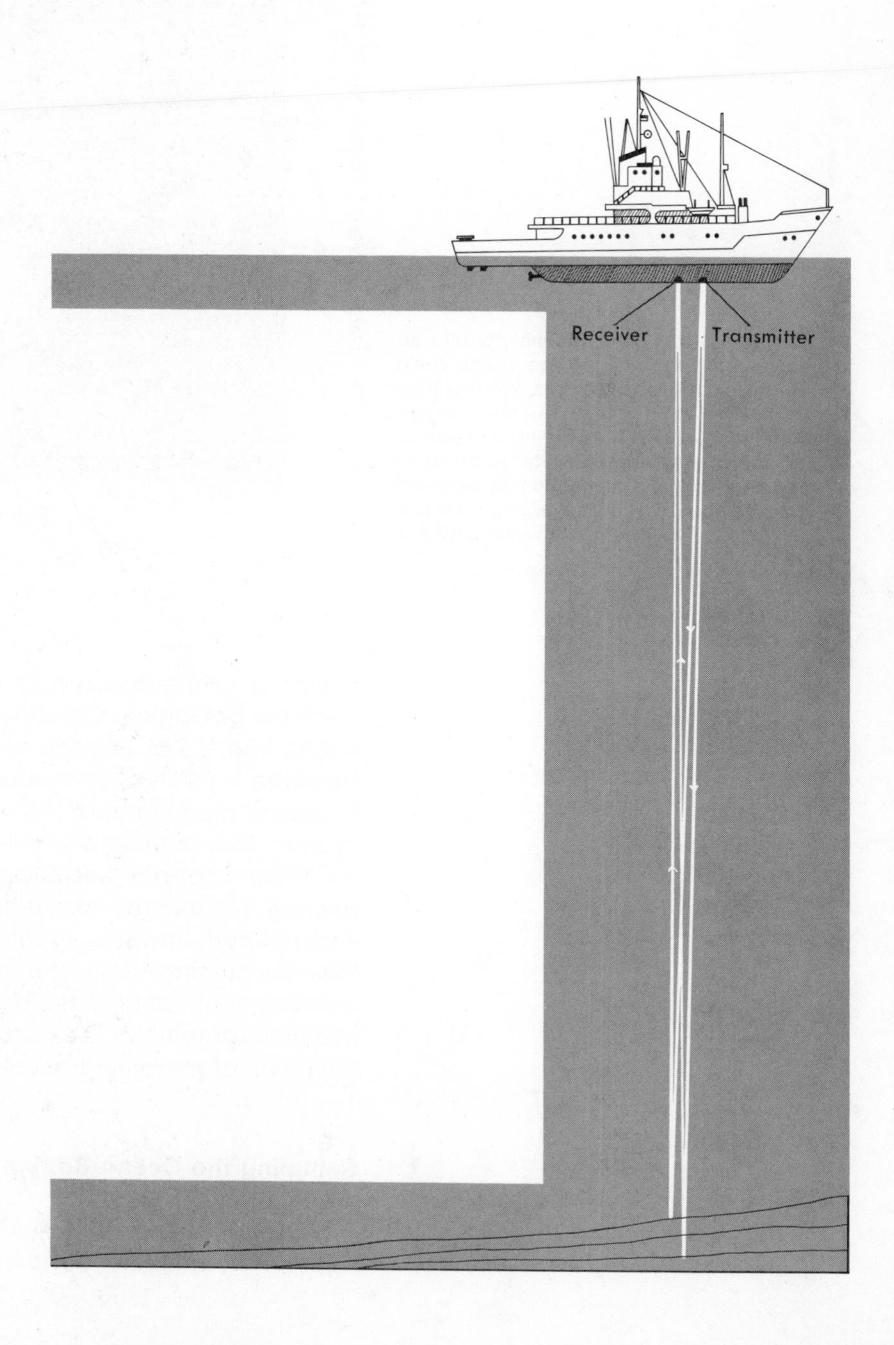

FIGURE 2.23 The principle of continuous seismic profiling. Part of the energy is reflected back by the ocean floor; the remainder continues into the ocean bottom and eventually is reflected back to the surface by different layers of sediments and rocks.

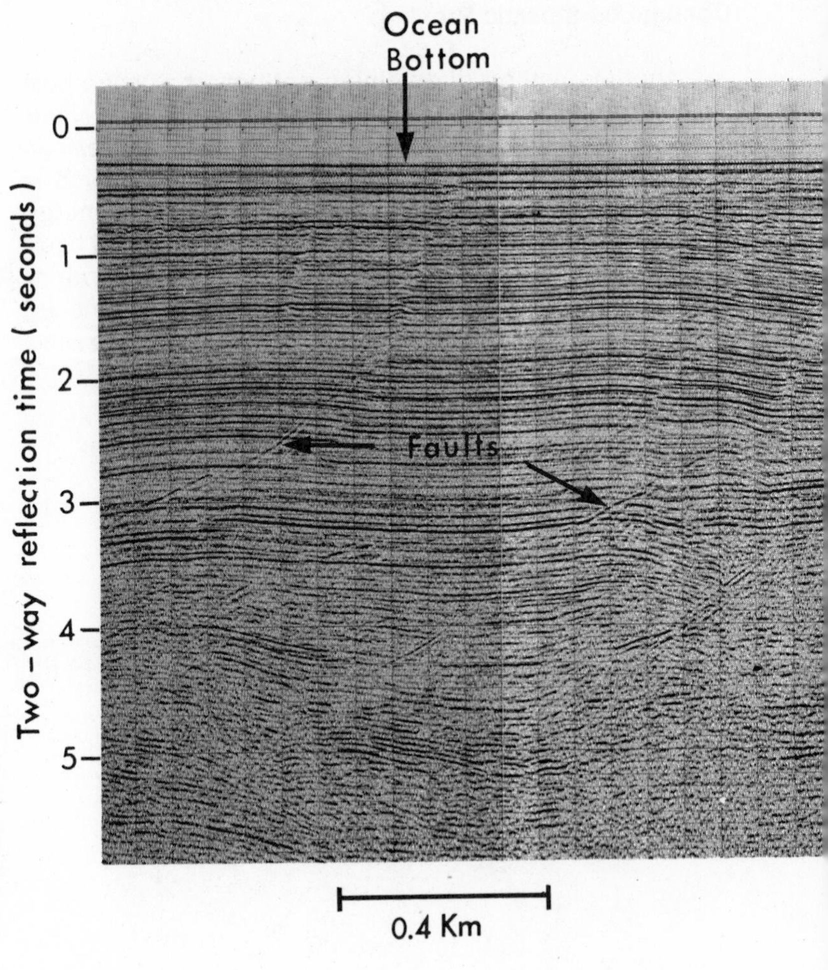

FIGURE 2.24 Continuous seismic profile obtained by a moving ship off the Texas coast. Salt deposited millions of years ago has been forced upward to form a dome, and the consequent tilting of the rocks helps to entrap oil. The speed of sound in seawater is about 1.5 km/sec, and that of unconsolidated sediment varies from about 1.7 to 2.5 km/sec. (Courtesy Carl H. Savit, Western Geophysical Corporation.)

moves, a continuous profile of the ocean bottom as well as of the geological structure below the ocean floor is obtained (Fig. 2.24). The use of continuous seismic profiling, however, is limited to the study of only a few hundred to a thousand meters below the ocean floor, but it can be used in even the deepest water. For deeper penetration below the ocean bottom, ***seismic refraction*** (Fig. 2.25) and ***reflection*** techniques requiring explosions are employed. This method, however, is considerably more expensive and time-consuming than the continuous seismic profiling technique. All of these techniques are extremely important in oceanographic studies as well as in detection of possible locations of petroleum-bearing rock structures offshore.

Sampling the Ocean Bottom

Ocean bottom samples can be obtained in various ways. The simplest devices are the ***grab*** and ***dredge***

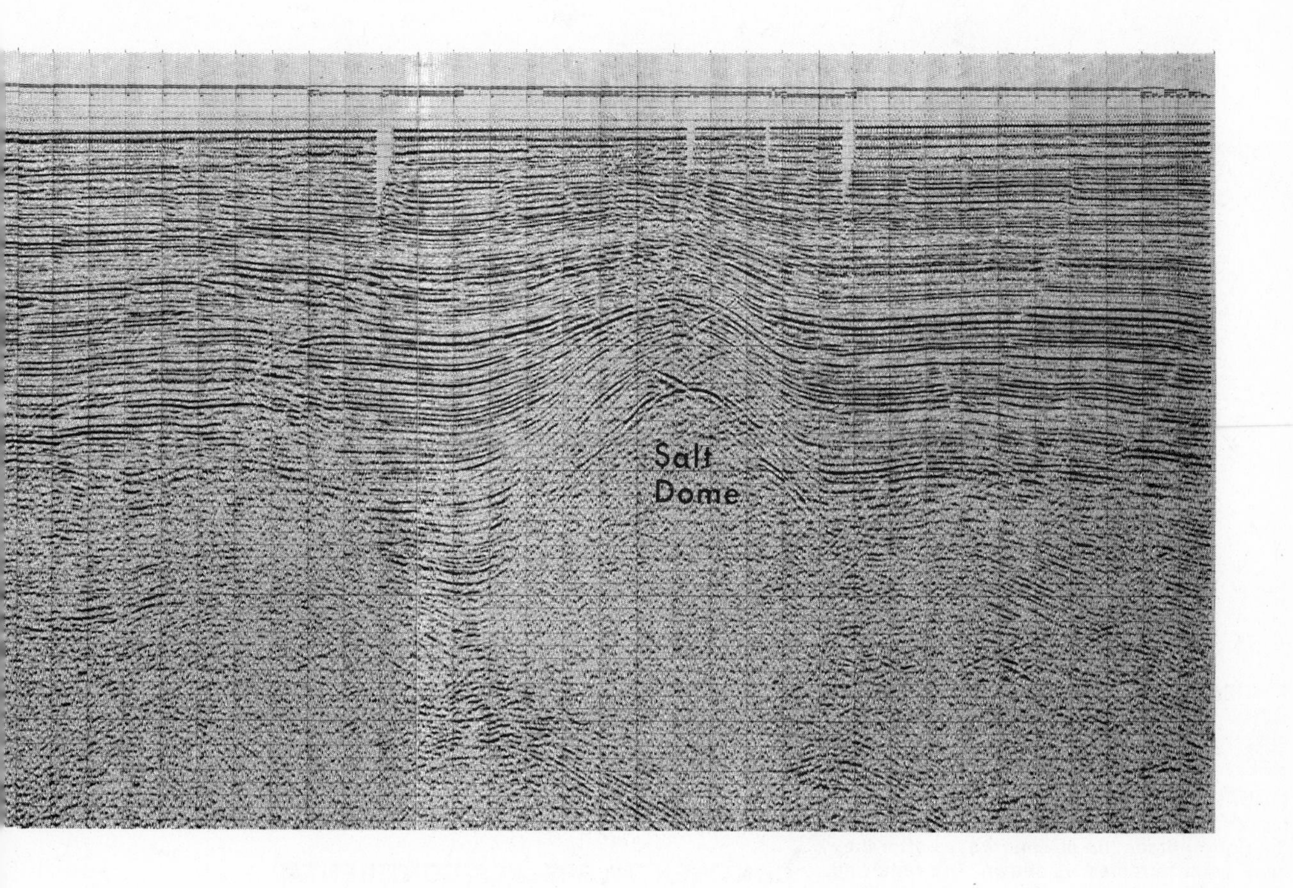

(drag) bottom samplers (Fig. 2.26). The latter is used by a moving ship, primarily to gather sediment and rock samples.

The best technique for deep sampling of the ocean bottom is by ***coring.*** Most coring devices involve the driving of a pipe into the ocean sediments (Fig. 2.27). In order to obtain deeper core samples, expecially in hard rock, drilling equipment very similar to that used by oil companies is needed. Unfortunately, drilling requires stable platforms such as those used in offshore oil drilling or on board a specially constructed drilling ship such as the *Glomar Challenger*[2] (Fig. 2.28), thereby making it the most expensive sampling technique.

[2]*Glomar Challenger* is the drilling ship of the Deep Sea Drilling Project. This project is managed by the Scripps Institution of Oceanography, supported by the US National Science Foundation as well as some foreign nations. It has been in operation since 1968 and has drilled several hundred holes, enabling scientists to study rocks from depths greater than 1,000 m below the ocean floor.

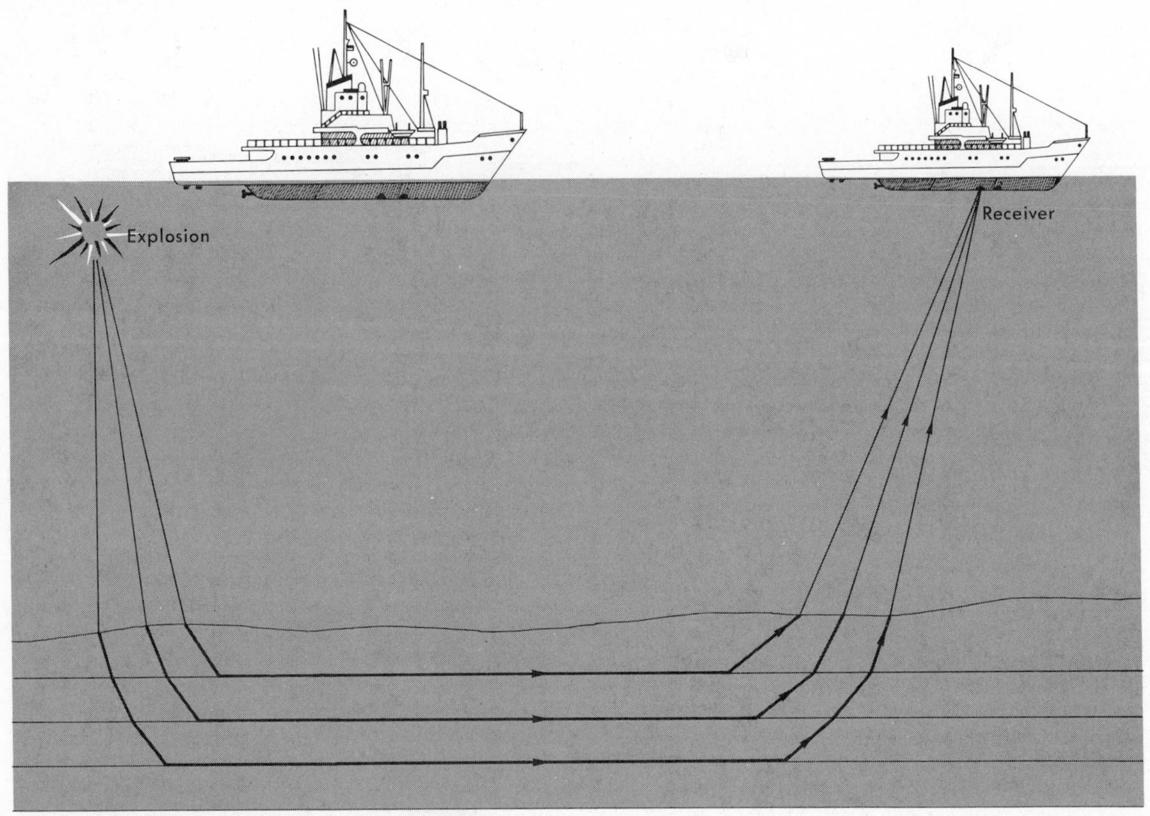

FIGURE 2.25 Seismic refraction technique for which two ships are required. The ship on the right records the seismic waves after they have been refracted as shown. The receiving ship may be miles away from the transmitting ship.

CONTINENTAL SHELVES, CONTINENTAL SLOPES, AND CONTINENTAL RISES

The ***continental shelf*** is the relatively shallow and flat portion of the sea floor (Fig. 2.29) that extends from the shoreline to the ***shelf break***, where an abrupt increase in slope occurs. The shelf break is located at an average depth of about 130 meters. The slope of an average shelf is so slight (a drop of about two meters every kilometer) that it is undetectable to the naked eye. The continental shelves constitute about 5 per cent of the earth's surface, or about three times the area of the Soviet Union. The average width of the continental shelves is about 70 km but varies considerably from place to place. For example, practically no continental shelf exists off Chile, whereas off northern Siberia its width is about 1500 km. The sediment cover of the continental shelf is made up mostly of sandy materials, except off the glaciated coasts of higher latitudes, where glacially derived gravel-like debris dominates.

Continental shelves are considered to be part of the continents. Many geologists believe that they were produced during glaciations when the sea level was lower, perhaps at the level of the shelf break. During the glacia-

FIGURE 2.26 Some ocean-bottom sampling equipment. *A,* Orange-peel grab; *B,* rock dredge; *C,* anchor bag dredge. (Photos courtesy of the Woods Hole Oceanographic Institution.)

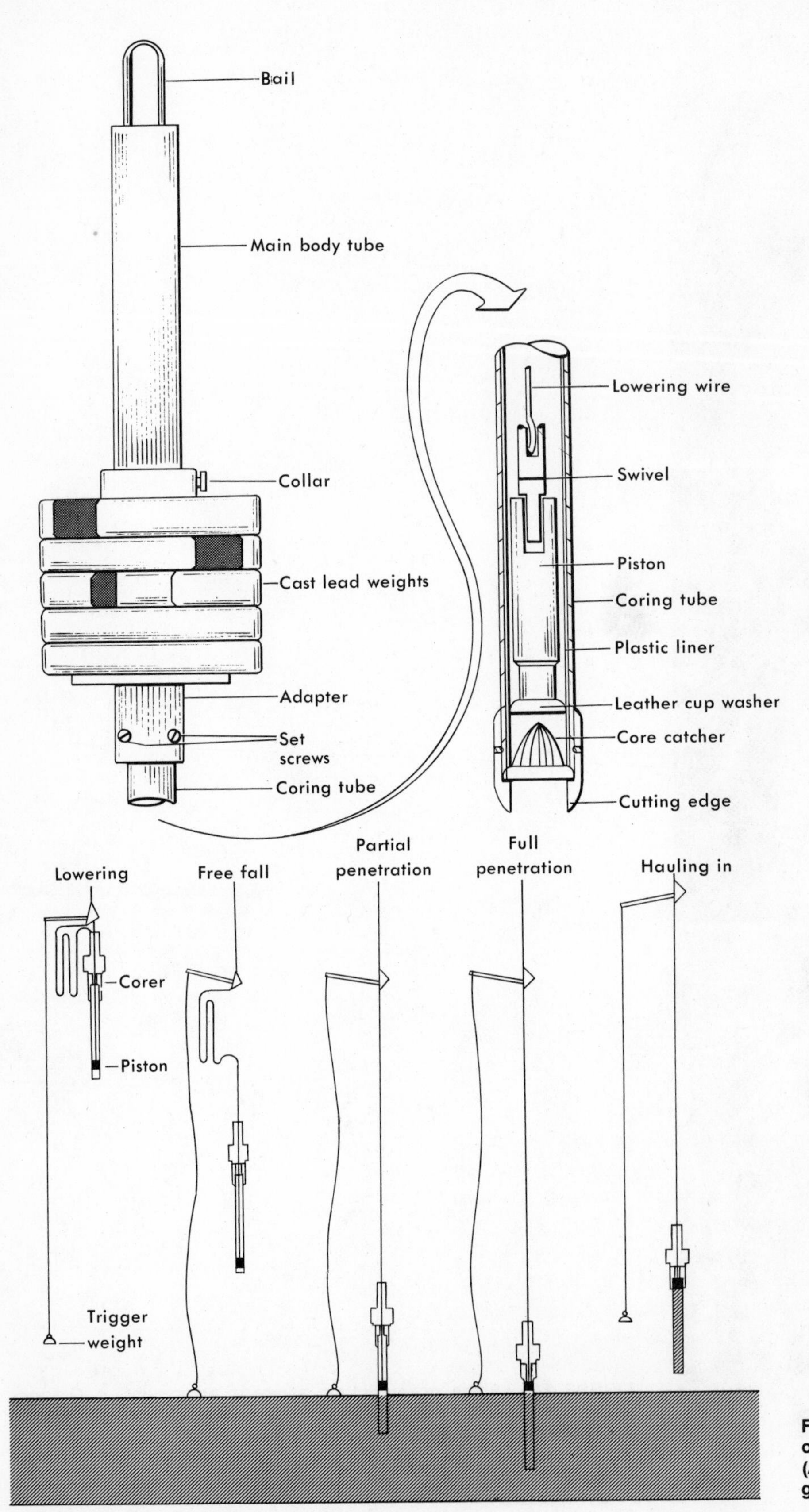

FIGURE 2.27 Structure and operation of the piston corer. (After U.S. Naval Oceanographic Office, 1968, Pub. 607.)

FIGURE 2.28 The drilling ship *Glomar Challenger*. *A*, The ship at sea; *B*, The drilling platform aboard the ship. (*A* courtesy of Deep Sea Drilling Project, Scripps Institution of Oceanography; *B* courtesy of Lamont-Doherty Geological Observatory of Columbia University).

tions of the past million years or more, coastal areas could have been worn down to this new lower sea level by stream and wave erosion. When the glaciers melted, about 15,000 years ago, these coasts submerged to produce the continental shelves. However, not all continental shelves of today need have been produced in this manner; some shelves may be the result of sediment accumulation from river run-off or coastal submergence.

The ***continental slope*** is the narrow band (average width, about 25 km) extending from the shelf break to the deep ocean floor. It has an average slope of about 4 degrees (a drop of about 70 meters per kilometer) but is highly variable. For example, off Santiago, Chile, the slope

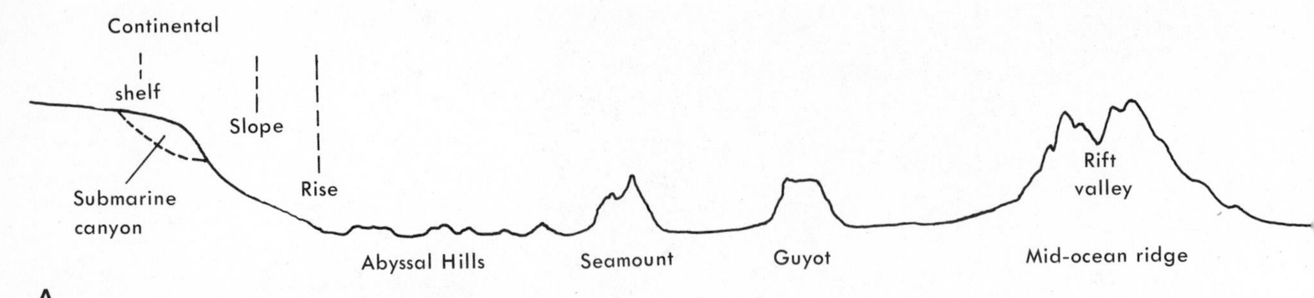

A

FIGURE 2.29 *A*, **Hypothetical ocean profile, illustrating the major features of the ocean floor.** *B*, **Artist's conception of the floor of the North Atlantic Ocean. (*B* from a painting by Heinrich Berann, Courtesy of ALCOA.)**

is approximately 45 degrees. About 60 per cent of continental slope sediment is mud, and the rest is composed of sand, gravel, rock and organic remains. The origin of the continental slopes is still not well understood, although many seem to be the result of large scale movements associated with plate tectonics.

Continental slopes do not always extend to the deep ocean floor. They are often interrupted by broad sedimentary wedges called ***continental rises***. Large quantities of sediments may be transported across the continental shelves and slopes and deposited into the deep ocean by ***turbidity currents*** (discussed in the next section) as well as by submarine landslides. They may eventually produce continental rises.

SUBMARINE CANYONS

In addition to the numerous submarine valleys found throughout the continental shelves and slopes, there are many V-shaped rocky canyons called ***submarine canyons***. They are confined mostly to the continental slopes, although many, such as the Hudson Canyon, extend far into the continental shelves. They are almost perpendicular to the coast and usually have many branches. The largest one, the Monterey Canyon off California, is deeper and wider than the Grand Canyon (Fig. 2.30). Some, such as the Hudson and the Congo Canyon, are found off river mouths.

The origin of submarine canyons is still unresolved. Undoubtedly, different canyons are produced by different means. Those found off rivers may be drowned portions of rivers that must have extended farther out to sea during glaciations, because sea levels were lower, and the rise in sea level after glaciation could have drowned their mouths. In some cases, sinking of river mouths could be due to large-scale earth movements. However, many submarine canyons are not associated with rivers.

Another explanation for the origin of submarine can-

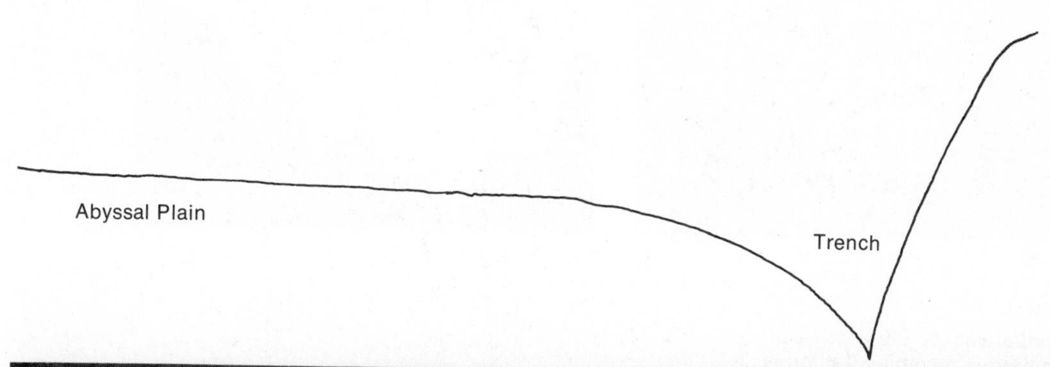

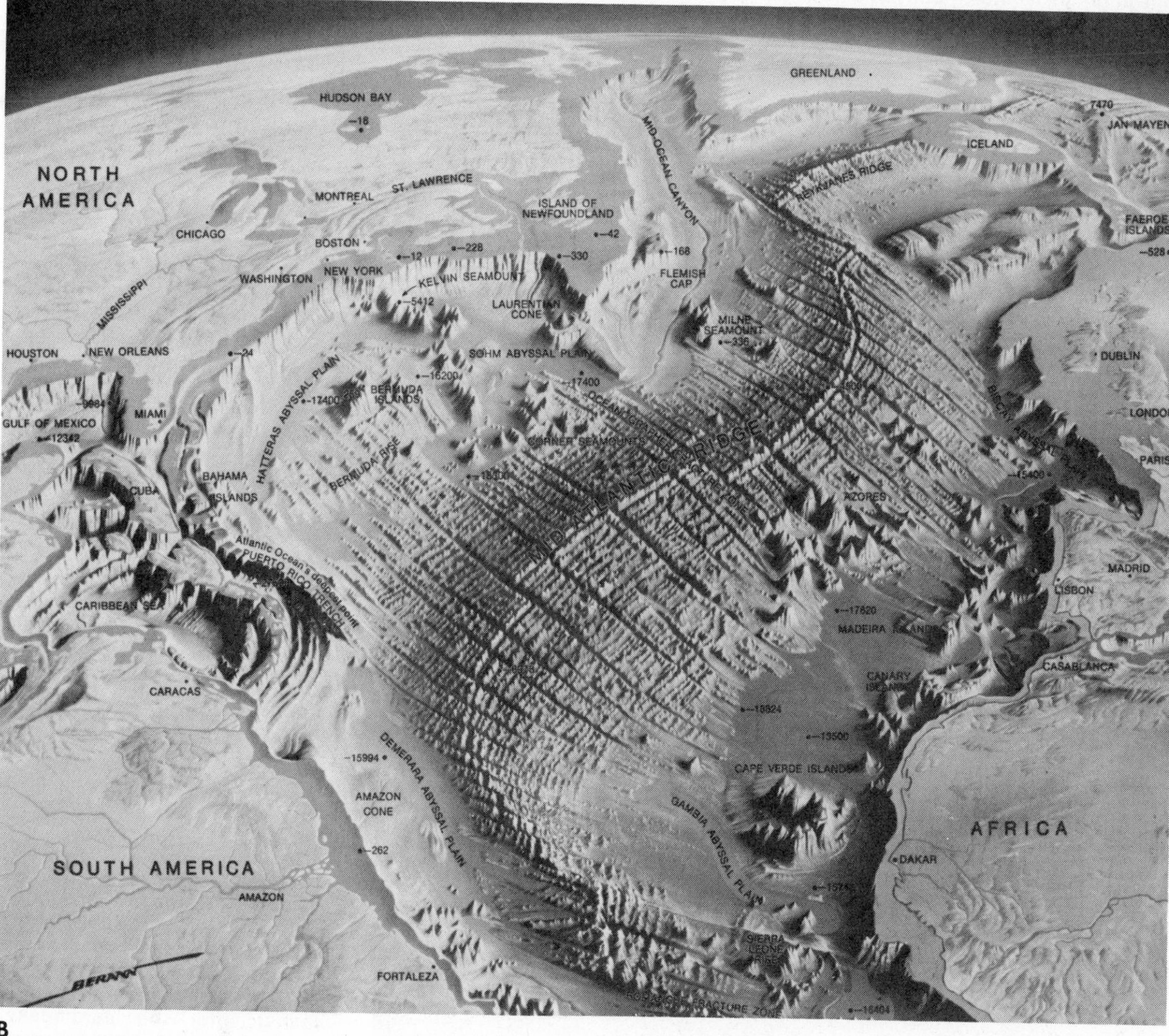

B

yons is erosion by ***turbidity currents***. Turbidity currents are dense fast-flowing bodies of sediment-laden water. Their high density results from the large quantities of mud and sand kept in suspension by the high velocity of the

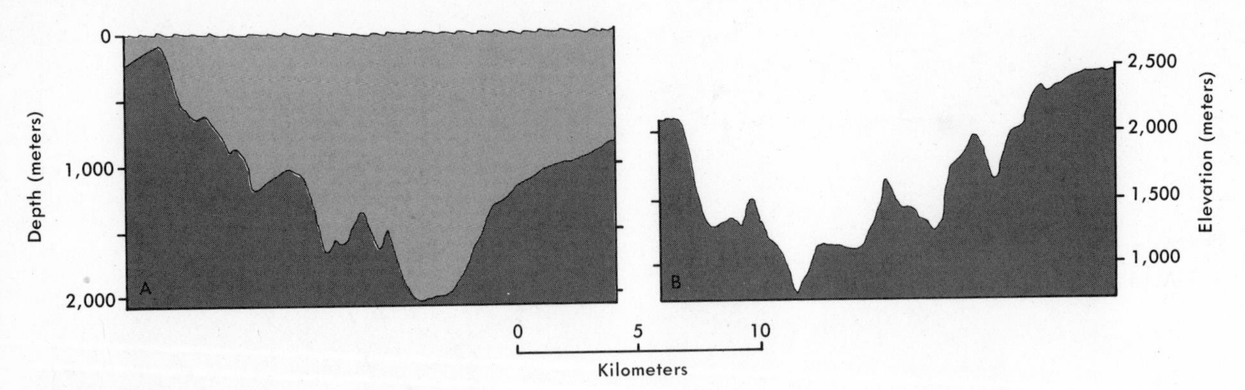

FIGURE 2.30 Comparison of the cross-sections of (*A*) the Monterey Canyon off California and (*B*) the Grand Canyon of the Colorado River. (After Shepard, 1973.)

current. Because of their greater density, the currents flow downslope along the bottom of the ocean, and the suspended particles, aided by the high speed of water, are believed to be capable of cutting canyons in hard rocks. It is likely that turbidity currents could maintain as well as deepen submarine canyons that were formed by the drowning of river mouths. Turbidity currents are easy to produce in laboratory experiments, but they have never been directly observed in the oceans. It is thought that earthquakes or other disturbances can dislodge large quantities of sediments resting precariously on many continental shelves and slopes. When the sediments are mixed with water as they tumble down the slopes, turbidity currents could result. The most compelling evi-

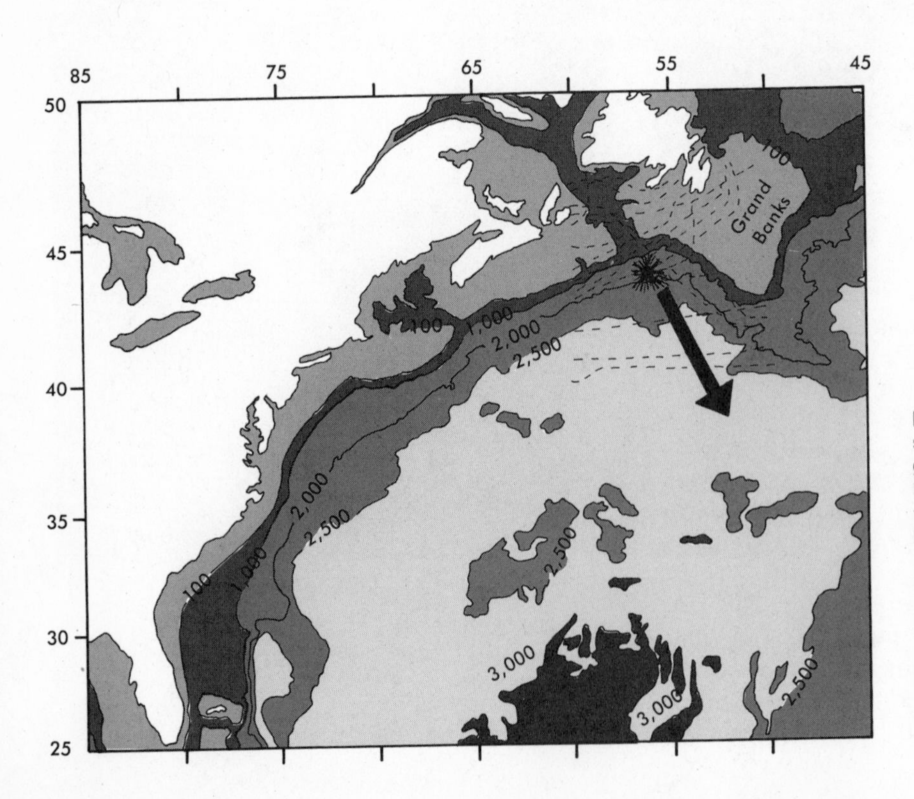

FIGURE 2.31 Breaking of the submarine telephone and telegraph cables after the 1929 Grand Banks earthquake near Newfoundland. The cables (dashed lines) were broken in a sequential manner after the earthquake, which is believed to have triggered a turbidity current, indicated by the arrow. (After Heezen and Ewing, 1952, and US Naval Oceanographic Office: Oceanographic Atlas of the North Atlantic Ocean, Publication 700, 1965.)

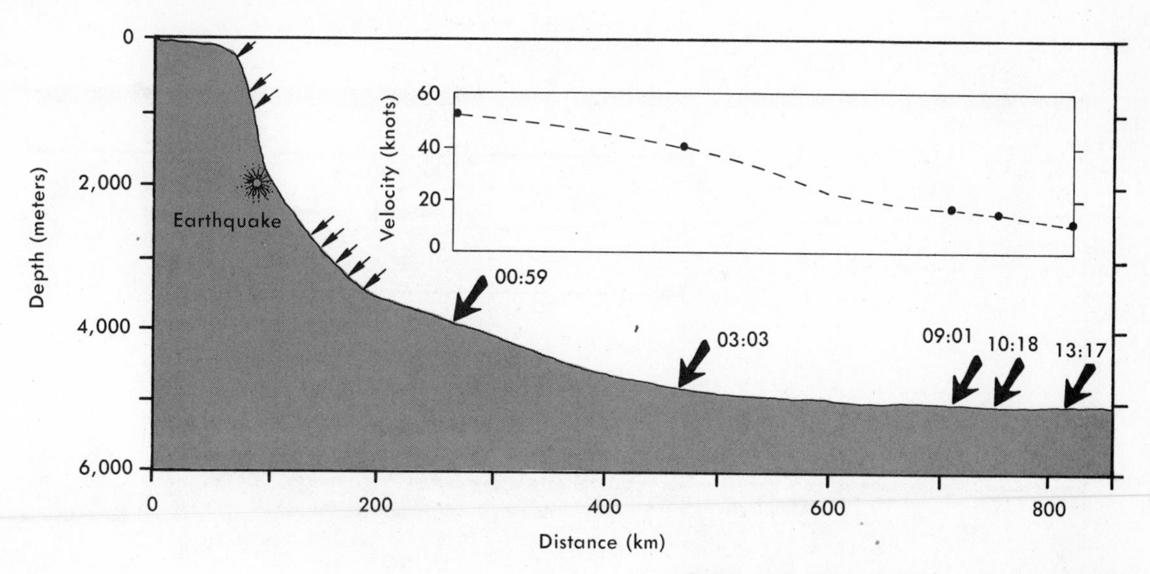

FIGURE 2.32 Cross-section of the Grand Banks area, showing times of cable breakage after the earthquake. Small arrows indicate cables broken at the time of the earthquake. Inset shows velocity of the turbidity current as interpreted from the times of cable breakage. (Modified after Heezen and Ewing, 1952.)

dence for turbidity currents comes from systematic suboceanic telephone cable breaks. In 1929, an earthquake in the Grand Banks area near Newfoundland resulted in the sequential breaking of several transatlantic telephone and telegraph cables (Fig. 2.31). The cables were believed to have been broken by a turbidity current caused by the earthquake. Analysis of the time of cable breakage (Fig. 2.32) indicates that the turbidity current itself traveled at speeds of up to about 55 knots (nautical miles per hour, 1 knot equals about 1.15 statute miles per hour). Drillings in the area confirmed that sand and other shallow water sediments were carried by the turbidity current more than 500 km into the deep ocean. Similar events have occurred in other oceans at other times, for which turbidity currents provide a plausible explanation.

ABYSSAL HILLS AND ABYSSAL PLAINS

Large portions of the deep sea, especially in the Pacific, are characterized by numerous small extinct volcanoes (usually only a fraction of a kilometer in height), known as ***abyssal hills***. In contrast, other regions of the ocean floor may be almost featureless or flat. These are called ***abyssal plains***. Seismic profiling studies indicate that abyssal plains are produced by the burial of abyssal hills by sediments, including sediment brought by turbidity currents. Biological sediments, if available in large quantities, may be sufficient to bury abyssal hills and produce abyssal plains. Obviously, abyssal hills exist only in areas where insufficient sediment has been available to cover them. Figure 2.33 illustrates some abyssal hills and abyssal plains.

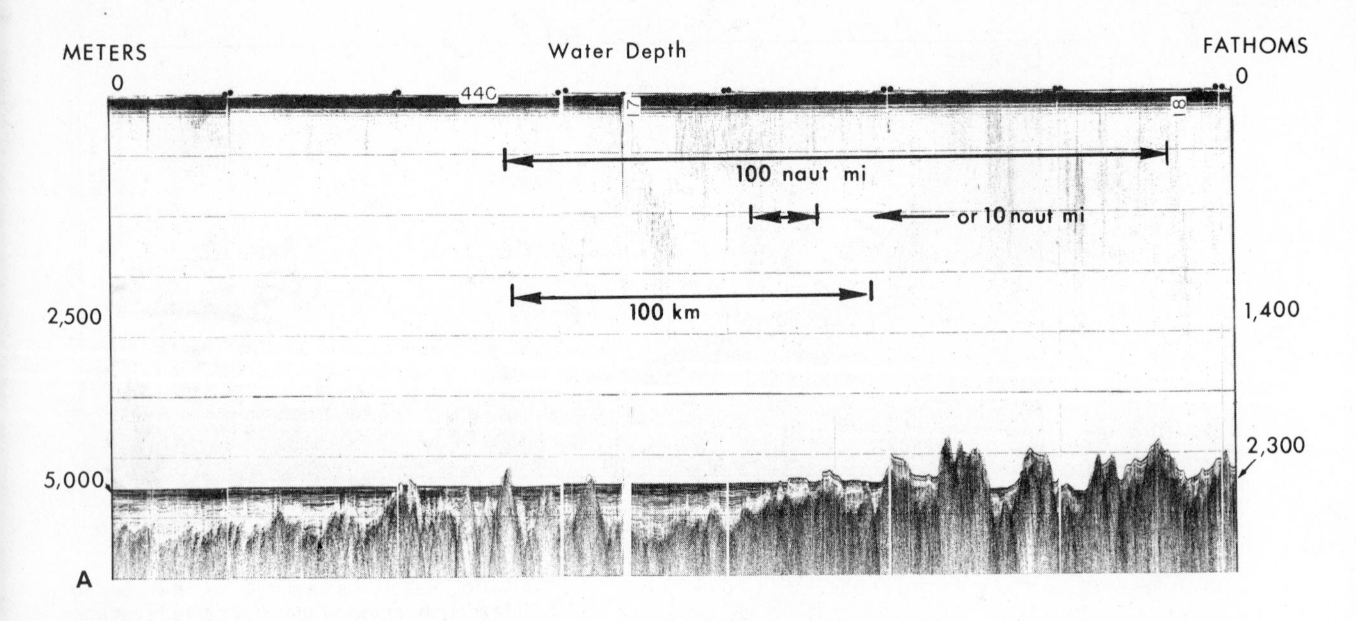

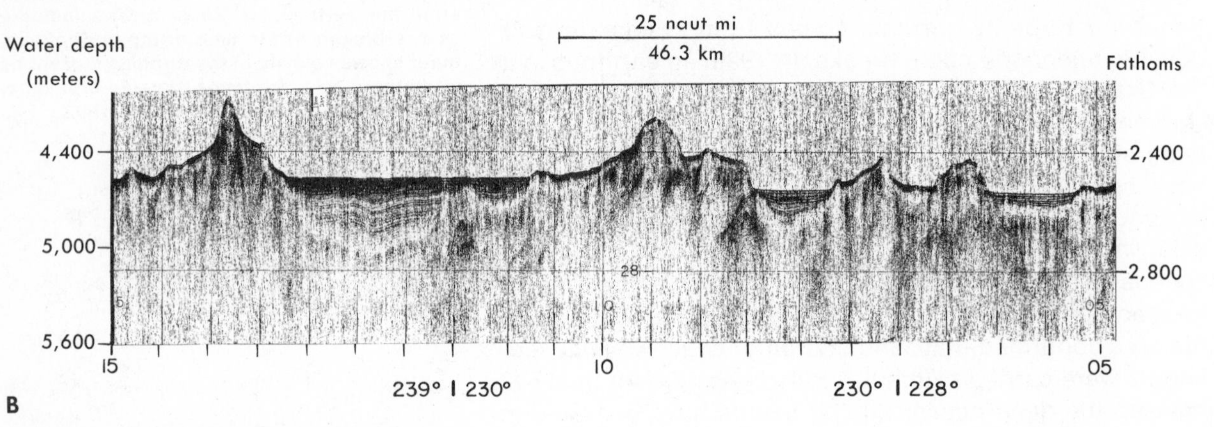

FIGURE 2.33 Continuous seismic profiles showing abyssal hills and abyssal plains along the edge of the Mid-Atlantic Ridge, in water about 5 nm deep. Note the burial of abyssal hills to form abyssal plains. (*A* courtesy of Lamont-Doherty Geological Observatory; *B* From Hayes and Pimm, 1972.)

MID-OCEAN RIDGES AND OCEANIC TRENCHES

Perhaps the most striking features of the ocean floor are the mid-ocean ridges (Fig. 2.34). They form a nearly continuous volcanic mountain range running through all the oceans, although only the Mid-Atlantic Ridge is located near the center of an ocean. The axes of mid-ocean ridges are sometimes characterized by longitudinal ***rift valleys***, whose depths may exceed that of the nearby ocean floor (Fig. 2.35). Mid-ocean ridges may have widths of as much as 3,000 km and may rise to more than 2 km above the surrounding sea floor. The ridges may rise above the sea surface to form islands such as the Azores, Tristan da Cunha, or Iceland. As seen earlier, the rocks along the axes of the mid-ocean ridges are youngest, and ages of rocks increase progressively away from them, as evidenced by drillings and magnetic measurements. The

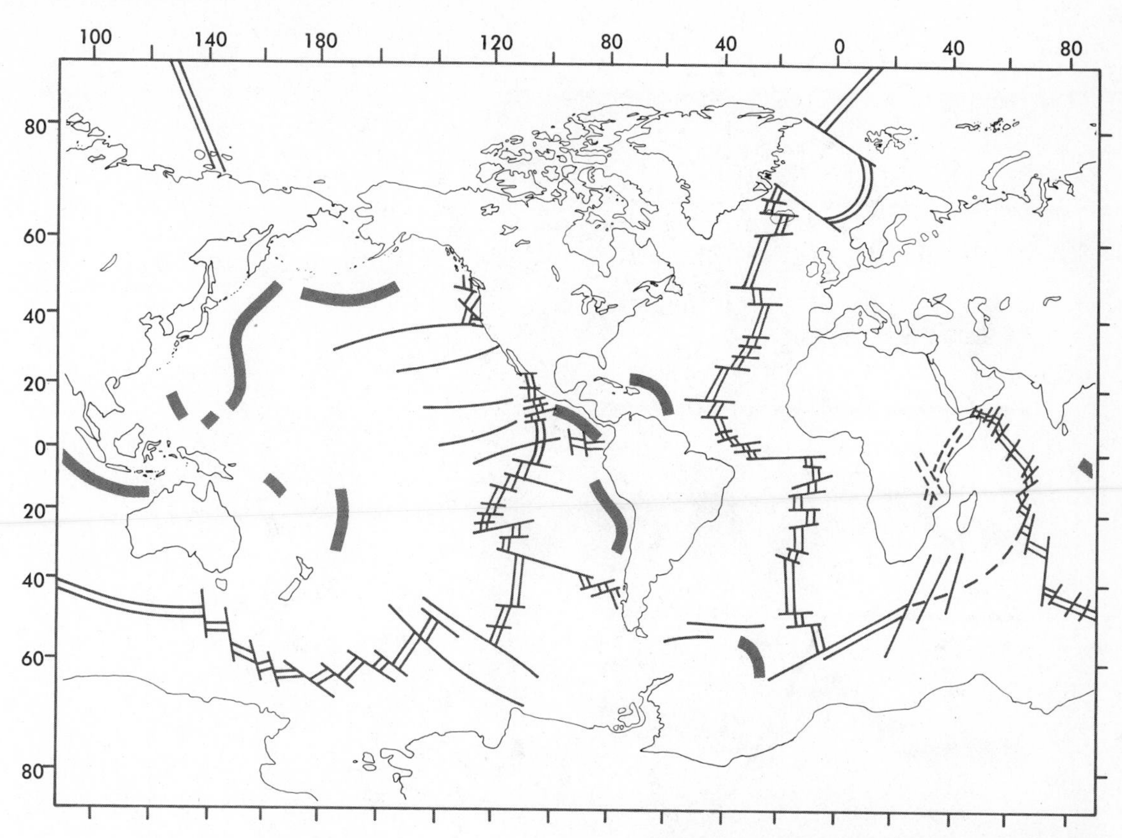

FIGURE 2.34 Locations of mid-ocean ridges and oceanic trenches. Heavy lines indicate trenches, and double lines indicate ridges. Note the fracture zones offsetting the ridges. The double-dashed line in Africa represents the East African Rift. (After Isacks et al., 1968, and Le Pichon, 1968.)

mid-ocean ridges are broken and offset by numerous fractures, and both ridges and offset regions (between rift valleys) are characterized by frequent shallow earthquakes (see Figs. 2.17 and 2.18).

The ***oceanic trenches*** (Figs. 2.34 and 2.35*b*) represent the deepest portions of the oceans (the greatest depth, the Challenger Deep, is in the Marianas Trench and is about 11,500 meters—deeper than Mount Everest is tall). They are never found in the centers of the oceans; rather they are located near and parallel to continents and island chains. Most of the trenches are located around the edges of the Pacific Ocean. Like the mid-ocean ridges, the trenches are characterized by volcanoes and earthquakes. However, unlike those of the mid-ocean ridges, trench earthquakes are among the deepest known (see Fig. 2.18). As discussed earlier, the origin of mid-ocean ridges and trenches is a vital link in our understanding of plate tectonics.

SEAMOUNTS, ATOLLS, AND GUYOTS

Seamounts (undersea volcanic peaks) are present in all oceans but are most common in the Pacific (Fig. 2.36).

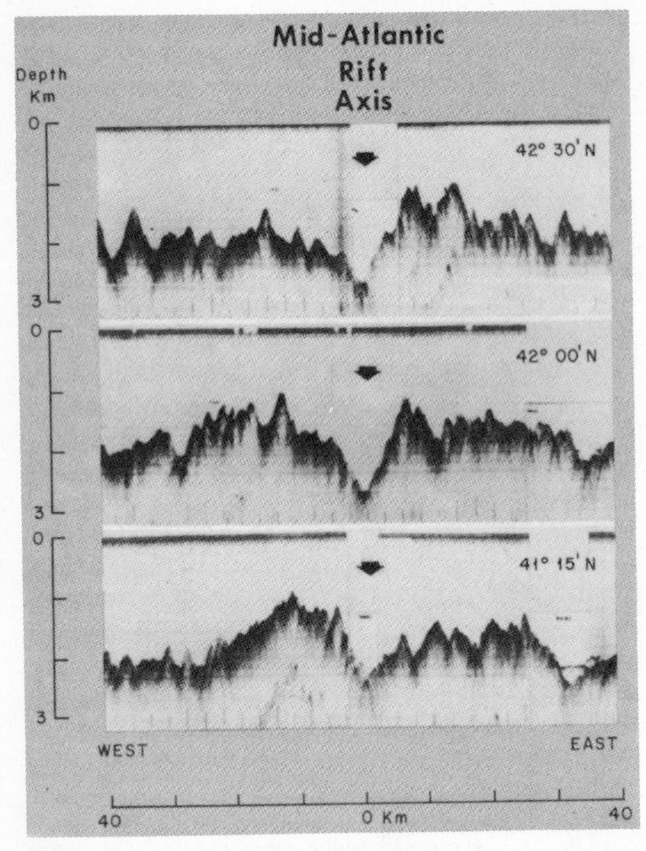

A

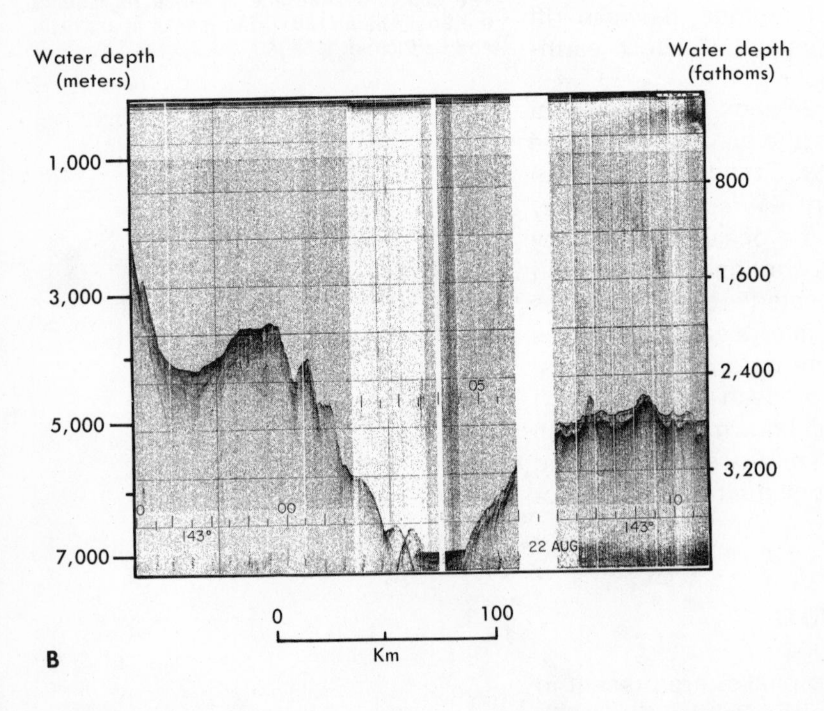

B

FIGURE 2.35 *A*, Continuous seismic profile across the rift valley of the Mid-Atlantic Ridge. *B*, A similar profile across the Aleutian Trench. (*A* from National Academy of Science, 1972; *B* courtesy of Lamont-Doherty Geological Observatory R/V, Conrad Cruise # 12, 1969.)

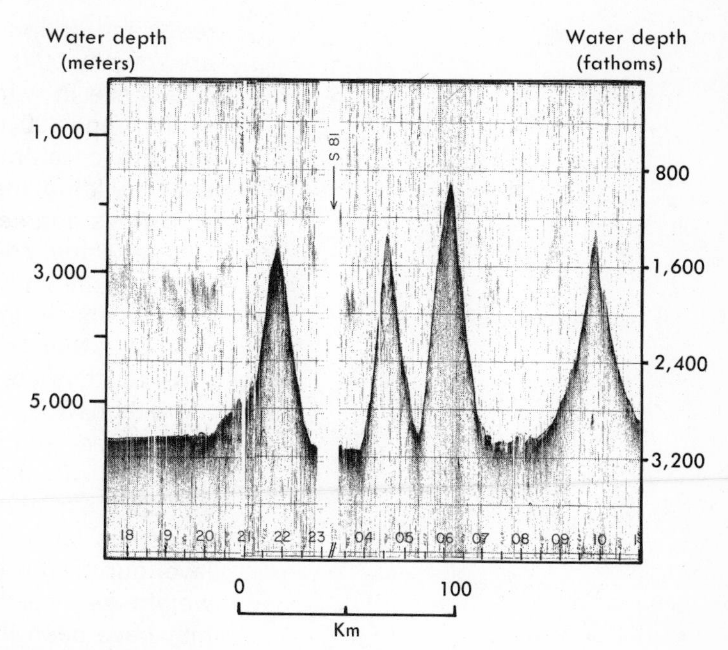

FIGURE 2.36 Continuous seismic profile across seamounts near the Mid-Atlantic Ridge. (From Hayes and Pimm, 1972.)

Although probably formed in the same manner, they are larger than abyssal hills, reaching heights of more than one kilometer. Some, like the Hawaiian Islands, are above sea level. Many seamounts, including the Hawaiian Islands, are found along chains, with the oldest at one end and the youngest at the other end.

Atolls (Fig. 2.37) are ringlike islands made up of coral

FIGURE 2.37 *A–C*, Origin of an atoll resulting from subsidence of a volcano and the upward growth of a coral reef. *D*, Formation of a guyot by the rapid subsidence of an atoll. (After Shepard, 1973.)

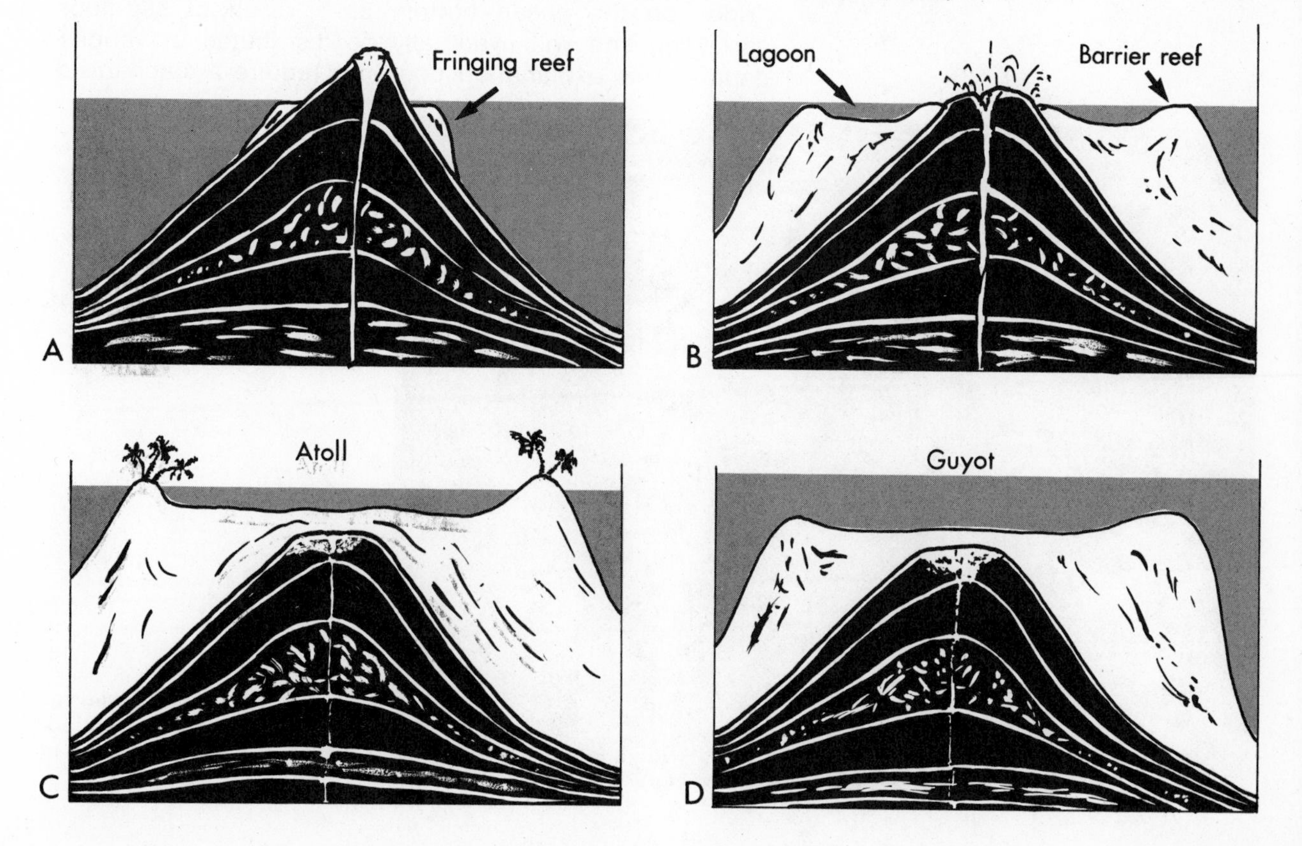

reefs, and drillings into them have indicated that the reefs are actually built over extinct volcanoes. The reef-building corals live in warm tropical waters (above 20° C), mostly in the upper 20 meters of the oceans (see also Chapter 10). If a coral-infested volcano subsides, the corals will keep building the reef upward. Atolls result when entire volcanoes submerge below sea level (Fig. 2.38). In 1842, Charles Darwin developed this theory of the origin of atolls, and it is still an accepted explanation.

Guyots are flat-topped seamounts. Although evidence indicates that the flat surfaces are produced by wave erosion, they are found at various depths, and some are more than two kilometers deep. Apparently, the guyots were submerged after the flattened surface was produced. Lowering of sea level during the last glaciation is not a sufficient explanation for most guyots, as their surfaces are at depths far greater than the postulated lowering of sea level during the glaciation. Subsidence due to their own weight may be the answer in some cases. Other guyots may have been formed when an atoll subsided so rapidly that the corals were killed (Fig. 2.37*d*).

Plate tectonics can explain the origin and distribution of many of the features discussed above. For instance, we have seen that the top surfaces of guyots are found at various depths. Because ocean depths generally increase away from the mid-ocean ridges, guyots which were formed near the mid-ocean ridges can be visualized to "ride" on the ocean bottom as a result of sea-floor spreading and will eventually be distributed at various depths. This explanation does not require a mechanism

FIGURE 2.38 Atolls of the Tuamoto Archipelago in the Central South Pacific. (Courtesy of NASA.)

for their sinking, and a similar theory can be extended to the origin of many atolls.

Volcanoes created near the mid-ocean ridges should move away owing to sea-floor spreading and should become inactive, and new ones should be created in their place. The chainlike arrangement of many seamounts and islands, with the youngest and active ones near the ridges and the oldest and inactive ones farther away, may be explained by sea-floor spreading. Unfortunately, not all such features follow this pattern. For example, the active volcanoes of the Hawaiian Islands are far away from the ridge in the Pacific, constituting a serious weakness in the theories of sea-floor spreading and plate tectonics. The formation of many such volcanoes may be explained in terms of ***hot spots,*** huge vertical pipelike regions within the mantle, where the materials are believed to be hotter, perhaps molten or partially molten, than in surrounding regions. Several hot spots are believed to exist within the mantle, one of them near the Hawaiian Islands. As the lithosphere moves on top of the hot spot, volcanoes are produced, and as sea-floor spreading continues, these are moved away from the hot spots, and new volcanoes may be produced in their places. Thus, the Hawaiian Islands will cease to be active, and new active volcanoes will be produced near the hot spot. Hot spots may be similar to mid-ocean ridges except that they are localized rather than extended.

OCEAN SEDIMENTS

Sediments are comprised of particles that settle onto the sea floor. Many sediments are derived from land and are brought to the oceans by streams, wind, and glaciers. Others are remains of organisms that live in the oceans, and a small fraction is even extra terrestrial in origin. Sediments are probably the most economically important feature of the sea floor. Ancient sediments are the source of petroleum and various mineral resources such as chalk, which is produced by marine organisms. Other mineral resources such as gold and diamonds are found on the continental shelves off land areas rich in these precious minerals. ***Manganese nodules,*** another resource of the sea, are just beginning to be mined. Their origin is poorly understood, but they are formed by successive coatings of manganese-rich minerals precipitated around objects such as rock particles and shark teeth.

Sediments provide important clues to the origin and history of the oceans. Fossils composing or found in sediments can be used to determine the age of the sediments, which in turn may tell the age of the ocean floor itself.

They also preserve a detailed record of past climatic and chemical history of seawater, a record which is just beginning to be investigated in detail.

Classification of Sediments

Ocean sediments may be classified in various ways, based on their origin (source), size, chemical composition, or place of deposition. Figure 2.39 shows the general distribution of sediments on the ocean floor. The following is a simple classification of ocean sediments:

(1) Terrigenous (derived from land)
(2) Pelagic (formed in or over the sea)
  (a) Biological (ooze)
  (b) Inorganic
    (i) Red clay
    (ii) Chemical precipitates

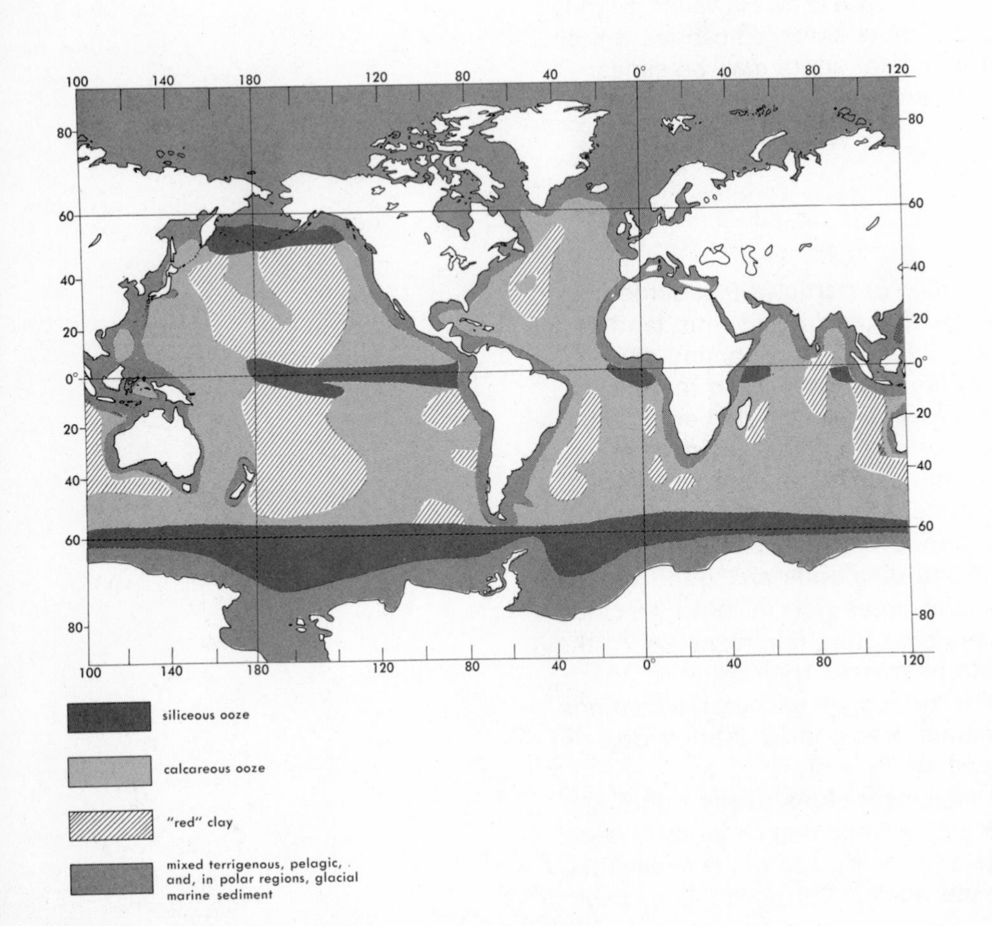

FIGURE 2.39 Distribution of sea-floor sediment. (After Heezen and Hollister, 1971.)

TERRIGENOUS SEDIMENT

Terrigenous or land-derived sediments are mostly the result of erosion, entering the sea via rivers, wind, or glaciers. Consequently, these deposits tend to be most abundant near shore, especially near major rivers. Land-derived sediments consist basically of the same materials found near the surface of the earth. They are mostly sand (1/16 to 2 mm in size, usually composed of quartz or feldspar), silt (1/16 down to 1/256 mm, same composition as sand), and clay (less than 1/256 mm, chiefly hydrous aluminum silicates, usually derived from the weathering of igneous rock). In addition, terrigenous sediments include airborne dust and volcanic ash. For example, airborne dust from the 1973 drought in north Africa was detected as far away as the Caribbean area. Volcanic ash may dominate the sediment near volcanic islands.

Because heavy (usually larger) particles sink faster than light ones, the heavier materials tend to settle to the bottom first and therefore are closer to the source. Thus, in general, terrigenous materials are coarser near shore and finer away from the shore. Of course, currents can change these distributions. Turbidity currents are capable of transporting even larger sediments great distances, and the fine sediments on the deep ocean floor may be disturbed by currents moving over them. A current flowing swiftly through a channel picks up the finer particles (leaving the coarser ones) and carries them until the current is slowed in less restricted water. Therefore, channels with swiftly flowing water tend to have coarser sediments or exposed rock, whereas sheltered coves frequently become mud flats, characterized by fine sediments. Sediments accumulate and erode more quickly along the beach than in any other part of the ocean environment, perhaps as much as two to three meters in a few months. (Beaches are discussed in greater detail in Chapter 8).

PELAGIC SEDIMENT

Pelagic sediments (those formed in or over the sea) are either biological or inorganic. Sediments containing 30 per cent or more organic remains are considered ***biological sediments***. They are very fine and are appropriately called ***oozes***. Some are made of calcium carbonate (calcareous) and others of a glassy form of hydrated silicon dioxide (siliceous).[3] Calcareous sediments are generally absent in depths greater than about 5,000 meters, owing to increased solubility of calcium carbonate under

[3]This type has the same composition as the gemstone opal.

FIGURE 2.40 Sediment-forming skeleta of minute marine organisms. *A*, Radiolaria; *B*, Coccoliths, broken diatom shells, and radiolarian "cages"; *C*, Chalk cliffs of Dover, composed of about 90 per cent coccoliths; *D*, Foraminifera. (*A* courtesy of Scripps Institution of Oceanography; *B* courtesy of Susumu Honjo, Woods Hole Oceanographic Institution; *C* courtesy of British Tourist Authority; *D* courtesy of Allan W. H. Bé, Lamont-Doherty Geological Observatory.)

high pressure and cold temperature. Figure 2.40 shows the skeletons of some organisms that constitute most of the biological sediments

All of the organisms in Table 2.1, except the pteropods, are single-celled and quite small. The plant members are microscopic. Radiolarians and foraminiferans are usually less than one millimeter in diameter. The pteropods are larger, about one centimeter in length, but their shells are thin and break into small pieces easily.

Table 2.1 MAJOR SOURCE AND CHEMICAL COMPOSITION OF BIOLOGICAL SEDIMENTS

SOURCE	*COMPOSITION* Siliceous (SiO_2)	Calcareous ($CaCO_3$)
Plants		
Diatoms	x	
Silicoflagellates	x	
Coccolithophores		x
Animals		
Radiolaria	x	
Foraminifera		x
Pteropods		x

Diatoms are among the most common of marine plants. They are so important as food for herbivores that they are sometimes called the "grass of the sea." Their glassy shells settle to the bottom as diatom ooze. In some areas of past productivity, they form deposits of ***diatomaceous earth*** that is hundreds of meters thick. This rock is sometimes used in filters such as those used to clarify beer, to strengthen concrete, and to give a flat finish to paint. Thick deposits of diatomaceous earth on land, such as at Lompoc, California, indicate that the area was once covered by the sea. Silicoflagellates are also silica-producing plants and have the same general distribution as diatoms, although they are much less abundant.

Coccolithophores may be considered plantlike because they contain chlorophyll. Coccolithophore sediments are composed of minute calcareous buttons (coccoliths) that once covered the organism's single-celled body and which volumetrically compose the bulk of most chalk sediments. The white cliffs of Dover, England, comprise more than 90 per cent coccoliths.

Radiolarians and foraminiferans, both protozoans (single-celled animals), are similar in that they feed by the use of thin projections or ***pseudopodia***, onto which dead organic matter (detritus) and bacteria get stuck. The animal then draws the pseudopodia into the cell and digests the food. Radiolarians have siliceous skeletons, whereas foraminifera generally have calcareous shells and sometimes produce thick deposits of limestone.

Pteropods are the only multicelled animals that occur in oozelike sediment. They are closely related to snails and have a calcareous shell. Pteropods drift in the water, suspended with the aid of winglike feet. They may be locally abundant, and they contribute, upon death, to calcareous sediment.

The ***inorganic pelagic sediments*** contain less than 30 per cent organic remains and are found in the deepest portions of the ocean floor (solution in the oceanic water column destroys most skeletons before they reach the bottom) and in areas of low biological productivity. They include ***red clays***, which in most cases are actually brown in color. These sediments are very fine, usually less than 0.05 mm in diameter. Their origin is not well understood. They probably include the finest fraction of wind-blown (terrigenous) materials plus meteorite dust. They may even contain very fine but unrecognizable biological sediments.

Another inorganic substance, although not a sediment in the strictest sense, is the manganese nodule. Although spectacular, manganese nodules form a very small fraction of ocean sediment volume. They are chemically precipitated from seawater as coatings of manga-

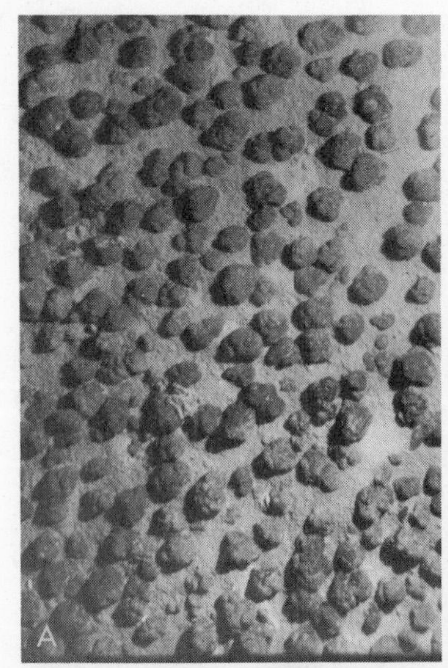

FIGURE 2.41 Manganese nodules. *A,* Nodules on the sea floor; *B,* The R. V. *Prospector,* a deep-sea mining ship; *C,* Emptying a dredgeload of manganese nodules on board the ship. (*A,* courtesy of Smithsonian Oceanographic Sorting Center; *B* and *C* Courtesy of Deep-Sea Ventures, Inc.)

nese minerals around objects such as rock particles and shark teeth. They are usually less than 25 cm in diameter (Fig. 2.41). Their exact mode of origin is not well understood. Associated with them are cobalt, nickel, and various other valuable substances. Many other minerals, not necessarily economically important, are also being precipitated on the sea floor. The legal aspects of mining of these resources in international waters are discussed in Chapter 12.

Sedimentation

Sediments are important to us because they tell much about the history of the oceans. Figure 2.42 illustrates a split core of ocean sediments. From the kinds of fossils

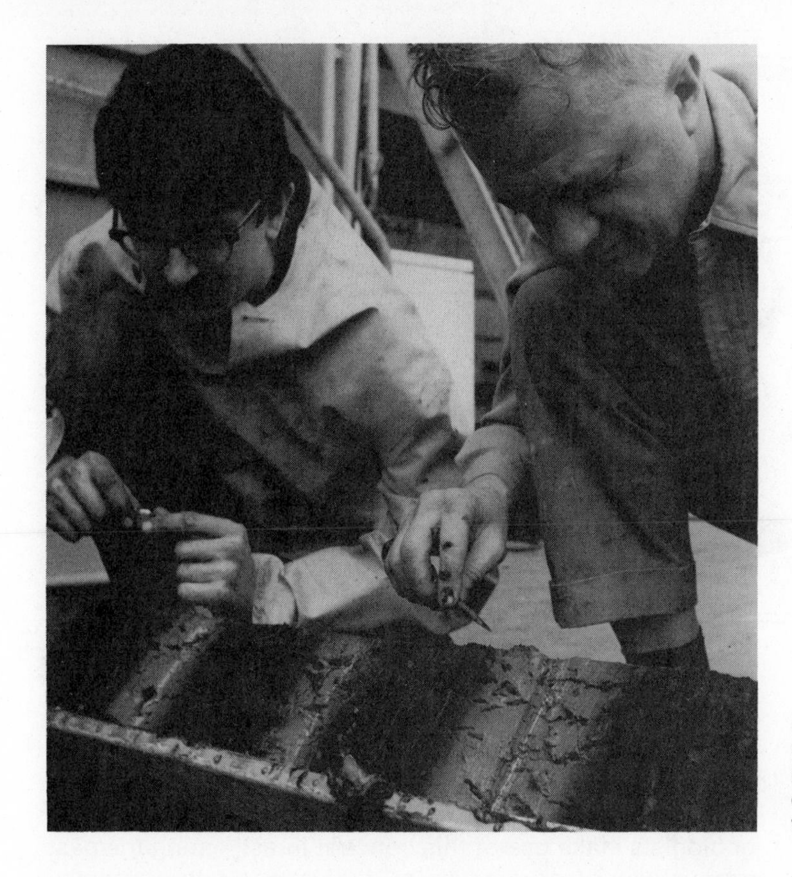

FIGURE 2.42 A sediment core being examined on board ship. (Photo by K. N. Sachs, courtesy of Woods Hole Oceanographic Institution.)

(skeletons) found in various layers, paleontologists can tell not only the age of the layer but also something about the environment (including climate) that existed when the sediments were deposited. It has also been found through the study of sediment cores that sediment thickness increases away from the axis of the mid-ocean ridges. This finding is consistent with the concept of sea-floor spreading, which states that the sea floor farther from the ridge is older than the sea floor near the ridge and has had more time to accumulate sediments. The oldest sediments in the oceans are younger than 200 m.y. and are found farther away from the mid-ocean ridges, implying that any older sediments which may have been present were carried deep into the oceanic trenches and recycled. Of course, many parts of the oceans are not bordered by trenches, which indicates that these ocean basins are younger than 200 m.y.

Sediments accumulate at rates ranging from less than 0.5 to more than 10 cm of thickness per 1,000 years. Many biological sediments tend to accumulate at a rate of about 2 cm per 1,000 years (Fig. 2.43), or 20 meters of sediments in a million years. However, higher rates of sedimentation may occur in areas of high biological productivity, and this rate of production of marine life may have varied

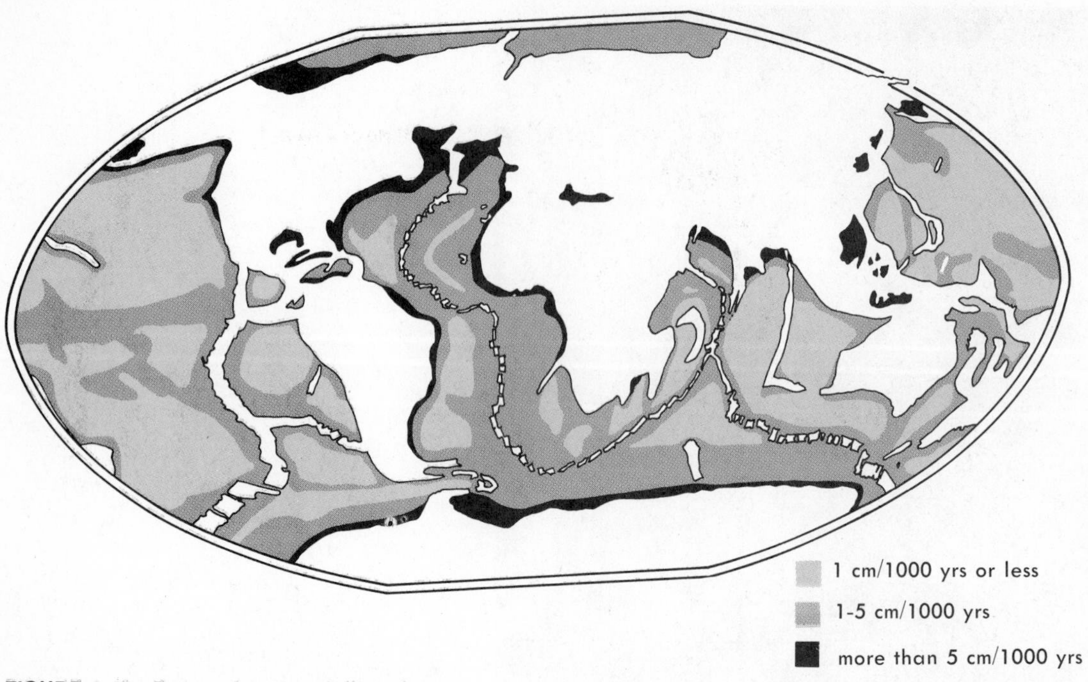

FIGURE 2.43 Rates of accumulation of deep sea marine sediment (in centimeters per 1,000 years). (After Heezen and Hollister, 1971.)

tremendously over periods of millions of years. Petroleum geologists make use of this concept in estimating the past productivity of an area as an indication of the potential accumulation of fossil fuels (oil and gas).

Oil comes mainly from animal protein, as indicated by the presence of foraminiferan remains at many oil sites. High concentrations of diatom shells in sedimentary rock may indicate the presence of natural gas, which is derived from animals and plants. The greatest potential for development of oil and gas reserves is found within sedimentary rocks with a low diversity of ooze species but great numbers of individual skeletons, as these two factors tend to indicate high productivity. Sedimentation in shallow areas with a high productivity and low species diversity will result in entrapment of the decaying organisms before their organic matter is completely broken down into inorganic salts by bacteria. A lack of oxygen near the sea floor also helps to retard decomposition of organisms. Many of the productive oil fields of today were environments such as mentioned above, millions of years ago. The production of oil indicates an inefficient utilization of organic matter by relatively few species. A greater diversity of species results in a more efficient utilization of organic matter and consequently in less production of petroleum. The energy we release by burning fossil fuels was initially trapped from sunlight by plants millions of years ago. By utilizing fossil fuels, we help to achieve the energy efficiency that ancient ecosystems were unable to effect.

SUGGESTED READINGS

Bullen, K. E. 1955 (September). The Interior of the Earth. *Scientific American.*

Calder, Nigel. 1972. *The Restless Earth.* New York, Viking/Compass.

Donn, William L. 1972. *The Earth: Our Physical Environment.* New York, Wiley.

Heezen, Bruce C. 1956 (August). The Origin of Submarine Canyons. *Scientific American.*

______, and Charles D. Hollister. 1971. *The Face of the Deep.* New York, Oxford University Press.

Shepard, Francis P. 1959. *The Earth Beneath the Sea.* Baltimore, Johns Hopkins Press.

______. 1973. *Submarine Geology,* 3rd ed. New York, Harper & Row.

Vine, Frederick J. 1966. Spreading of the Ocean Floor: New Evidence. *Science,* ***154***: 1405–1415.

Watkins, J. S., M. L. Bottino, and M. Morisawa. 1975. *Our Geological Environment.* Philadelphia, W. B. Saunders Co.

Wegener, Alfred. 1966. *Origin of Continents and Oceans.* New York, Dover Publications. (Translation of the original German edition of 1929.)

A Nansen bottle, used to collect water samples from the depths of the sea. (Photo courtesy of University of Rhode Island)

CHAPTER 3

Chemistry of the Oceans

Knowledge of chemistry of seawater is important for an understanding of factors that control the distribution and production of ocean life, movement of ocean water, and pollution. In addition to the resources below the sea floor, seawater itself contains a valuable supply of chemicals and, of course, fresh water which may be obtained through desalination processes.

ORIGIN OF SEAWATER

There are many hypotheses regarding the origin of seawater, and all are related to the origin of the atmosphere. These explanations are highly speculative, however, since a time scale of up to five billion years is involved.

One clue to the origin of seawater lies in the study of the distribution of certain inert (noble) gases (neon, argon, krypton, and xenon). The weight of one molecule of water is less than that of these gases. An argon atom, for example, weighs more than twice as much as a water molecule. These gases are found in very small proportions in the earth's atmosphere as compared to amounts present in the atmosphere of the sun and of other stars. If one assumes that the sun and the earth evolved from a similar source, it is implied that large quantities of these inert gases must have escaped from the earth's atmosphere. The ease with which a gas will leave the atmosphere depends on the weight of its molecules, the gravitational attraction of the earth, and the atmospheric temperature. The tremendous gravitational attraction of the sun is capable of retaining essentially all of its atmosphere, in spite of its very high temperature. The relatively low concentrations of inert gases in the earth's atmosphere imply that the early earth's atmosphere must have been very hot, causing a great loss of gases. Because the weight of a water molecule is less than that of these inert gases, most of the water vapor in the early atmosphere was probably also lost. Thus, the present waters of the oceans must have accumulated only after the atmosphere had cooled considerably.

Essentially all of the oceans' waters must have had

SALT AND THE AGE OF THE OCEANS

It has been suggested that the age of the oceans can be determined by knowing the total salts dissolved in the oceans and by knowing the amount of salts added to the oceans each year through river run-off. The total amount of sodium in the oceans, for example, is about 1.5×10^{19} kilograms, and the rivers of the world are presently estimated to supply about 1.7×10^{10} kilograms of sodium each year to the oceans. If all the sodium in the oceans were added through rivers at the present rate, the oceans are about 900 million years old. It is believed, however, that about half the sodium entering the oceans via rivers is recycled from the oceans as a result of the wind's picking up and transporting sea spray. This would double our estimate of the age of the oceans to 1800 m.y. Assuming that about half the amount of sodium in seawater has been trapped in sedimentary rocks, the age of the oceans can be extended to about 2700 m.y. (2.7 billion years), a very reasonable figure when one considers the many inaccuracies and uncertainties inherent in the method.

their origin within the earth,[1] coming to the surface via hot springs and volcanoes and being released as rocks crystallized from a molten state. Because this process is continuing today, in the long run (hundreds of millions of years), the volume of the oceans should continue to increase. All of the waters of the oceans (1.4 billion cubic kilometers) could have been derived in this manner at an average rate of only about 1/2 cubic kilometer of water per year during the last three billion years.

How and in what form the dissolved materials in seawater were added to the oceans is not well understood. Most of the dissolved materials were probably initially derived from within the earth through volcanic and hot spring emanations as well as from the weathering (decomposition) of igneous rocks. For example, most of the sodium in seawater could easily be accounted for by rock weathering. On the other hand, the chlorine in seawater could be derived from volcanoes and hot springs. Sodium and chlorine make up about 86 per cent of all the dissolved materials in seawater.

Has the salt concentration of the oceans remained much the same as it is today, or has it been drastically different during past geological time? One can only speculate about this question. A precise answer would require knowledge of the rate of increase of the ocean volume, the rate of chemical addition through volcanic and hot spring activities, the rate of rock weathering, and the rate of removal of these substances from seawater by deposition. Rough estimates for some of these factors can be made, based on the study of ancient rocks and from a knowledge of the physical and chemical properties of water. For example, no matter what its history might be, we are fairly certain that the concentration of sea salts could never have been double that of today's oceans, because direct precipitation of salts would occur before such a concentration were reached. This suggests that the saltiness of the oceans probably was not drastically different than it is today, throughout all but its earliest history.

Although the oldest known salt deposit on earth (proof of a saline ocean), is only about 800 m.y., we know that at least slightly salty oceans probably existed as far back as three billion years ago, because fossils of lime-secreting algae of that age have been found. Lime ($CaCO_3$) is one of the salts of today's ocean. However, this does not indicate the actual degree of saltiness of the ancient oceans. This problem remains a mystery.

---

[1]It is estimated that the present atmosphere of the earth could hold only a fraction of 1 per cent of the total volume of the oceans; also, a great deal of water apparently escaped before the earth cooled.

ORIGIN OF LIFE

The early (more than four billion years ago) atmosphere of the earth probably contained no free oxygen, as that gas would have been consumed in the oxidation of exposed materials such as iron. It is thought that most of the gases present came from volcanic activity and included methane (CH_4), ammonia (NH_3), water (H_2O), nitrogen (N_2), and some carbon dioxide (CO_2) and carbon monoxide (CO). It is known that a mixture of hydrogen, methane, ammonia, and water vapor, when exposed to electric sparks, will form amino acids, the building blocks of proteins. Other organic molecules have also been synthesized from simple gases in the presence of sparks, ultrasonics, light, heat, and ultraviolet radiation. These include many of the components of living organisms, such as sugars, proteins, and even the constituents of DNA (deoxyribonucleic acid) and RNA (ribonucleic acid). DNA and RNA are the chemical carriers of the genetic information that is necessary for life.

The early atmosphere, subjected to high-energy ultraviolet radiation, heat, and possibly lightning, would have produced many organic compounds that dissolved in the water collecting at the surface of the earth. The first living organisms were certainly very simple, dropletlike affairs but were capable of self-duplication (probably using RNA as a genetic material). The sea probably was like a dilute salty broth, with the organic molecules serving as food for the first anaerobic bacterialike organisms.

As organic food became relatively scarce, there was an environmental pressure toward evolution of the process of photosynthesis, including the production of oxygen. The next stage, perhaps, was the evolution of animals which could eat the photosynthetic plants.

It has been estimated that life has existed on earth for as much as four billion years and photosynthetic processes about 2 to 3.5 billion years. The development of the sexual reproduction process may have occurred as many as 1 to 2 billion years ago. As oxygen continued to accumulate in the atmosphere, it finally produced a high-atmosphere ozone (O_3) layer, which shielded the earth from much ultraviolet radiation, the very radiation that was probably responsible for creating the first life. The reduction of ultraviolet light intensity eliminated the need for shielding of organisms by seawater (which permits only very limited penetration of ultraviolet light) and eventually allowed the invasion of land by living organisms. However, the oldest known marine organisms (three billion years) are 6 to 7 times older than the oldest known land organisms (400 to 500 m.y.), suggesting that colonization of the land had to await the evolution of more complex organisms.

COMPOSITION OF SEAWATER

Seawater is only about 96.5 per cent water; the rest (3.5 per cent) consists of dissolved solids such as common salt (sodium chloride) and relatively small amounts of dissolved gases such as oxygen. Actual concentrations of dissolved substances, however, may vary from place to place, even within the same ocean.

Dissolved Solids

Water dissolves everything, at least in minute amounts. However, because some substances are more soluble than others, certain elements are more abundant than others in the sea. For example, sodium chloride is very soluble, calcium carbonate (limestone, clam shells, and so forth) is much less soluble, and silicon dioxide (as in quartz, and glass) is almost insoluble.

Most of the dissolved material in seawater is composed of relatively few elements. Table 3.1 shows that about 95 per cent of dissolved solids are made up of only six ***major elements*** (chlorine, sodium, magnesium, sulfur, calcium, and potassium). In fact, 86 per cent of all dissolved salts consist of only two elements, sodium and chlorine. While all of the major elements are important to marine life, sodium chloride and magnesium are economically important as well. For convenience, concentrations are reported in parts per thousand by weight (grams per kilogram or ‰) of seawater. Note that the major elements are found in concentrations greater than 0.3‰. Although actual concentrations of the major elements vary from place to place, they are found in constant ratios.[2] For example, the ratio of magnesium to chlorine (Mg/Cl) is practically the same (0.067) in all oceans. This is due to the fact that the world oceans have been mixing with each other for millions of years.

Actually, some of the elements in seawater are bound to other elements in the form of complex ***ions*** (electrically charged atoms or groups of atoms). Such ions behave as if they were composed of single atoms. For example, sulfur is usually bound with oxygen to form the sulfate ($SO_4^{=}$) ion, which has two negative charges. Dissolved carbon adheres to hydrogen and oxygen to form the bicarbonate (HCO_3^{-}) ion, having a negative charge.

[2]A. Marcet (in 1819) and J. G. Forchammer (in 1865) reported that the relative proportions of elements in seawater vary only slightly, based on analysis of numerous water samples. In 1884, W. Dittmar came to the same conclusion based on analysis of seventy-seven water samples from various depths collected around the world during the *Challenger* Expedition. The ratios obtained by Dittmar are very close to those obtained today.

Table 3.1 ELEMENTAL COMPOSITION OF DISSOLVED SOLIDS IN SEAWATER*

ELEMENT		AMOUNT
Major Element		***In Parts Per Thousand (‰)***
Chlorine	(Cl)	19.35
Sodium	(Na)	10.76
Magnesium	(Mg)	1.29
Sulfur	(S)	0.86
Calcium	(Ca)	0.41
Potassium	(K)	0.39
Minor Element		***In Parts Per Million***
Bromine	(Br)	67
Carbon	(C)	27
Strontium	(Sr)	8
Boron	(B)	4
Silicon	(Si)	3
Fluorine	(F)	1
Important Trace Element		***In Parts Per Billion***
Nitrogen	(N)	500
Lithium	(Li)	170
Rubidium	(Rb)	120
Phosphorus	(P)	70
Iodine	(I)	60
Iron	(Fe)	10
Zinc	(Zn)	10
Molybdenum	(Mb)	10
Copper	(Cu)	3
Uranium	(U)	3
Cobalt	(Co)	0.1
Mercury	(Hg)	0.1
Silver	(Ag)	0.04
Gold	(Au)	0.004

*Assuming a salt concentration of about 35.12 ‰ and excluding hydrogen and oxygen, which are mostly in the form of water. Only 26 elements are shown here, because most of the others are present in minute amounts. (After Goldberg, 1965; Culkin, 1965; and Fitzgerald et al., 1974.)

Table 3.2 shows seven major ionic constituents of seawater. Thus, hydrogen and oxygen can be chemically bound with other elements in forms other than water (H_2O). The contribution of hydrogen and oxygen toward the salt content of seawater may be about 4 per cent or more of the total, when they are bound with other elements.

The ***minor elements*** (bromine, carbon, strontium, boron, silicon, and fluorine) collectively constitute only about 0.3 per cent of dissolved solids and are generally found in concentrations of 1 to 65 parts per million (0.001 to 0.065‰). Carbon, bromine, and silicon, although present only in small amounts, are essential for the growth of marine plants. Carbon in the sea is mostly in the

Table 3.2 MAJOR CONSTITUENTS OF SEAWATER*

CONSTITUENT		*AMOUNT (‰)*	*CUMULATIVE PER CENT OF TOTAL*
Chloride	(Cl^-)	19.35	55.1
Sodium	(Na^+)	10.76	85.7
Sulfate	($SO_4^=$)	2.71	93.4
Magnesium	(Mg^{++})	1.29	97.1
Calcium	(Ca^{++})	0.41	98.3
Potassium	(K^+)	0.39	99.4
Bicarbonate	(HCO_3^-)	0.14	99.8

*Assuming a total salt concentration of about 35.12 ‰ and taking into account the fact that sulfur and carbon are generally combined with other elements in the sea. (After Culkin, 1965.)

form of bicarbonate (see Table 3.2), but may also be present as carbonate ($CO_3^=$), carbon dioxide (CO_2), or carbonic acid (H_2CO_3). Silicon as silica (SiO_2) in the form of opal makes up the skeletons of some marine plants and animals, including certain sponges. Bromine, which may be extracted from seaweeds, seawater, and terrestrial salt deposits (formed in former seas that are now dry land), has a number of industrial uses such as in photography, and in the manufacture of certain drugs and detergent gasoline.

Dissolved solids that are found in concentrations of less than one part per million are termed ***trace elements***. These include ions of nitrogen such as nitrate (NO_3^-), nitrite (NO_2^-), and ammonia (NH_4^+); iron; copper; and iodine, all of which are essential for the growth of plants. Other elements such as mercury, silver, gold, and uranium are valuable, but they are present in concentrations too small to be extracted economically. Excessive amounts of some elements such as copper and mercury may be toxic to marine organisms or can be concentrated by certain forms of marine life and in turn be toxic to man.

Measurement of Salinity

Salinity refers to the total amount of dissolved solids in water and is usually measured in grams per kilogram (‰) of seawater. Ocean water generally varies from about 30 to 36‰, with an average of about 34.7‰ salinity. However, some briny areas of the Red Sea have salinities of as high as 270‰. Estuarine waters (where rivers enter bays and the sea) have a range of salinity from almost fresh water to about 30‰, depending on river run-off. The distribution of salinity and factors affecting it are discussed in detail in the next chapter.

Accurate determination of salinity is of great importance to biologists and chemists studying the ocean as well as to physical oceanographers, who are concerned with the relationship of salinity to ocean circulation. Patterns of salinity variations in the oceans may reveal the paths of deep ocean currents and of nutrients and the amount of river run-off. These are discussed in the next chapter and in Chapter 7.

A simple but highly inaccurate method of salinity determination is the evaporation of a known amount of seawater and the weighing of the resulting dry salt. This procedure is inaccurate because some of the salts escape into the atmosphere in bubbles and because any inorganic matter which was not actually in solution (suspended material, for instance) remains along with the dry salt. In addition, water itself may be bound with certain salts in the process of simple drying.

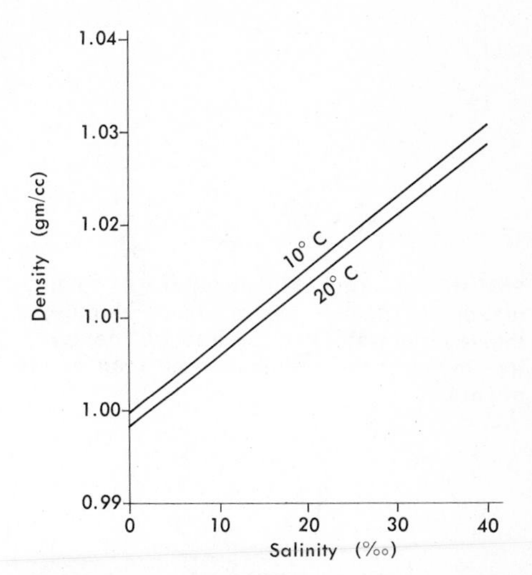

FIGURE 3.1 Relationship between salinity and density of seawater at 10° C and 20° C and atmospheric pressure.

Chemical analysis of seawater, although time-consuming, has been used to obtain an accurate estimate of salinity. As stated earlier, the ratios of major constitutents of seawater remain virtually constant. This enables one to calculate salinity by knowing the concentration of only one of the major constitutents. Before the advent of electronic measuring devices now in use, salinity was in fact calculated by measuring the concentration of a single ion. Since the chloride ion (Cl^-) is the most abundant dissolved constituent of seawater and is relatively easy to measure chemically, the chloride content or ***chlorinity***[3] (Cl‰) was most useful in estimating salinity, as expressed in the following equation:

$$S‰ = (1.80655) \times Cl‰$$

The density of seawater is directly related to its pressure, temperature, and salinity. Therefore, if the temperature and pressure are held constant, only salinity affects density (Fig. 3.1), and thus salinity may be estimated by measuring density. This procedure requires a simple hydrometer (similar to those used in testing batteries), a small weighted glass tube that floats higher in the water as salinity increases. Rough estimates of salinity may be achieved quickly in this way, but accurate estimates by this method are very time-consuming and consequently are not usually used in oceanographic work.

The index of refraction[4] of seawater is directly proportional to the salinity for a particular temperature and pressure (Fig. 3.2). Light is bent (refracted) as it enters a water

FIGURE 3.2 The effect of salinity on the refractive index of seawater at 10° C and 20° C.

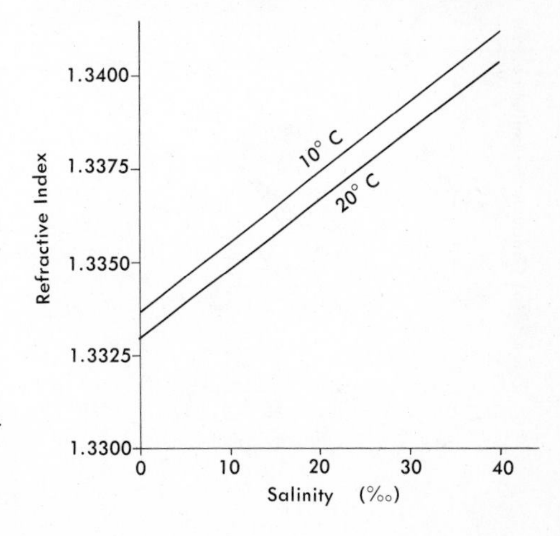

[3]Chlorinity is now defined as the weight in grams of silver required to precipitate the chloride, bromide, fluoride, and iodide in 0.3285233 kg of seawater.

[4]The index of refraction of seawater is the ratio of the speed of light in vacuum to the speed of light in seawater.

sample, and this bending is measured using a refractometer (Fig. 3.3), which may be calibrated to give a direct reading of salinity.

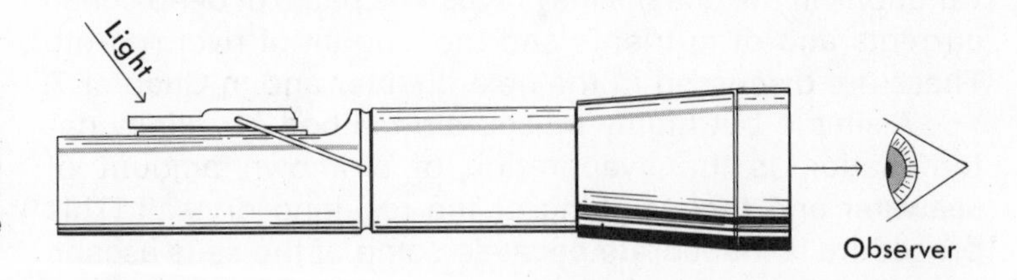

FIGURE 3.3 The refractometer, a salinity-measuring device. Light passes through a thin layer of water and is refracted. The refractive index or the salinity can be read by the observer.

The methods described above are either too costly or too inaccurate for widespread use; consequently, most salinity determinations made today are based on measurements of electrical conductivity of seawater. The conductivity of water increases with increases in salinity, temperature, and pressure. Therefore, if the temperature, pressure, and conductivity are known, salinity may be accurately and quickly estimated (Fig. 3.4).

FIGURE 3.4 The effect of salinity on the electrical conductivity of seawater at 10° C and 20° C, at atmospheric pressure.

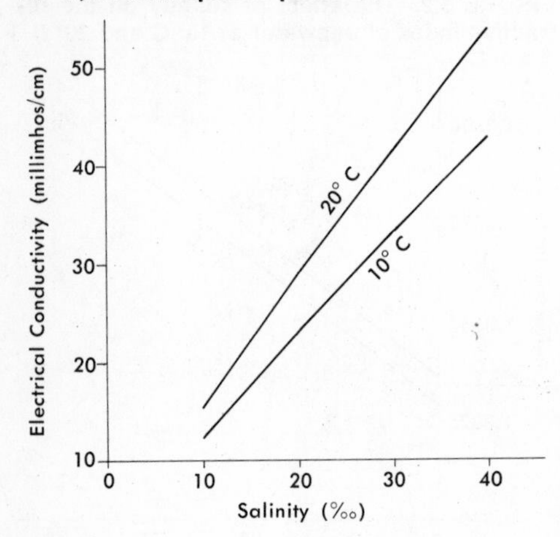

Dissolved Gases

Seawater also contains several dissolved gases, including oxygen, carbon dioxide, nitrogen, and hydrogen sulfide, and the inert gases helium, argon, neon, krypton, and xenon. The inert gases do not ordinarily combine with other elements and are not used by marine organisms. The other gases can be produced or consumed by organisms in addition to being dissolved from the atmosphere. The absolute concentration of each gas in seawater depends to a certain extent on its relative concentration in the atmosphere, its solubility, and the temperature and salinity of the water.

NITROGEN

The most abundant dissolved gas in the sea is nitrogen (about 15 ppm). However, nitrogen as a gas is apparently of little importance to marine life except to a few nitrogen-fixing bacteria and plants. Most of the nitrogen essential to the growth of marine organisms is derived solely from nitrogen-containing salts (about 0.5 ppm in seawater) such as nitrates, which are so rare that they tend to limit the amount of life the ocean can support.

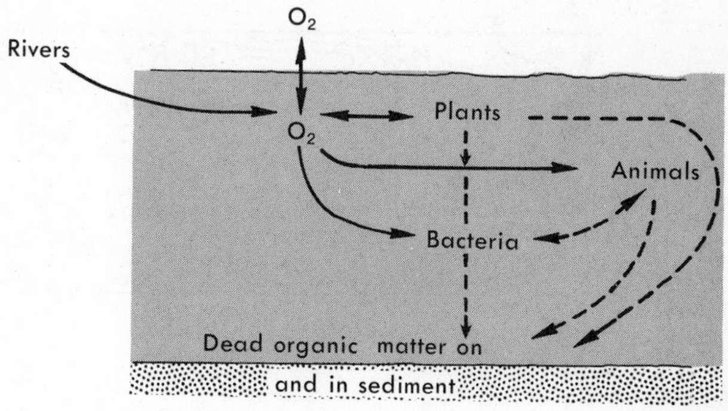

FIGURE 3.5 Oxygen cycle in the sea. Solid arrows indicate flow of elemental oxygen; dashed arrows indicate flow of oxygen in organic matter. Oxygen in CO_2 is supplied to the plants from the atmosphere and water. CO_2 is produced by bacteria, plants, and animals during respiration.

OXYGEN

Oxygen is produced by the following reaction as a result of the photosynthesis of organic matter by plants in the sea as well as on land:

$$CO_2 + H_2O + \text{light (in the presence of chlorophyll)} \longrightarrow \text{organic matter} + O_2$$

Oxygen also enters the sea via rivers and the atmosphere. It may leave the seawater by being consumed by marine organisms (animals, bacteria, and plants), as well as through loss to the atmosphere. All organisms, including plants, utilize oxygen during respiration, the process by which they release energy used for their growth and activity. In general, plants produce more oxygen than they consume by respiration, and the remainder is used by animals and bacteria. In contrast to photosynthesis, which requires sunlight, respiration can take place day and night. The reaction that takes place during respiration is shown below:

$$\text{Organic matter} + O_2 \longrightarrow CO_2 + H_2O + \text{energy}$$

Figure 3.5 illustrates the pathways of oxygen in the sea.

A typical vertical distribution of O_2 in the sea (excluding polar regions) is shown in Figure 3.6. Oxygen concentrations are generally at a maximum near the sea surface, as a result of intense photosynthetic activity of plants and the atmospheric exchange of oxygen. This corresponds to the wind-mixed, lighted surface layer. Beneath this layer is the oxygen minimum zone (at about 700 m in the figure), in which bacteria and animals consume oxygen by respiration, in the absence of light and plant life. In addition, the lowered oxygen concentrations at these depths is caused by decay of dead organic matter that has fallen from the productive surface layer. Curiously, at greater depths, the oxygen concentration tends to increase with depth. This effect is due to transport of oxygen by deep ocean cur-

FIGURE 3.6 Typical variation of oxygen (O_2) and total dissolved carbon (as CO_2) in the North Central Pacific Ocean.

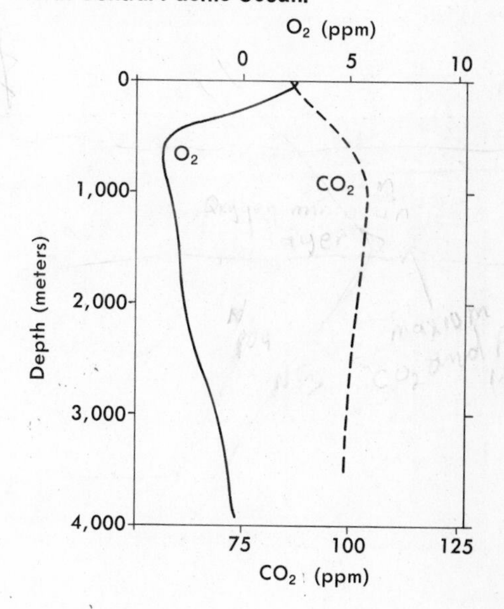

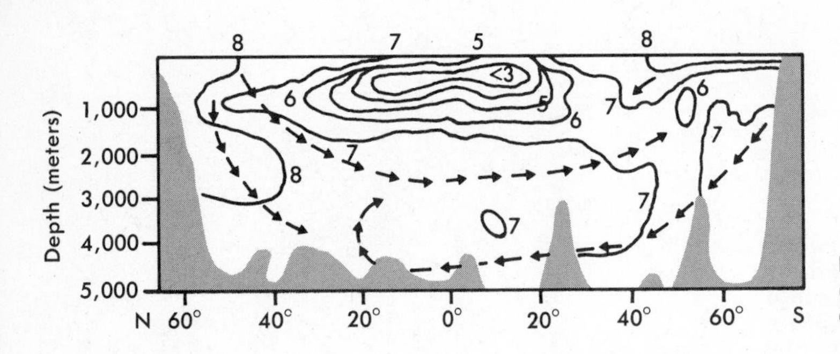

FIGURE 3.7 Vertical distribution of oxygen (in ppm) in the Atlantic Ocean. Arrows indicate currents. (Modified after Wattenberg, 1933.)

rents into these regions (Fig. 3.7) from subpolar areas, where cold dense oxygen-rich water sinks. (Oxygen is more soluble in cold water than in warm water.)

Carbon Dioxide

Carbon dioxide (CO_2) is released when plants, animals, and bacteria respire. Plants, however, consume it during photosynthesis and in general consume more carbon dioxide than they produce. In addition, carbon dioxide enters the sea from the atmosphere, and rivers carry it into the sea in the form of bicarbonates of sodium, calcium, and potassium, in the form of carbonates, and as carbonic acid. The bicarbonate and carbonate salts are produced by weathering of rocks on land. Carbonate is removed from the water by plants and animals which use it in the construction of their shells and skeletons. Some of these hard remains settle to the bottom and become part of the sediment. In cold deep ocean water (greater than about 5,000 m), the carbonates are more soluble than

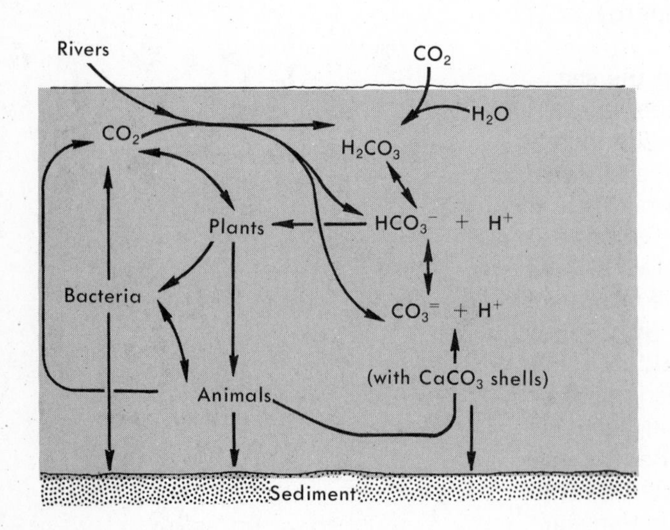

FIGURE 3.8 Carbon cycle in the sea. (H_2CO_3 = carbonic acid, HCO_3^- = bicarbonate, $CO_3^=$ = carbonate, CO_2 = carbon dioxide.)

near the surface, and they dissolve but are eventually recycled back to the surface waters. Consequently, carbonate sediments are restricted to warm, shallow water.

Figure 3.8 illustrates the pathways of the carbon dioxide system, discussed above. Notice that the carbon is cycled and recycled through the system. Actually, there is a net input of carbon via the rivers and atmosphere, which is balanced by the deposition of carbonate shells and skeleta onto the sea floor.

The concentration of total dissolved carbon dioxide (including the bicarbonates and carbonates) increases down to about 1,000 meters, corresponding to the oxygen minimum zone (see Fig. 3.6). The lower carbon dioxide concentration near the surface is mostly due to consumption by plants. The increase of dissolved carbon dioxide is related to the decomposition of dead organisms, which results in the conversion of particulate organic matter into nutrients and carbon dioxide.

Hydrogen Sulfide

Hydrogen sulfide (H_2S) is a gas that has an odor very much like rotten eggs. It is produced by anaerobic bacteria (bacteria that live in the absence of dissolved oxygen), which may obtain oxygen from sulfur compounds and ions, and produce hydrogen sulfide as waste during respiration. Other anaerobic bacteria use nitrate (NO_3^-) or nitrite (NO_2^-) as an oxygen source and may produce ammonia (NH_4^+) as a by-product instead of hydrogen sulfide. Anaerobic conditions are found in areas where free oxygen is consumed faster than it can be supplied, as in the deeper water of stagnant basins such as fiords and in the Black Sea, and in marine sediments such as mud flats. These sediments are characteristically black as a result of the conversion of hydrogen sulfide to other sulfur-containing compounds. When the black sediment comes in contact with oxygen (as when a shovelful of mud is turned over or when a marine animal burrows through the mud), oxygen becomes available to it and the black color slowly disappears as the sulfur compounds are consumed by aerobic bacteria (those that require oxygen).

In addition to the preceding rather localized conditions, hydrogen sulfide may be produced in the open ocean when excessive amounts of organic matter are decomposed, producing anaerobic conditions. Such a condition exists periodically along the Peruvian coast, in a phenomenon known as El Niño (see Chap. 7). During these times, changes in the current patterns result in warmer nutrient-poor water entering an area normally fed by cooler nutrient-rich water. These changes result in extensive fish kills. Decay of the dead fish uses up all the available oxygen, and anaerobic bacteria proliferate, causing hydrogen sulfide build-up. Sulfur compounds tend to coat buildings and ships, giving them a black appearance. Hence, El Niño is sometimes called the "Callao Painter," after a Peruvian port city.

NUTRIENT CYCLES IN THE SEA

In addition to contributing to the salinity of seawater, many of the chemical constituents of seawater are required as raw materials for the production of ocean life. All of the major constituents listed in Table 3.2 are required for plant growth and are used in the living tissues of organisms, as are many minor and trace elements, including nitrogen, phosphorus, iron, copper, and iodine. Some organisms, such as diatoms, silicoflagellates, radiolarians, and certain sponges, have skeleta containing silicon dioxide (opal); however, it is not required as a constituent of their tissues (see also Chap. 11).

Plants require nitrogen and phosphorus in rather large quantities. Living tissues contain about 100 atoms of carbon to about 15 atoms of nitrogen and 1 atom of phophorus. The ratio of nitrogen to phosphorus (N/P) by weight in living tissue is about 7 grams nitrogen to 1 gram phosphorus. This is nearly the same ratio as is found in the open sea, indicating that plants tend to absorb these

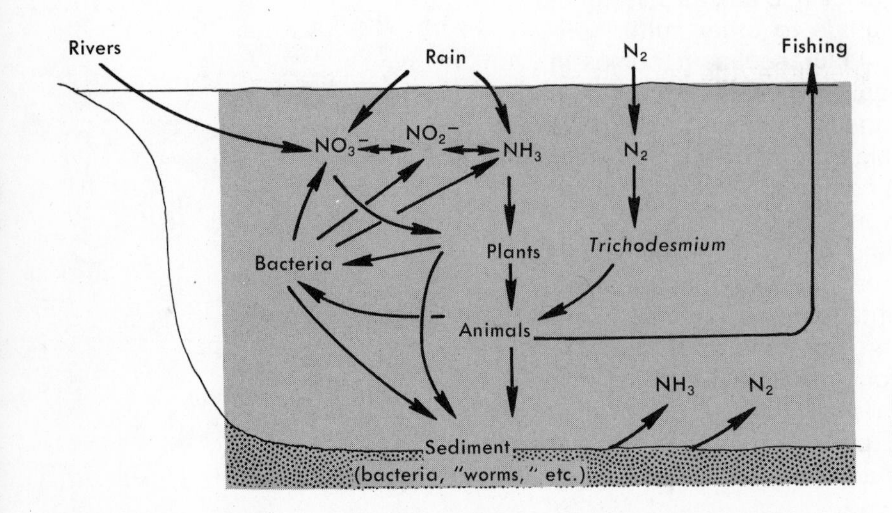

FIGURE 3.9 The nitrogen cycle in the sea. (NO_3^- = nitrate; NO_2^- = nitrite; and NH_3 = ammonia, which forms the ammonium ion NH_4^+.)

elements in the same relative proportions. The similarity between the N/P ratios of organic matter and of seawater may be due to evolutionary adaptation of plants to the chemical environment, or it may indicate a largely biological source of nitrogen and phosphorus in the sea. When organisms die and decompose, their nitrogen and phosphorus are released back into solution. Near shore and in estuaries, the N/P ratio may vary considerably, owing to differences in composition of nutrient-rich run-off and pollution. Pollution of the coastal ocean is discussed in detail in Chapter 8.

Nitrogen compounds that flow into the sea through river run-off and rain enter the nitrogen cycle (Fig. 3.9) through plants, mainly as the nitrate ion. Plants are consumed by animals and bacteria, which release nitrogen compounds back into the water, where they become available to the plants again. Various bacteria release nitrates, nitrites, and ammonia and thus are of great importance in regenerating these nutrients. In warm surface water, a blue-green alga, *Trichodesmium*, is important in converting dissolved nitrogen into organic matter in areas where other nitrogen sources have been depleted.

Not all nutrients are recycled within the sea. Some dead organisms settle onto the sea floor and are buried by rapidly thickening sediment, and some nitrogen compounds as well as other potential nutrients are lost from the sea. Some of this organic matter may eventually be converted to fossil fuels. In addition, some organic matter is carried from the sea as a result of fishing by birds and humans. If the total nutrient content of the sea is to remain stable, potential nutrients lost through sedimentation and fishing must be balanced by input from sources such as rivers and the atmosphere.

The tendency for dead organisms to sink from the well-lit surface water results in transport of nitrogen compounds into the deep sea, where a very small fraction is lost to sediments, and the rest remains until brought near the surface by ocean currents. Therefore, the concentration of nitrogen compounds, especially nitrates, usually increases with depth (Fig. 3.10), frequently forming a maximum at about 1,000 meters, due to decomposition of the dead organisms that have fallen from the surface waters. This nutrient maximum is correlated with the oxygen minimum (see Fig. 3.6), which is caused by bacterial metabolism. The nutrient depletion near the surface is most pronounced in tropical waters (Fig. 3.11), where surface heating results in density stratification and a lack of vertical mixing. These waters tend to be low in productivity as compared to polar water, where vertical mixing brings back nutrients to the surface.

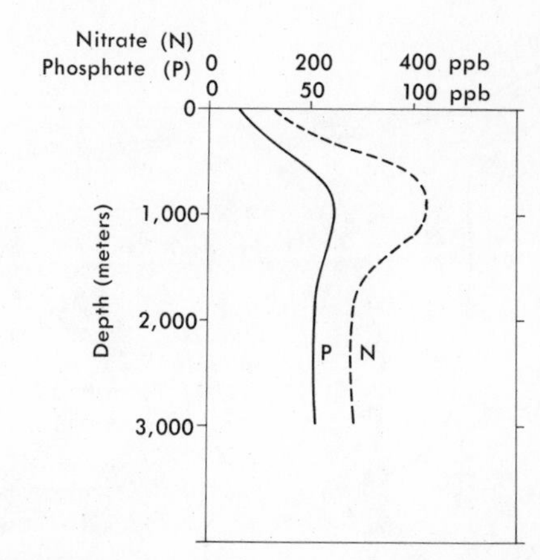

FIGURE 3.10 Typical variations in nitrate and phosphate with depth in the Atlantic Ocean (ppb = parts per million). (After Sverdrup et al., 1942.)

FIGURE 3.11 Vertical distribution of nitrate-nitrogen (in ppb) in the South Central Atlantic. Arrows indicate deep-ocean currents. (Modified after Sverdrup et al., 1942.)

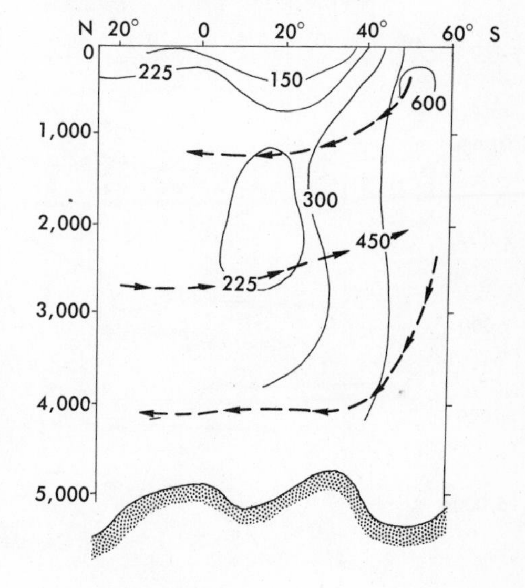

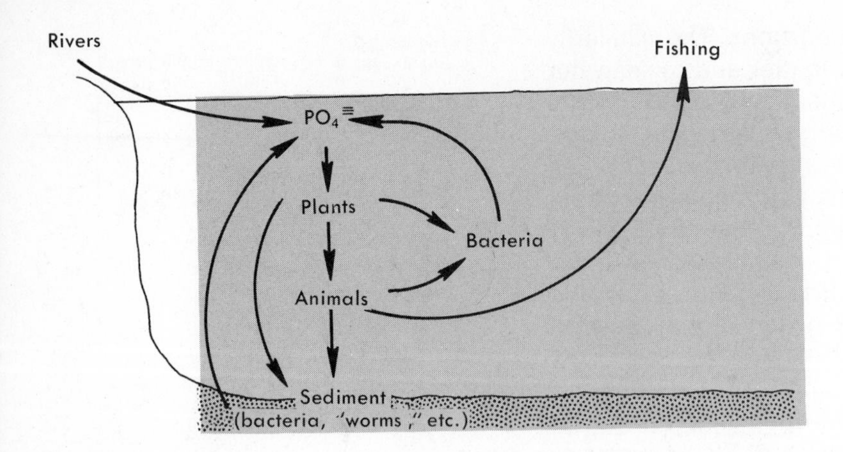

FIGURE 3.12 The phosphate cycle in the sea.

Phosphates ($PO_4^{\equiv}$) are cycled through the biological system in a manner similar to nitrates. The phosphate cycle is shown in Figure 3.12. Phosphates, like nitrates, enter the sea via rivers and run-off, including rainwater running over rocks covered with guano (bird feces). They leave the water through the fishing activities of birds and humans. In addition, some phosphates, such as in fish teeth, are trapped in sediment. The input and output of nutrients is nearly in balance for the oceans as a whole, but nutrient pollution (an overabundance of nutrients) and overfishing may upset the equilibrium locally. This may have adverse effects on the biota, especially in promoting the growth of certain undesirable species, such as red-tide organisms.

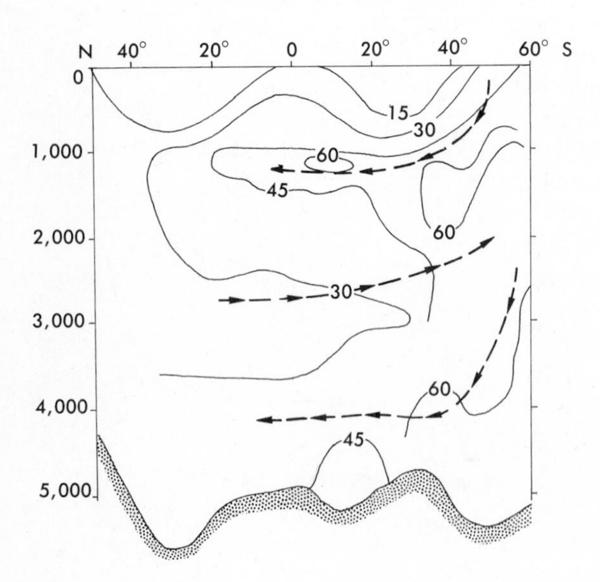

FIGURE 3.13 Vertical distribution of phosphate-phosphorus (in ppb) in the Central Atlantic. Arrows indicate deep-sea currents. (Modified after Sverdrup et al., 1942.)

The distribution of phosphates in the sea is very similar to that of nitrates, both vertically (see Fig. 3.10) and horizontally (Fig. 3.13). The distribution of nutrients reflects their depletion in warm surface water, regeneration in deep water, and transport by currents.

DESALINATION

The fresh water need of the United States in 1960 was about 300 billion gallons a day, and it is expected to double by 1980 and triple by the year 2000. In recent years, the quest for more fresh water has turned our attention to seawater, which is about 96.5 per cent pure water. More than 1,000 desalination plants (Fig. 3.14), with a combined capacity of producing over 500 million gallons of water per day, are in operation or under construction throughout the world. The plants in the United States alone have a capacity of over 60 million gallons per day. The largest desalination plant in the world, under construction in 1975, is in Hong Kong and will have a capacity of about 48 million gallons per day.

There are many methods of desalinating seawater. The oldest technique is distillation, whereby seawater is evaporated using solar energy and fresh water is produced by condensing the resulting water vapor. Julius Caesar is said to have used this method to obtain fresh water for his army. Such solar stills are used on a small-scale basis even today, especially in desert areas. Other more sophisticated

FIGURE 3.14 A modern desalination plant at St. Thomas, Virgin Islands. (Photo courtesy of U.S. Department of the Interior.)

techniques requiring great amounts of energy but capable of desalinating large quantities of seawater are in operation throughout the world.

In spite of the economic considerations, desalination may be an answer to the world's water needs. But it may also create other problems, especially the disposal of the warm and highly concentrated brine, which at the present time is sent back to the sea. This undoubtedly could cause great damage to the ecology of the coastal ocean, especially when many desalination plants are concentrated in one general area, say, along most of the Atlantic coast of the United States. No one can foretell what the consequences would be, but research is being conducted on this problem. It has been proposed that the brine be mixed with sewage in such a way that the resulting mixture would sink to greater depths. This proposal, although economically advantageous, warrants further research as it involves many complex oceanographical and environmental factors. Some use for most or all of the brine may be found in the future. For example, metals and other salts might be extracted from it. At present, most of the magnesium and bromine in the United States come from the sea. Although sodium chloride is also extracted from seawater, most of the United States' supply currently comes from terrestrial salt deposits.

Perhaps we should consider safer alternate sources of fresh water, in addition to conservation and recycling of presently available water supply. Fresh water can be extracted from the atmosphere. Large cooling towers could be set up in tropical islands to obtain the moisture in the air. Cooling can be accomplished by pumping cold deep water from the ocean. Because this water is rich in nutrients, it can be reused for fish farming. Another alternative is to tow polar icebergs to populated regions and obtain fresh water by melting them. Both of these alternatives may be competitive with desalination of seawater and, most importantly, may be less damaging to the environment.

SUGGESTED READINGS

Broeker, Wallace S. 1974. *Chemical Oceanography.* New York, Harcourt Brace Jovanovich.

Harvey, H. W. 1957. *Chemistry and Fertility of Sea Water.* New York, Cambridge University Press.

Hunt, Cynthia A., and Robert M. Garrels. 1972. *Water: the Web of Life.* New York, W. W. Norton.

Keosian, John. 1964. *The Origin of Life.* New York, Reinhold.

Popkin, Roy. 1968. *Desalination, Water for the World's Future.* New York, Praeger.

Riley, J. P., and G. Skirrow (eds.). 1965. *Chemical Oceanography.* New York, Academic Press.

Rubey, William M. 1951. Geologic History of Seawater. *Bulletin of the Geological Society of America,* **62**:1111–1148.

Sverdrup, H. U., M. W. Johnson, and R. H. Fleming. 1942. *The Oceans.* Englewood Cliffs, New Jersey, Prentice-Hall.

This strange ship is actually a floating instrument platform. It is towed to sea and flipped into the very stable upright position. *A, Flip* flipping; *B, Flip* flipped. (Photos courtesy of Scripps Institution of Oceanography.)

CHAPTER 4

Physical Nature of Seawater

Most of the natural processes that take place in the oceans depend on the characteristics of their water. It is because of the unique nature of water that the sea transmits and absorbs light energy, that it is salty, that ice floats, waves erode the shore, and marine life flourishes. An understanding of these properties is essential to a study of the processes that take place in the sea. In addition, the large-scale distribution of temperature and salinity control the deep circulation in the sea. The interaction of air and sea has a major impact on terrestrial weather and climate. In fact, most of the rain that falls on land enters the air through evaporation at sea.

LIGHT IN THE SEA

As light reaches the surface of the sea, part of it is reflected, and the rest penetrates the water. Reflection is least when the sun is directly overhead (about 2 per cent on a calm sea), increasing as the sun approaches the horizon, until essentially 100 per cent reflection occurs. The amount of light that actually enters the sea varies considerably, depending on a number of factors, including sun

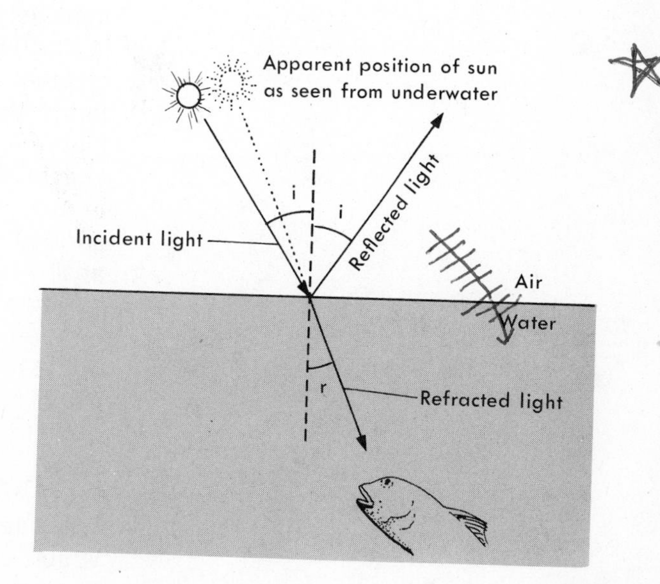

FIGURE 4.1 Reflection and refraction of lights at the air-sea interface. Note that because of refraction, the sun appears to be closer to the vertical than it really is.

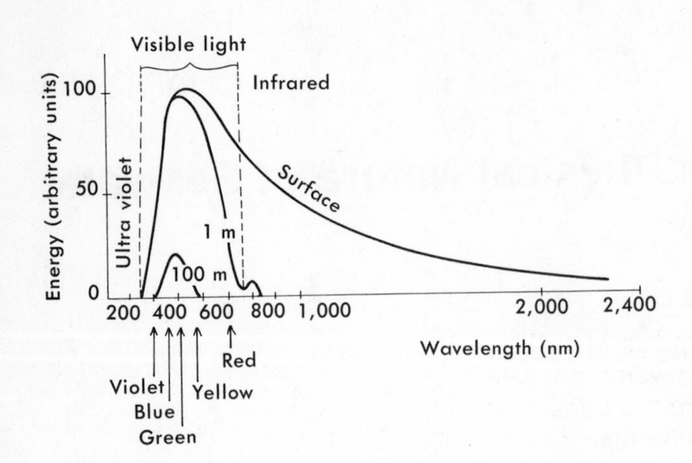

FIGURE 4.2 The spectrum of light reaching various depths in the sea. Wavelengths are given in nanometers (nm) or billionths of a meter. (After Sverdrup et al., 1942.)

angle, sky conditions, sea surface condition, and clarity of water.

The velocity of light is greater in air than in water. Therefore, when light enters the sea, it is refracted or bent downward (Fig. 4.1). Refraction of light follows Snell's Law and is discussed in detail in Appendix B. Thus, to a fish or a diver, the sun appears to be closer to the vertical than it really is. As light penetrates into deeper denser water, it is refracted even more toward the vertical.

As light travels through water, it becomes progressively dimmer, owing to absorption and scattering by tiny particles in the water and absorption by the seawater itself. Sunlight is actually composed of a spectrum of different wavelengths or colors (Fig. 4.2), and not all colors of light penetrate the same distance into the sea. Most of the red light is filtered out in the upper few meters. Many fish and shrimps that live below the depth to which red light penetrates are themselves red. In the absence of red light, they appear black and are thus camouflaged. In the clearest oceanic water, blue light penetrates the deepest (more than 500 meters), and green light penetrates less. Figure 4.3 shows the total light penetration in oceanic and coastal waters. At depths of about 600 meters, the intensity of light may be equivalent to that of starlight reaching the surface of the earth. In fairly clear coastal water, selective absorption of blue light by particulate matter and dissolved yellow substances results in relatively greater penetration of green light, and about 99 per cent of the total light is filtered out in the upper 50 meters. In very ***turbid water*** (water that is heavily laden with suspended particulate matter), most of the light is filtered out in the first meter.

As light energy is absorbed in the sea, it is converted into heat, unless the light is absorbed by a living plant, in which case some of the light energy is converted into

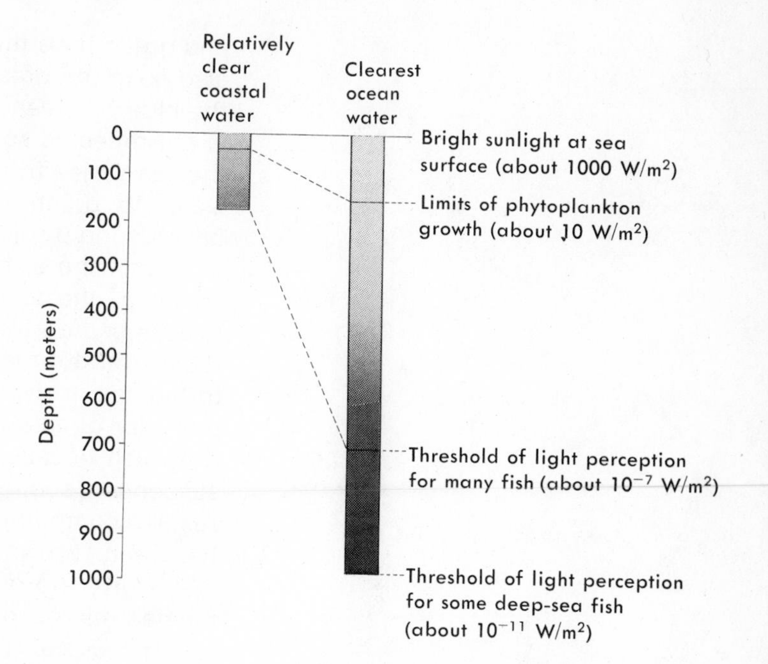

FIGURE 4.3 Total light penetration in clear coastal water and the clearest ocean water on a bright sunny day. The intensity of starlight reaching the sea surface is equivalent to that of sunlight reaching a depth of about 600 meters in the clearest ocean water on a bright day. (After Blaxter, 1970, and Clarke and Denton, 1962.)

chemical energy used in the growth of the plant. Since most light energy is absorbed near the surface, this is the region where the greatest warming occurs. Warm water is lighter (less dense) than cold water and tends to stay at the surface, resulting in stratification of warm surface water over cold deeper water. Between these two layers is the transition zone of rapid temperature change with depth known as the ***thermocline*** (Fig. 4.4).

The observed color of the sea is due to the color of the light that is reflected from the surface and the light that is back-scattered from within the sea. Because blue

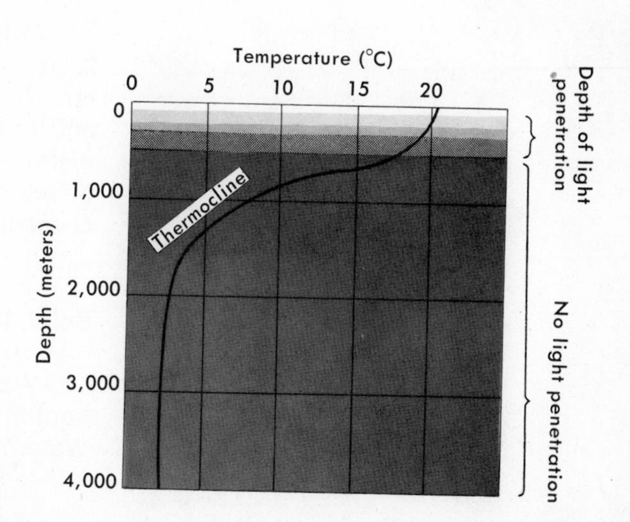

FIGURE 4.4 A thermocline in relation to light penetration.

and green have the greatest penetration in seawater, they also have the greatest chance for back-scattering; hence the blue or green color of the sea. Light that is absorbed is not reflected so it is not seen from above. The color of the sea varies from deep blue in the clearest open ocean water to green in clear coastal areas to yellowish or brownish in turbid coastal water.

Near shore, the water tends to look more greenish owing to the additional scattering of light by yellowish substances derived from the decomposition of plants in the sea and on land. These yellow substances that wash to the sea in rivers may even tinge some estuarine water the color of pale beer. Highly turbid water tends to appear brownish because of the reflection of light by brownish suspended sediments, which in turn owe their color to rustlike compounds of iron oxide. Nutrient-rich water may be green, brownish, or even a tomato-soup color as a result of the production of great amounts of highly pigmented microscopic plants.

The overall transparency of water may be measured with a ***Secchi disc.*** This is an old technique, which usually involves lowering of a black-and-white disc into the water and noting the depth at which it disappears from view. The depth thus measured indicates the point to which about 18 per cent surface light penetrates.

Most ocean light measurements today are made with ***photometers,*** light meters similar to those used in photography, which measure the light energy (in watts) reaching a particular point. Frequently, light reaching a particular depth is compared with the light at the surface, giving an indication of the transparency of the water. This technique, although it operates on the same principle as the simple Secchi disc method, is much more sensitive.

HEAT IN THE SEA

Solar energy and the thermal properties of water largely determine the state in which water exists (solid, liquid, or vapor), as well as the temperature distribution and movement of ocean water. The vertical temperature distribution and the process of evaporation may someday be an important source of electrical power and will be discussed later in this chapter.

Solid, Liquid, or Vapor

Water is the only substance that may exist as a solid, liquid, or vapor under ordinary temperatures on earth. Water may be found as ice, liquid, or water vapor (steam).

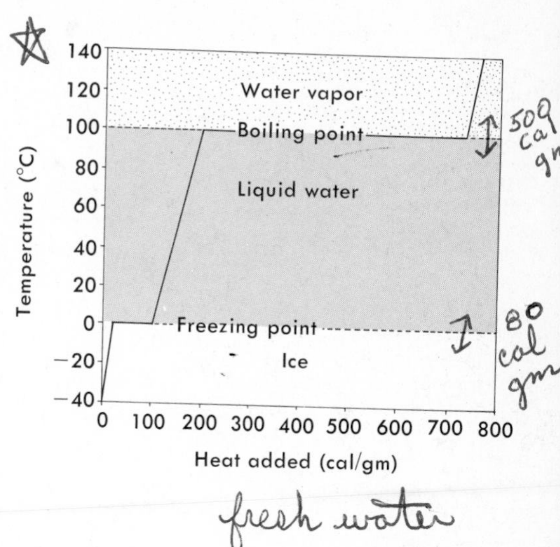

FIGURE 4.5 The changes of temperature and state of pure water as heat is added, starting with ice at −40°C.

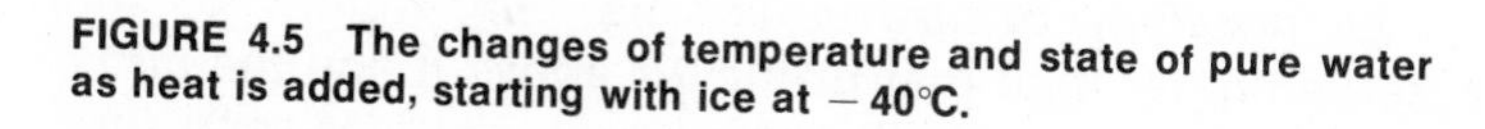

Water has one of the highest known ***heat capacities***, the amount of heat in calories[1] required to raise the temperature of one gram of a substance by 1° C. Therefore, it has a built-in capacity to resist changes in temperature. For example, the heat capacity of pure water is one calorie/gram,whereas that of iron is only about 0.1 cal/gm, and that of aluminum, nitrogen, and oxygen is about 0.2 cal/gm. Seawater, because of its dissolved salts, has a heat capacity of about 0.93 rather than 1.0 cal/gm but still warms and cools much less rapidly than land or air.

As heat is added or removed, the state in which water is found may be changed (Fig. 4.5). It takes 80 cal/gm to melt ice at 0° C, and 540 cal/gm to boil water at 100° C. Note that, in this case, as calories are added, no change in temperature occurs. However, water may also ***evaporate*** (be converted from liquid to vapor) at temperatures lower than the boiling point, but more heat is required. For example, at 20° C, 585 calories are required to evaporate one gram of water.

As salts are added to water, the freezing point is lowered (Fig. 4.6). Seawater with a salinity of 37‰ freezes at about −2° C. Water is one of the only substances known that becomes lighter and floats when frozen; most others contract as they freeze. The fact that ice floats is of great importance to aquatic life. If ice sank, the sea floor might be covered with ice, and the polar seas could freeze solid. Similarly, some lakes in high latitudes would freeze solid in winter, and many would never completely thaw in summer. The distribution of life, especially of bottom dwellers, would be greatly restricted.

FIGURE 4.6 The relationship between salinity and the temperature of freezing.

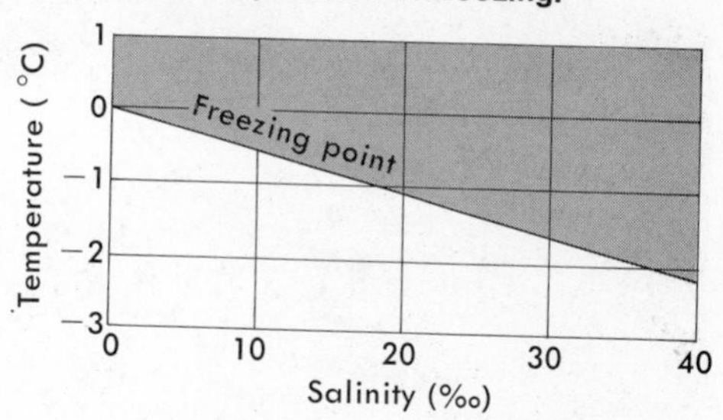

[1]A calorie is the amount of heat required to raise the temperature of one gram (about 1/5 teaspoon) of liquid water by 1°C.

Heat Budget

As we have seen earlier in this chapter, ***solar radiation*** is the most important contributor of heat to the sea. In fact, absorption of light energy accounts for more than 99 per cent of the heat entering the sea. In addition, minute (but measurable) amounts of heat reach the sea from the earth's hot interior. The amount of heat entering the sea is essentially in balance with that leaving the sea. Heat leaves the sea by a variety of means, the most important of which is evaporation.

Evaporation occurs when the temperature of the water is higher than that of the air above it and the air is not already saturated with water vapor. As mentioned earlier, a great amount of heat is transferred into the atmosphere when liquid water is converted into vapor. The water heats the air and increases its capacity to retain water. Because water temperatures vary much less than those of the atmosphere, there may be considerable difference in the rate of evaporation during the day and at night and from one location to another. An average of about one meter of water evaporates from the sea each year. The greatest evaporation occurs at about 20° north and south of the equator, the latitudinal belts in which

FIGURE 4.7 When moist air rises and cools, condensation of water vapor occurs, resulting in precipitation.

most of the great deserts of the world occur. The reverse process, ***precipitation,*** is greatest near the equator. These relationships may be due to air-sea temperature differences as well as to trade winds, which carry water vapor toward the equator. When the water is colder than the air, ***condensation*** of water vapor onto the sea occurs, because the air is chilled and its capacity to hold vapor is decreased. However, this process occurs at a much slower rate than evaporation. In reality, most water is returned to the sea when light, warm, moisture-laden air rises. When air rises and cools, water vapor condenses onto cold dust or other particles in the upper atmosphere, forming clouds, and the resulting water drops may fall back to the sea or land as precipitation (Fig. 4.7). The rising and cooling of moist air near the equator and along steep coastal hills and mountains account for the increased precipitation in these areas.

Condensation also carries some heat into the sea, but this source is insignificant compared to the loss of heat through evaporation.

Radiation of heat from the sea back into the atmosphere (***back-radiation***) ranks second to evaporation as a means of heat loss from the sea. Loss of heat through radiation depends only on the temperature of the water near the sea surface.[2] Consequently, back-radiation occurs at all latitudes.

Conduction may carry heat into or out of the sea, depending on the temperature of the air and the water. Heat leaves the sea by direct conduction from warmer water, or enters it from a warmer atmosphere. Most of the heat thus transferred flows from the sea to the atmosphere. Thus, there is a net loss of heat from the sea due to conduction.

The most important means of heat transfer are as follows:

(1) solar radiation (99 + % of incoming heat)
(2) evaporation (55% of outgoing heat)
(3) back-radiation (40% of outgoing heat)
(4) conduction (5% of outgoing heat)

For the oceans as a whole, these factors are related in the following ***heat budget equation***:

$$\text{solar radiation} = \text{evaporation} + \text{back-radiation} + \text{conduction}$$

Thus the amount of heat entering the oceans balances the

[2]Radiation of heat from an object (or the sea) increases with the ***absolute temperature*** of the object, expressed in degrees Kelvin: °K = °C + 273. Radiation occurs at temperatures greater than absolute zero (0°K).

FIGURE 4.8 The heat budget of the oceans.

heat that leaves them (Fig. 4.8), except for minor imperceptible changes from one year to the next. Such changes in the heat content of the oceans have certainly occurred since the origin of the oceans and have probably been related to changes in climate on earth throughout geological time.

More solar radiation reaches a square meter of equatorial ocean than reaches a square meter in polar regions (Fig. 4.9). Consequently, a greater amount of heat enters equatorial water than polar water. Because of the overall oceanic heat balance, a net poleward flow of heat must occur. This is accomplished primarily by ocean currents. A similar poleward transport of heat also occurs in the atmosphere.

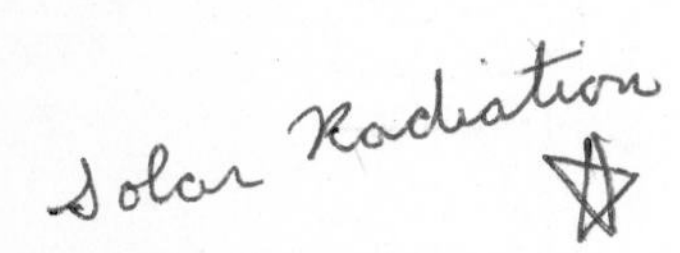

FIGURE 4.9 The amount of sunlight falling on equal areas of earth decreases from the equator toward the poles. The shaded areas on the earth are equal; the light energy reaching these areas is not.

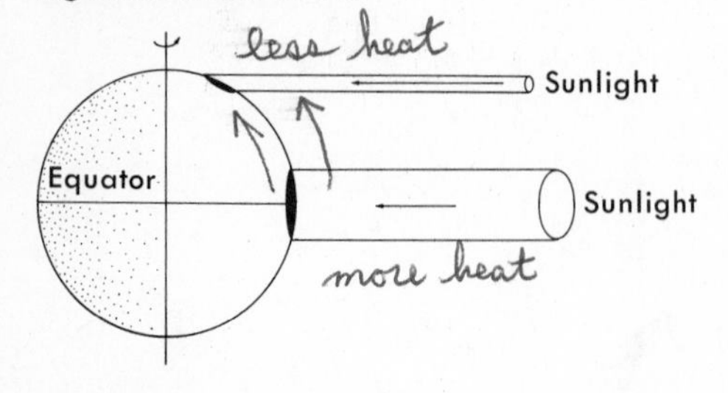

TEMPERATURE OF THE OCEANS

The distribution of temperature in the oceans is one of the factors that controls the distribution of marine organisms and the density of seawater. Because of the great heat capacity of water, the surface temperature of the sea varies far less than land temperatures. Consequently, the sea provides a rather stable environment for marine life and a moderating influence on the coastal climate.

Distribution of Temperature in the Sea

Sea surface temperatures are highest near the equator and lowest near the poles (Fig. 4.10). The average surface temperature is about 16° C, but it varies from about −2° C to about 30° C. Sea ice, however, may have a temperature far below −2° C. The distribution of temperature is partly dependent on current patterns, because some currents transport warm water to high latitudes, and others transport cold water toward the equator. It is because of such surface ocean currents that the surface ***isotherms*** (lines of equal temperature) do not necessarily parallel lines of latitudes but are displaced poleward in the western Pacific and Atlantic oceans and are closer to the equator in the eastern parts of the oceans.

In the equatorial Pacific, cooler water extending westward from the coast of South America indicates a westward-moving current in this area. Seasonal changes in temperature patterns may also be seen. The isotherms in February are generally south of where they lie in August. Seasonal temperature changes tend to be greater at mid-latitudes than in either polar or equatorial water, especially along the western Pacific and Atlantic oceans. Such variations are at least partly due to the prevailing westerly winds that blow off the land, which is subject to much greater seasonal temperature changes than are found at sea. The eastern Pacific and Atlantic oceans are more constant in their temperature, since the westerly winds blow from the sea with its moderating influence.

The temperature of deep water varies far less than the temperature of surface water. In some mid-latitude regions, a thermocline (a sharp temperature decline with depth) may be present near the surface during the summer and fall, owing to seasonal variations in the amount of solar heat reaching the area. In addition, a deeper permanent thermocline may be present in many regions (Fig. 4.11).

In a north-south section through the Atlantic (Fig. 4.12), it can be observed that the isotherms are located at greater depths under the equator as compared to polar regions. The 4° C isotherm (representing the overall average temperature of the oceans) occurs at about 1,600 meters at the equator, but reaches the surface at about 55° N and S latitudes. Water deeper than 3,000 meters is cold at all latitudes. Even at the equator, water near the bottom is very cold, about 2° C at 4,000 meters. However, the temperature near the ocean bottom progressively increases toward the North Atlantic, indicating the presence of a northward-flowing deep ocean current. At the poles, there is very little change in temperature with depth. East-west temperature patterns in the oceans are

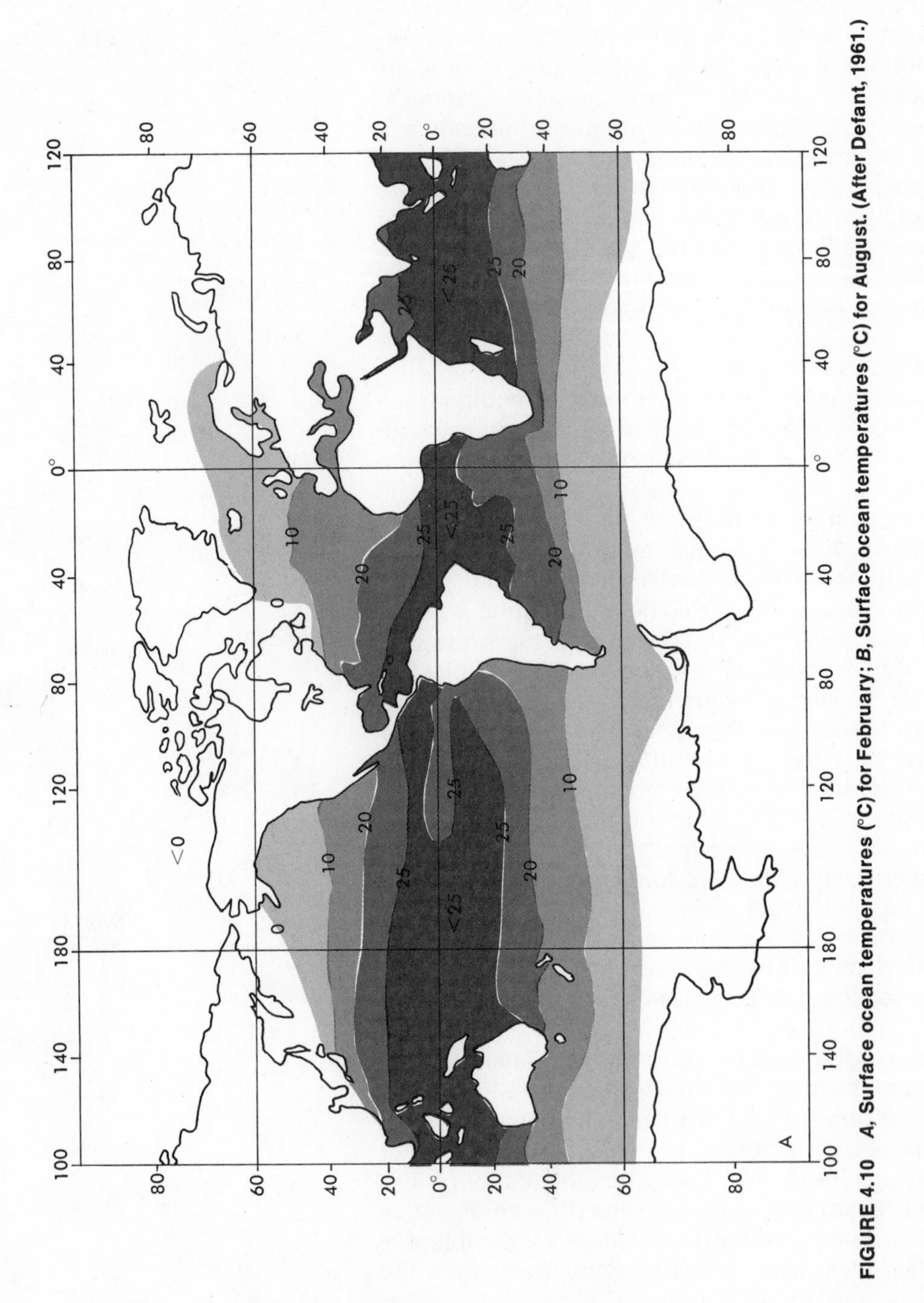

FIGURE 4.10 *A*, Surface ocean temperatures (°C) for February; *B*, Surface ocean temperatures (°C) for August. (After Defant, 1961.)

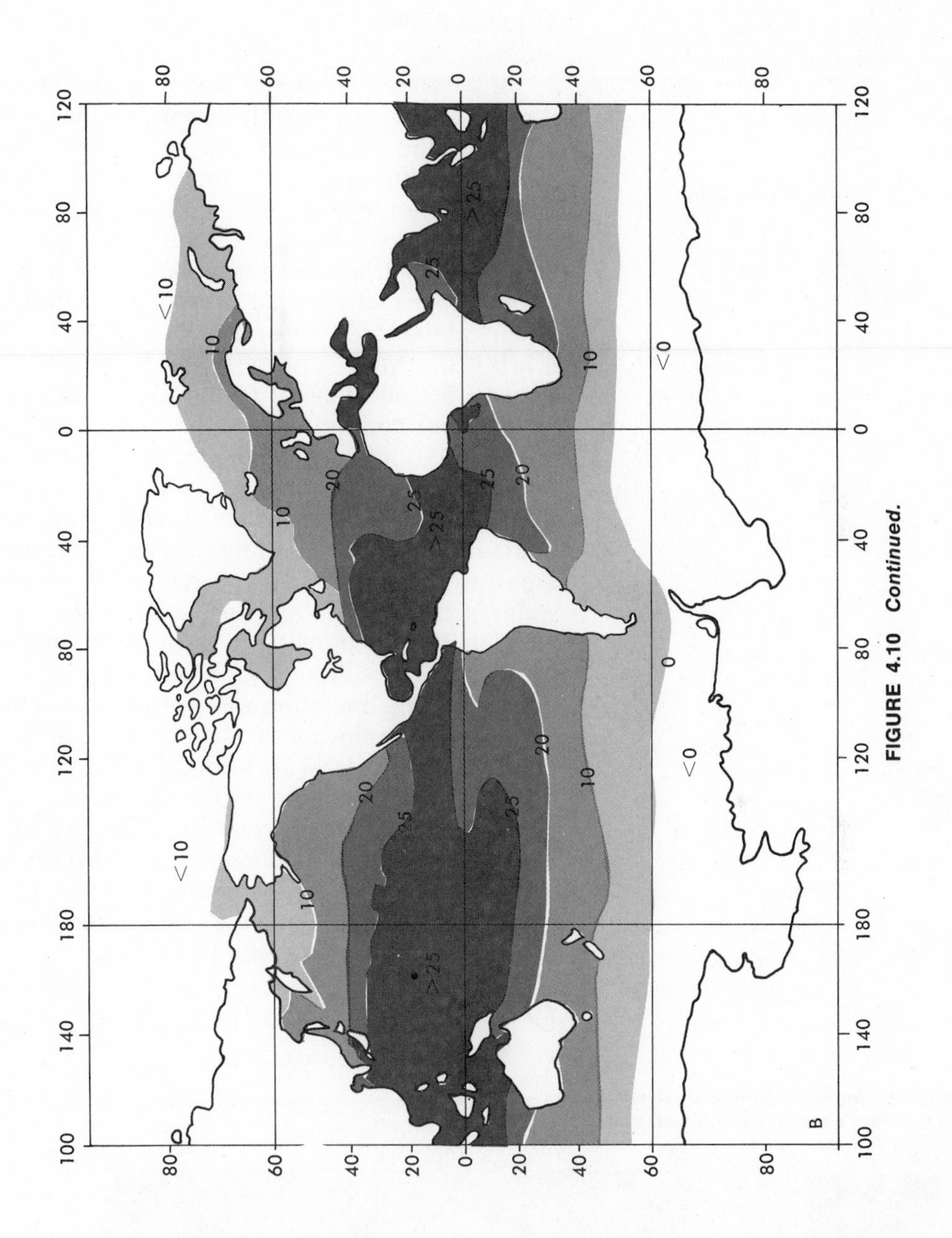

FIGURE 4.10 *Continued.*

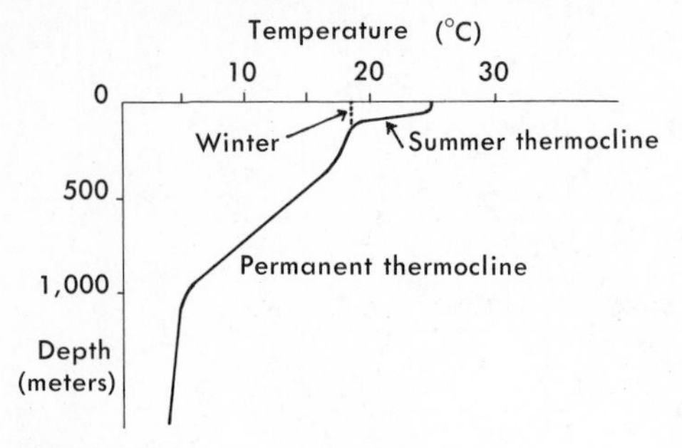

FIGURE 4.11 A double thermocline.

indicated by rather flat isotherms, as shown in Figure 4.13.

Thermal Power

It has been estimated that if all the heat in the Gulf Stream (which flows northward along the southeast coast of the United States and may be responsible for keeping much of northern Europe from becoming icebound) could be transferred to water that is colder by 25° C, tremendous power could be generated (about 75 times the power consumed by the United States in 1970). There are many parts of the oceans where a 20 to 25° C temperature difference occurs in the top 1,000 meters of water (see Figs. 4.11 and 4.12). If the warm surface water could be evaporated by cooling the air above it (especially under a vacuum), the water vapor could drive a turbine and produce electricity (Fig. 4.14). Cold ocean water pumped to the surface could be used to cool the air. After being warmed, this nutrient-rich water from the deep sea could then be transferred to a mariculture pond. The water vapor, after driving the turbines, would condense on the cold surface above the warm water, providing valuable fresh water. In fact, a small power plant using this principle was built by George Claude in Cuba in 1929. Although 22 kilowatts of power were thus generated, the operation was not an economic success. If ammonia, which evaporates much more easily than water, could be used to drive the turbines, a more efficient process could be achieved. In order for this concept to be productive, a tremendous amount of water would have to be pumped, and many millions of dollars worth of equipment would be needed. But the energy is there, and it is renewable. It may not be too many years before we derive some of our energy from the heat in the sea.

FIGURE 4.12 Longitudinal temperature structure in the Western Atlantic. (After Wüst, 1936.)

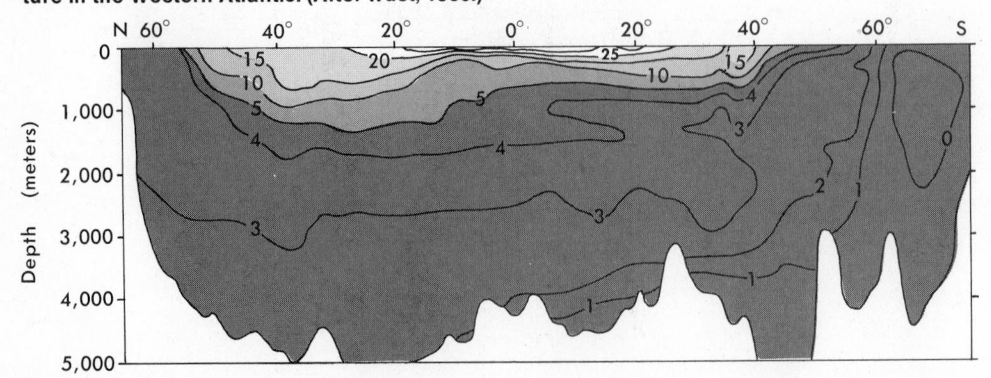

FIGURE 4.13 Latitudinal (east-west) temperature structure in the South Pacific. Note the gentle downward tilt of the isotherms toward the west. (After Stommel et al., 1973.)

Ocean Temperature Measurements

The temperature of surface water may be measured using an ordinary thermometer placed in a sample of water collected with a bucket or a water sampling bottle such as a Nansen bottle (Fig. 4.15). However, the temperature of a deep-water sample will change before it is brought to the surface, so it is necessary to measure its temperature *in situ* (in place). Special thermometers have been developed which may be attached to a Nansen bottle and lowered into the water upside-down and then reversed at the desired depth, automatically registering the temperature at that depth (Fig. 4.16). These thermometers have been devised so that the mercury column is intact on the way down, and then, when the thermometer is reversed, the mercury column breaks, isolating a smaller column of mercury, which is proportional to the temperature at that depth. After the temperature is read on board the ship, the thermometer is tipped back to the original position. The mercury column is rejoined and is ready to be used again. The water collected by the Nansen bottle may be used for various chemical analyses.

The bathythermograph is another temperature-measuring device (Fig. 4.17). It records the temperature and pressure continuously on a coated glass plate, from the surface to the desired depth. Since pressure is closely related to depth, this instrument produces a graph of temperature changes with depth on the glass plate.

Temperature may also be measured electronically with a temperature probe, which is a temperature-sensitive device connected to a meter (Fig. 4.18) on the ship. An electric current is set up in the probe and transmitted to the meter. The current is proportional to the temperature at the depth of the probe. These temperature-measuring units are frequently combined with devices that measure other parameters, such as electrical conductivity, salinity, depth, and dissolved oxygen, all at the same time.

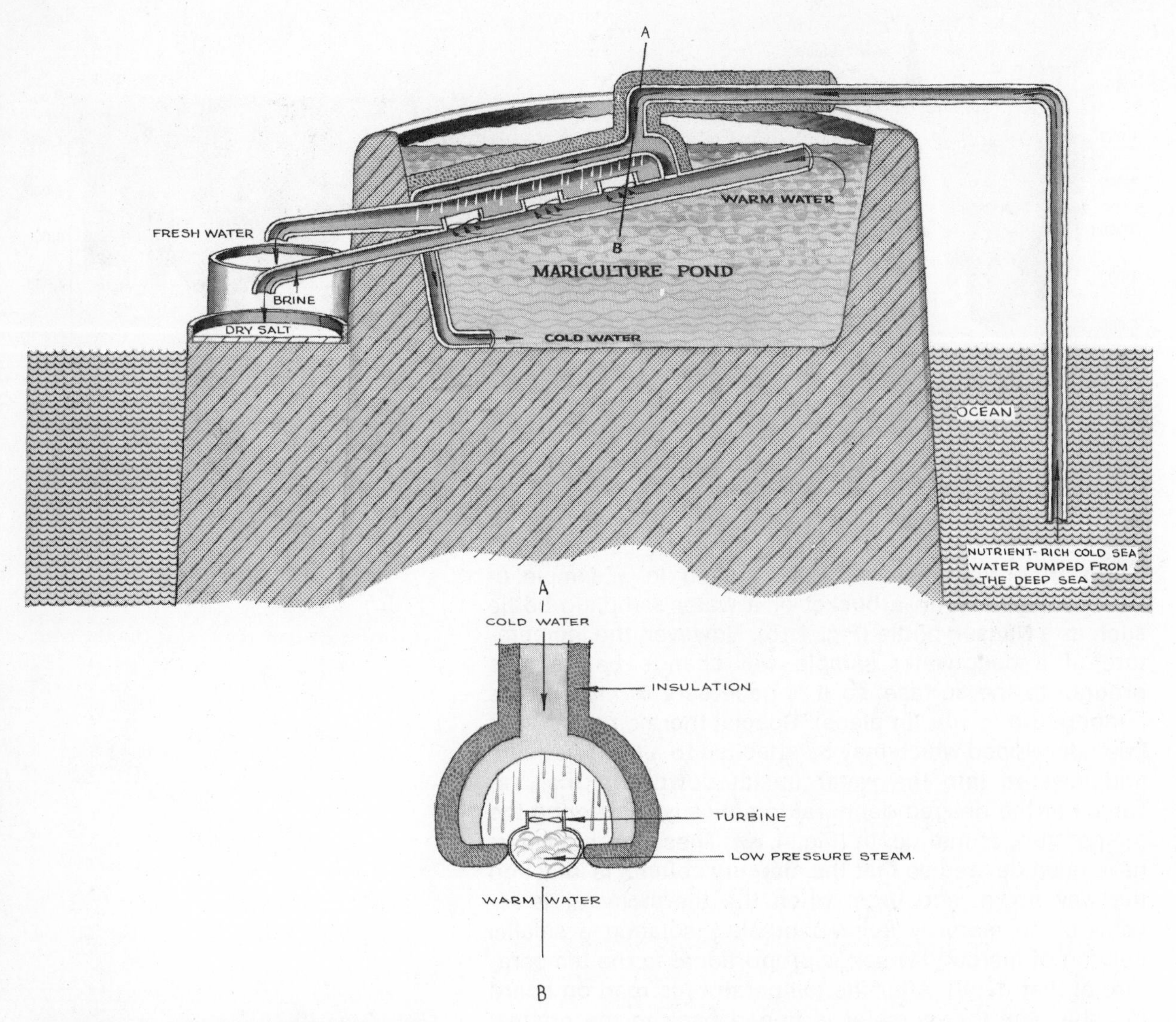

FIGURE 4.14 Diagram of a hypothetical thermal power station, which would use the power of evaporation to drive steam turbines.

SALINITY OF THE OCEANS

Salinity, the amount of dissolved solids in grams per kilogram of seawater (‰ or parts per thousand), depends on evaporation, precipitation, fresh water input from rivers, and mixing by currents. Salinity in the oceans generally varies from about 32 to 37‰, except in the Arctic and near shore, where it may be less than 30‰. The surface salinity in the Red Sea may be greater than 40‰, due to high rates of evaporation, but the average salinity in the oceans is about 35‰ or 3.5 per cent. Figure 4.19 shows the distribution of surface salinity in the oceans.

The surface salinity of the oceans depends on the difference between evaporation (E) and precipitation (P). Figure 4.20 shows the relationship between surface salinity, evaporation, and precipitation in the oceans. Notice that the areas of highest salinity, at 25° north and south latitudes, coincide with the regions where evaporation exceeds precipitation by the greatest amount. The mechanics of evaporation has been discussed earlier in this chapter. For the oceans as a whole, evaporation exceeds precipitation by about 10 cm/year. Because the ocean-atmosphere-land network is a closed system, approximately this amount of water re-enters the sea via the rivers each year.

Surface salinity, evaporation, and precipitation have a well-defined relationship for the oceans as a whole. This relationship is expressed by the equation

$$S = 34.6 + 0.0175\,(E-P)$$

where S is salinity (‰), and E and P are evaporation and precipitation, respectively, expressed in centimeters per year. It is interesting to note that 34.6 represents the average salinity at about 500 meters, implying that the surface waters may be mixed to that depth. Locally, salinity may be raised by freezing of seawater or lowered by river run-off and melting of ice. The lowered salinity observed in the Bay of Bengal, for example, is due to the tremendous run-off of fresh water from the Ganges River.

Figure 4.21 shows the north-south vertical distribution of salinity in the Atlantic. No general pattern of increase or decrease in salinity with depth is observed. However, there are regions of salinity maxima and minima, which indicate the presence of deep ocean currents (see Chapter 7) originating in high latitudes. These currents are caused by increases in density, which are mainly the result of low temperatures rather than of high salinity.

FIGURE 4.15 Nansen bottle being removed from the wire. (Courtesy of Woods Oceanographic Institution.)

DENSITY OF SEAWATER

Almost all movements in the deep ocean are caused by density gradients analogous to the atmospheric movements that create winds which drive surface currents in the ocean. Therefore, density is of prime interest to oceanographers in the study of currents. In addition, density is important to life in the sea. Water comprises the bulk of plant and animal cells. Consequently, the density of the biota is nearly the same as the water in which they live, which enables many organisms to be supported in the water and to be more or less neutrally buoyant. However, some plants and animals float on the surface or regulate their buoyancy with the aid of gas-filled bags or sacs.

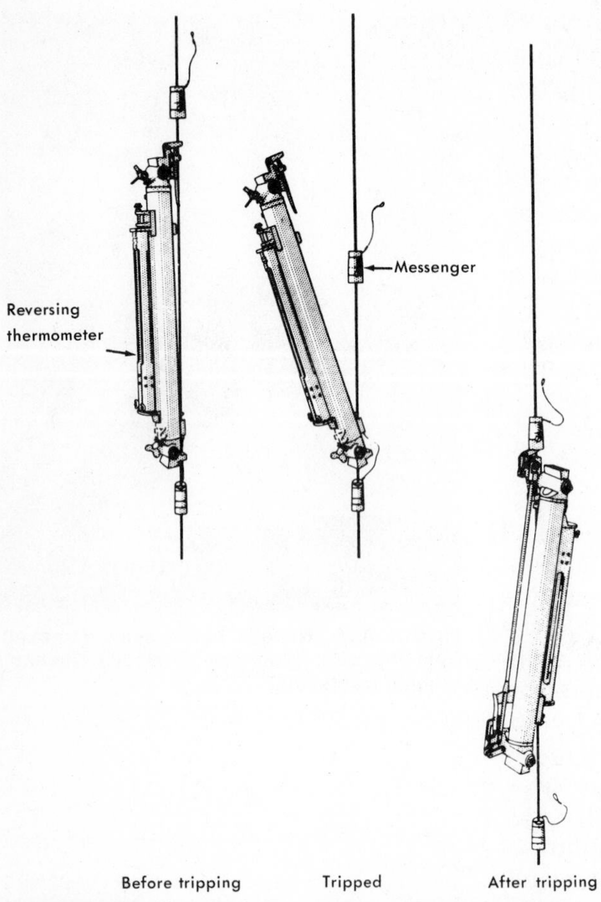

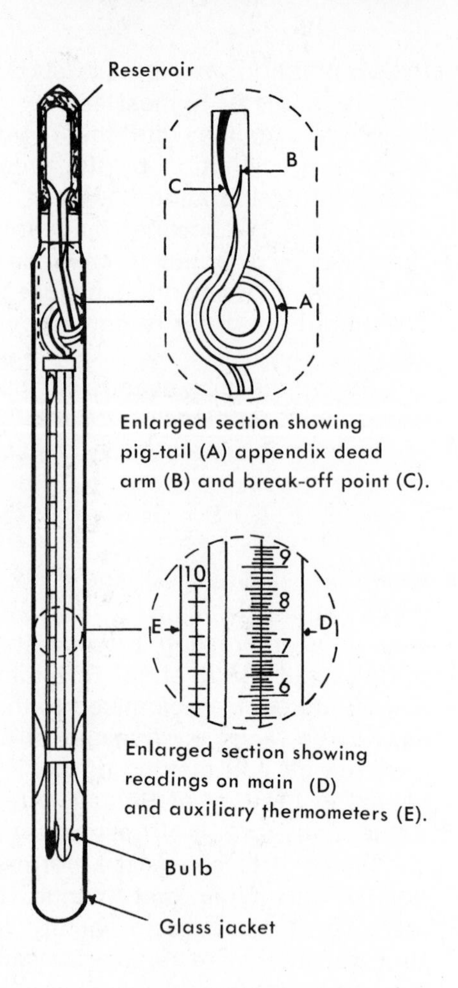

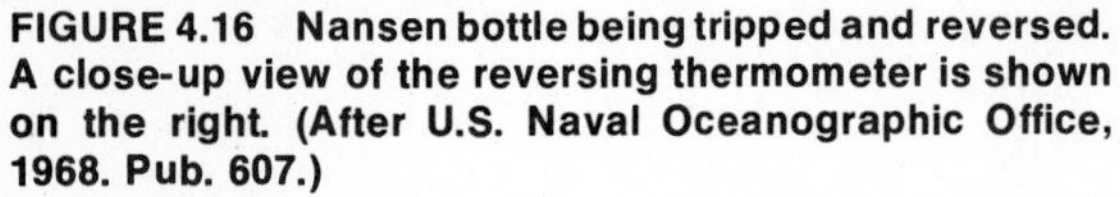

FIGURE 4.16 Nansen bottle being tripped and reversed. A close-up view of the reversing thermometer is shown on the right. (After U.S. Naval Oceanographic Office, 1968. Pub. 607.)

Factors Controlling Density

The density of seawater is a function of temperature, salinity, and pressure. In the open ocean, density increases (from about 1.02 to 1.05 gm/cc between the surface and 5,000 meters) with decreasing temperatures and increasing salinity and pressure. Although density at any depth in the ocean can be accurately calculated, measurements of density *in situ* are very difficult. The density of water is usually determined after bringing the water sample to the surface, discounting the effect of pressure. This method is justifiable for the study of horizontal currents, since we usually compare densities at the same depth (nearly the same pressure) at different localities.

Unlike the uniform effects of pressure on density as a function of depth (one atmosphere per 10 meters of water), the effects of temperature and salinity are localized and form the main driving mechanism of deep-ocean currents. Figure 4.22*a* illustrates the variation of density with temperature, at constant salinity and atmospheric

A

B

FIGURE 4.17 The bathythermograph (B-T), an instrument that automatically traces a record of the temperature changes with depth (pressure) as it is lowered into the sea. *A*, Close-up showing B-T, slides, and reader. *B*, B-T being retrieved. (Courtesy of NOAA.)

pressure. Figure 4.22*b* shows the variation in density with salinity, assuming constant temperature and atmospheric pressure. If we know the temperature and salinity of a water sample, its density at the surface (at atmospheric pressure) can be obtained by using a ***temperature-salinity graph*** (Fig. 4.23). Ordinarily, oceanographers use this graph to plot temperature and salinity data from various depths at a particular locality. This data can be used to study the deep-ocean currents, because we can identify waters from various localities by their temperature and salinity characteristics.

A uniform increase in density with depth has little effect on organisms. However, in areas where a sharp change (***pycnocline***) occurs over a small vertical distance—as in regions where very warm and therefore light water overlies a dense, cold layer—several striking phenomena may occur. The sudden increase in density may cause a slowing in the sinking rate of dead organic matter, which will accumulate at this level. An oxygen minimum zone may be formed owing to the decomposition of this accumulated organic matter by bacteria, which use up oxygen in the process and which may in turn drive away most of the oxygen-requiring organisms that ordinarily would eat the organic matter.

Temperature of Maximum Density

When ice melts, the resulting fresh water is more dense than the ice from which it was formed (about 0.08 gm/cc greater). The density continues to increase to a maximum at about 4° C, above which the density decreases (Fig. 4.24). For this reason almost all deep lakes in temperate regions have a dense 4° C layer of water near bottom no matter how hot summer becomes as long as the temperature drops to 4° C for substantial periods in winter. As water is warmed, the molecular movement increases, tending to expand the volume. However, at the same time, the structure of the water changes. Water exists as a mixture of single molecules and ***polymers*** (two or more water molecules bound together as a single unit). The polymers occur in groups of 2, 4, or 8 united molecules (Fig. 4.25), the size and shape of which change with

FIGURE 4.18 An electronic salinity, temperature, and depth profiling instrument, with remote-tripping water sampling bottles and reversing thermometers. *A*, Instrument being lowered from ship; *B*, Measurements being made aboard the ship. (Photos courtesy of Lamont-Doherty Geological Observatory of Columbia University.)

temperature. The lower density (between 0° C and 4° C) can be credited to the increased concentration of large bulky polymers.

The addition of salt to water lowers the temperature of maximum density (Fig. 4.26) until at 24.7‰ salinity, the water freezes at the temperature of maximum density. Therefore at higher salinities, seawater freezes before the theoretical temperature of maximum density is reached. As seawater approaches the freezing point, it becomes more dense and tends to sink, causing the water to be well mixed. The oceans do not have a 4° C bottom layer as do lakes but rather have water that may be as cold as 0° C or less. Consequently, the sea surface freezes only when the water is near the freezing point at all levels, from surface to bottom. As the initial ice is formed, much of the salt is excluded, causing the salinity of the water below the ice to increase, further lowering the freezing point. Sea ice (Fig. 4.27), therefore, forms less readily than lake ice.

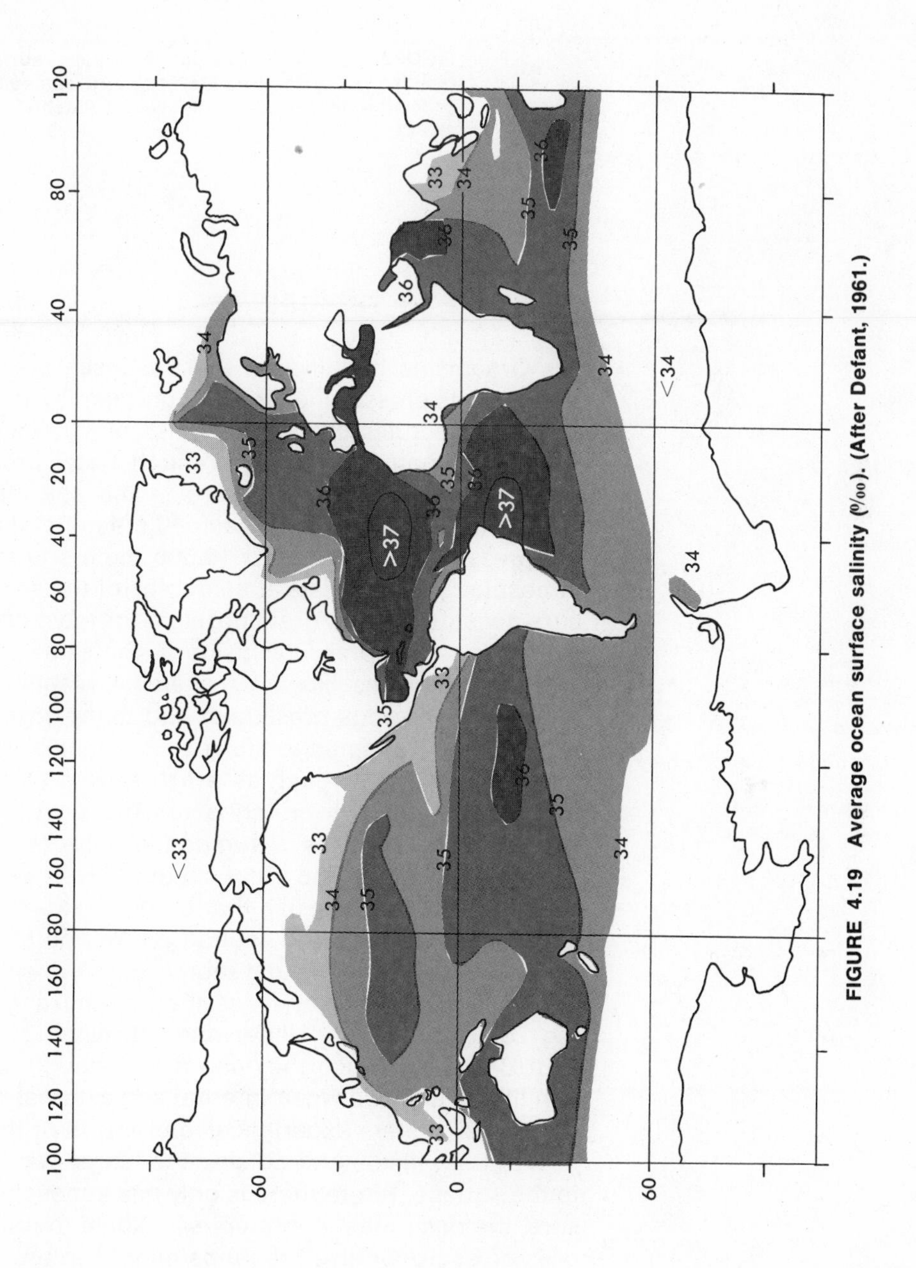

FIGURE 4.19 Average ocean surface salinity (‰). (After Defant, 1961.)

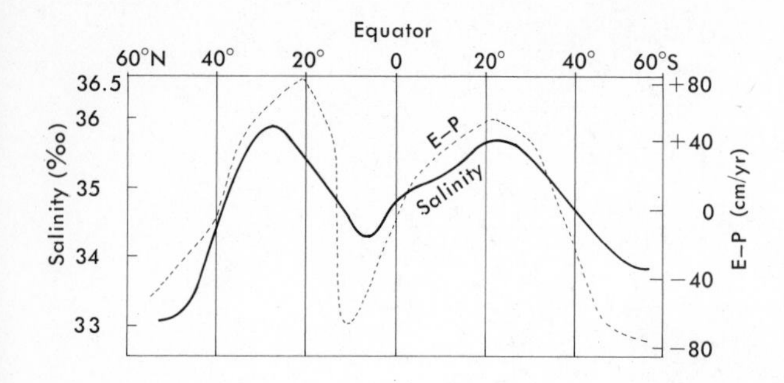

FIGURE 4.20 Relationship between surface salinity (solid) and the annual difference between evaporation and precipitation (E-P) for all oceans (dashed). (After Wüst, 1954.)

PRESSURE IN THE OCEANS

Organisms that live in the deep sea are subject to tremendous pressure, exerted upon them by the weight of the water that lies over them. A fish at 4,000 meters must withstand about 400 times as much water pressure as a fish near the surface. Pressure in the sea increases by about one atmosphere for each 10 meters of depth. Thus, the pressure in a trench at 10,000 meters is about 1,000 atmospheres. Life exists at all depths of the sea, and pressure does not pose a great problem for most marine organisms since the pressure inside is the same as that outside the organisms. However, chemical reaction rates may differ under various pressures, and many organisms may be chemically adapted to life at a particular range of depth.

A gas-filled sac such as a fish's swim bladder or the lung of a human diver tends to shrink with increasing depth unless more gas is pumped into the sac to equalize the pressure inside and out. Fish that have swim bladders, for example the many fish that carry on extensive vertical migrations (see also the section on Marine Migrations in Chapter 9), have glands that secrete gas during the descent. Similarly, divers may increase their depth range by carrying Self-Contained Underwater Breathing Apparatus (SCUBA), from which they can draw enough air to maintain the balance between internal and external pressure. A diver at 20 meters experiences a pressure of three atmospheres both inside and outside the lungs. When returning to the surface, where there is only one atmosphere of pressure, the diver must continuously exhale to relieve the excess pressure, or else the lungs might rupture.

SOUND IN THE SEA

The propagation of sound waves in the sea is of great importance in oceanography. Sound waves are used to determine ocean depths, thickness of sediment, and loca-

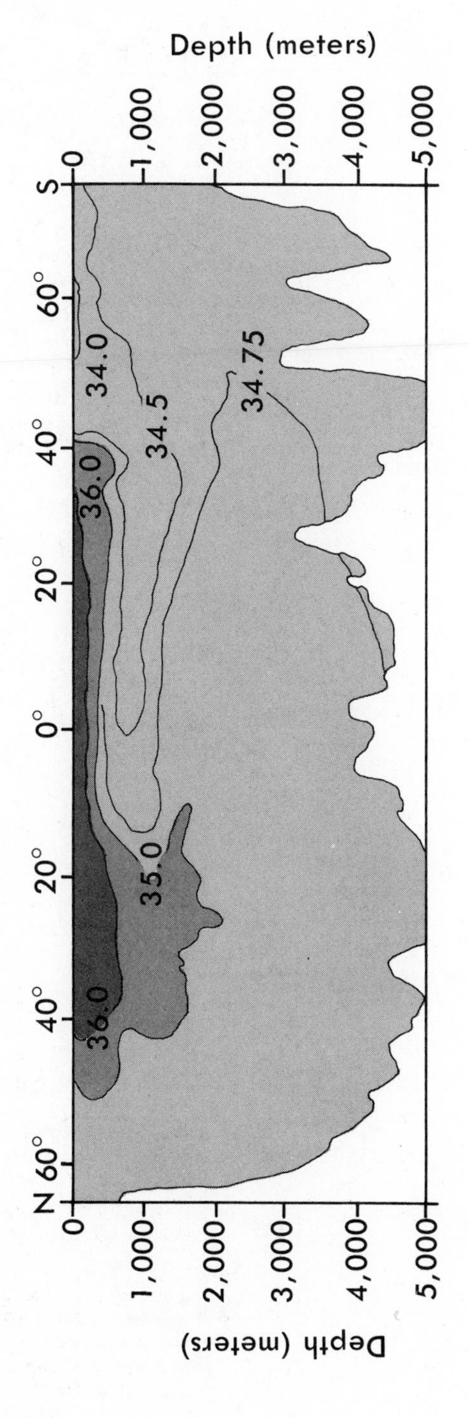

FIGURE 4.21 Longitudinal salinity (‰) structure in the Western Atlantic. (After Wust, 1936.)

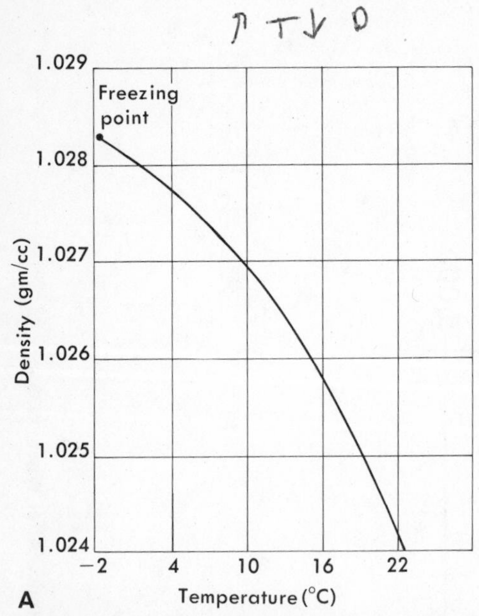

1.030
1.025
1.020
1.015
1.010
1.005
1
0.998
Density (gm/cc)
0
10
20
30
40
Salinity (‰)
B

FIGURE 4.22 *A*, **Density temperature relationship, at a salinity of 35‰ and at atmospheric pressure.** *B*, **Density-salinity relationship at 20° C and atmospheric pressure.**

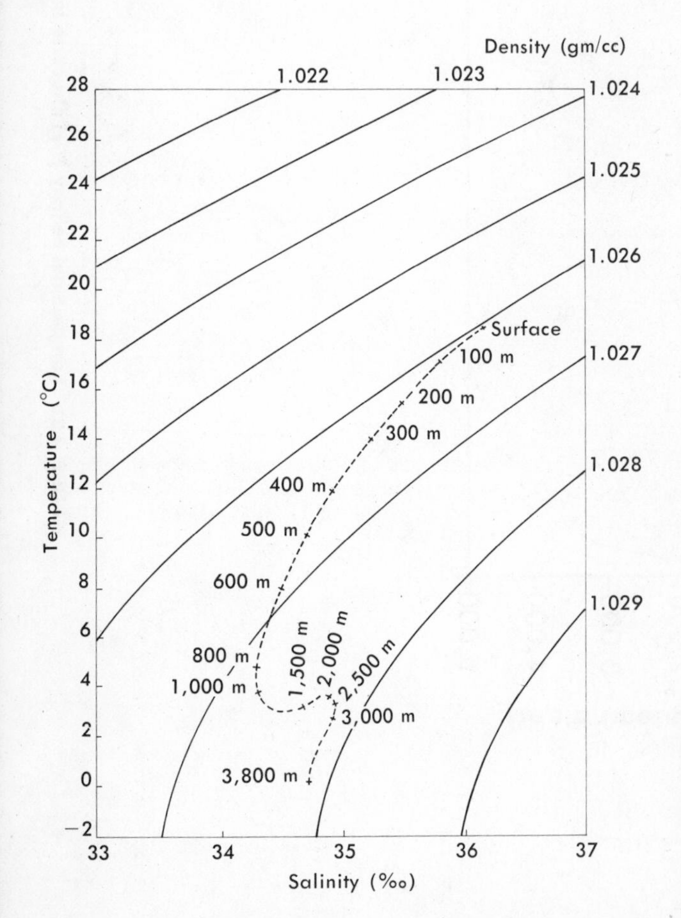

FIGURE 4.23 Temperature-salinity graph. Density values at the ends of the diagonal curved lines refer to the density of the water at atmospheric pressure. In this case, temperature and salinity values have been plotted for 14 water samples, which were collected at various depths at a sampling station located at about 30° S latitude in the Atlantic Ocean. Depths (in meters) are indicated on the dashed line connecting the data points. (After data of Wüst, 1936.)

tion of underwater objects such as fish and submarines. Sound transmission in the sea may also someday provide a new means of communication.

The velocity of sound in the sea varies from about 1,450 to 1,550 meters per second, as compared to about 1,100 meters per second in air, increasing with increasing temperature, salinity, and pressure. Figure 4.28 shows a typical profile of sound velocity and temperature with depth in the ocean. A sound velocity minimum is generally found at about 1,000 meters. Near the poles, however, the minimum is nearer the surface, and in equatorial regions, it may be found at a greater depth. The increase in sound velocity in the upper 1,000 meters is associated with the increase in temperature toward the surface (the thermocline). We know that the gradual increase in velocity below 1,000 meters is mainly due to the increase in pressure with depth, because variations of temperature and salinity are generally not appreciable in deep water. However, near the surface, where temperature and salinity variations with depth are significant, irregularities in this pattern may occur, depending on location, and a sound velocity maximum may be reached at about 100 meters. This effect is of considerable importance in the detection of underwater objects such as submarines. Figure 4.29 shows the refraction of sound waves near the sound velocity maximum. (See Appendix B for a discussion of the refraction of waves.) Refraction of the waves upward above the sound velocity maximum and downward below it produces a zone into which the sound waves do not penetrate. This ***shadow zone*** enables a submarine to avoid detection by a surface ship.

The sound velocity minimum shown in Figure 4.28 can aid ships in distress. A depth charge can be made to explode at the depth of minimum velocity. Refraction causes most of the energy associated with the explosion to be propagated through this zone (Fig. 4.30), known as the ***sound channel*** or ***SOFAR*** (sound fixing and ranging) ***channel***. Experiments have indicated that sound waves can be propagated in this manner for thousands of kilometers. The signals can be picked up by listening devices on ships, and the position of the ship in distress can be determined by triangulation, since the signals will arrive at different times at different listening stations. Although this technique is not fully utilized today, it may someday have great practical significance.

Echo sounding (see Chapter 2) often indicates a "false bottom," called the deep scattering layer, above the true sea floor. This is caused by reflection of sound waves off fish and other animals in the sea (see Chapter 9 for a more complete discussion of the deep scattering layer).

Animals in the sea sometimes make noises that can

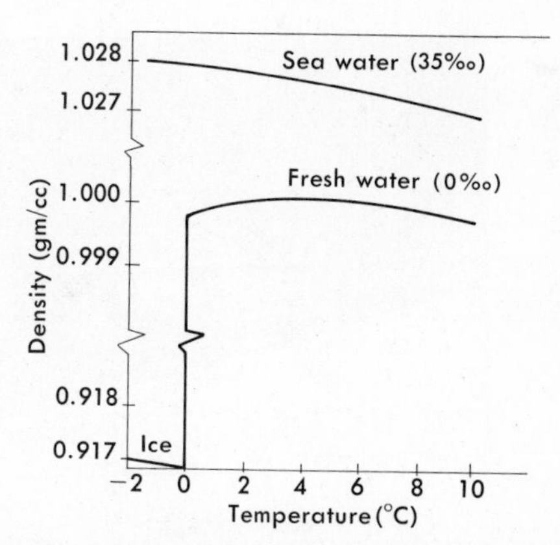

FIGURE 4.24 Temperature-density relationships for fresh water and for seawater of 35‰ salinity.

FIGURE 4.25 Relative composition of single water molecules and 2, 4, and 8 unit polymers of water at various temperatures. (After Euken, 1948.)

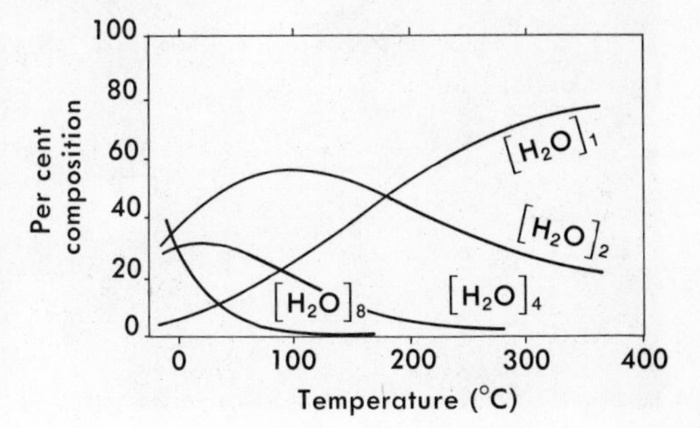

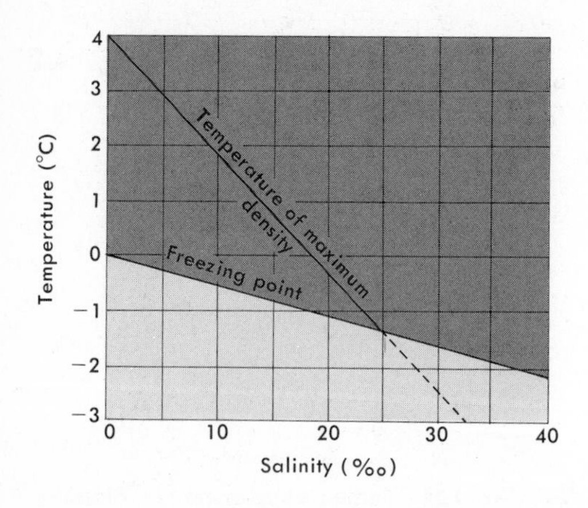

FIGURE 4.26 The relationship of freezing point and the temperature of maximum density to temperature and salinity. Note that above 24.7‰, the water freezes before the temperature of maximum density is reached.

FIGURE 4.27 Ice in the sea. (*A* courtesy of NOAA; B–E courtesy of US Coast Guard.)

A, Arctic *pack ice* (which rarely exceeds 4 meters in thickness) north of Alaska. The instrument shown is a buoy that transmits data about temperature, air pressure, and ice movements via orbiting satellites.

B, *Pancake ice*, a variety of sea ice. Rounded shapes are produced when ice crystals collide against each other owing to motions in the water. An average pancake is about one meter in diameter. Continued cooling may result in the formation of pack ice. The salinity of sea ice rarely exceeds 10‰.

C, A *pinnacled iceberg* in the Arctic. These icebergs are formed when glaciers reach the sea and are broken.

D, The Ross ice shelf, attached to the Antarctic continent. *Shelf ice* is formed in protected bays such as the Ross Sea from a combination of sea ice, snow, and land ice, and may reach about 300 meters in thickness.

E, *Tabular iceberg* in the Antarctic. Broken pieces of shelf ice result in tabular icebergs. Similar icebergs called *ice islands* are formed in the Arctic north of Canada and Greenland and may exceed 50 meters in thickness, but may have areas of hundreds of square kilometers.

FIGURE 4.27 *Continued.*

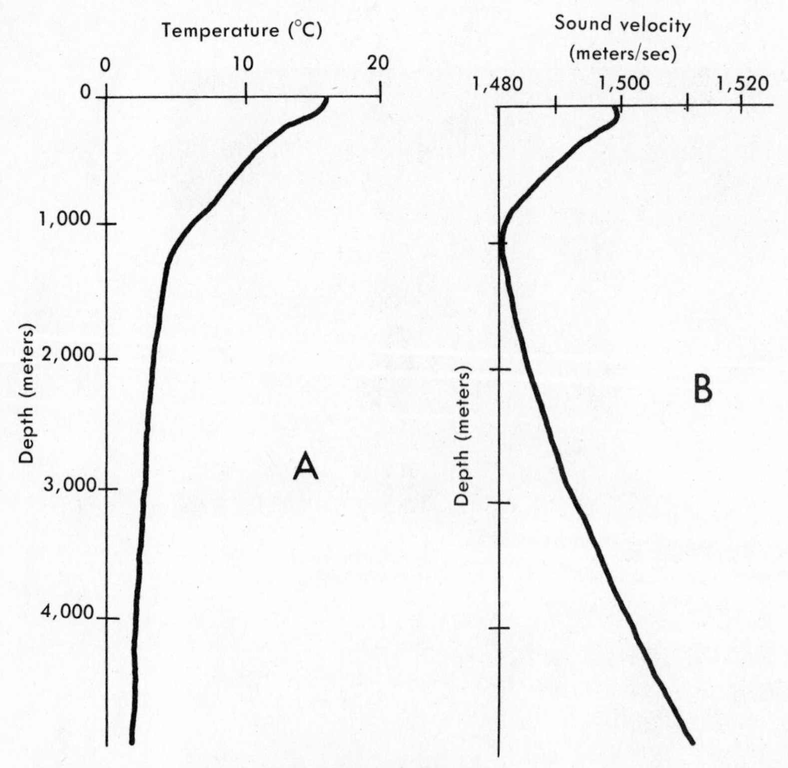

FIGURE 4.28 Typical temperature (*A*) and sound velocity (*B*) variations with depth in the sea.

FIGURE 4.29 A sound velocity maximum (*A*), usually located at a depth of about 100 meters, results in a sound shadow zone (*B*) when the sound source is near the surface. Sound waves are refracted upward above the sound velocity maximum and downward below it.

be heard by other animals. Croakers croak, snapping shrimps snap, and even fish and squids swimming through the water set up vibrations in the water that may be sensed by other organisms, usually through their skin (in the case of many fish, a tubular fluid-filled "lateral line" aids in the perception of vibration). Sounds produced by marine animals, including porpoises, are being studied in the hope that someday we may be able to com-

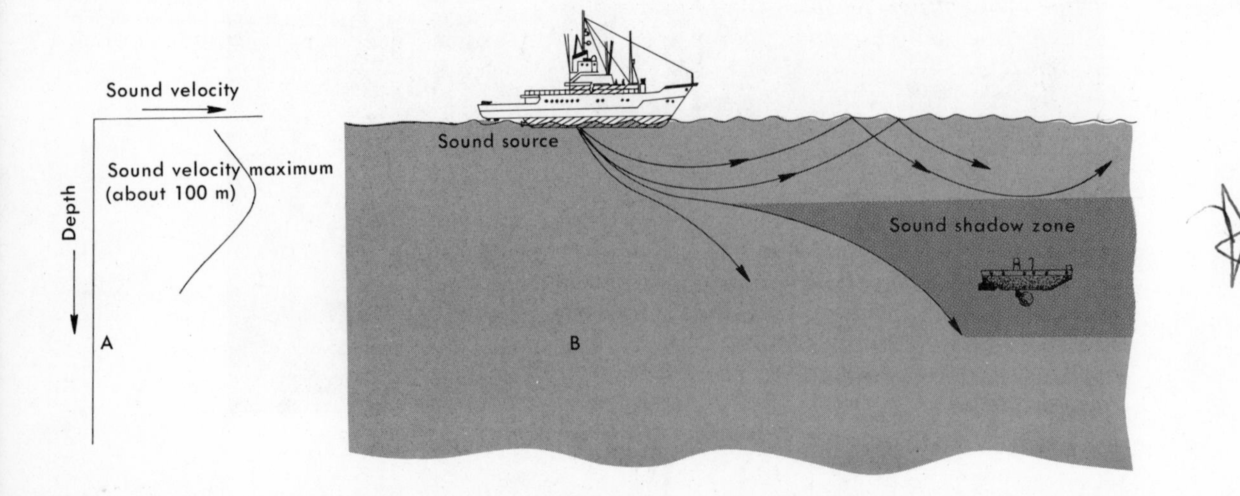

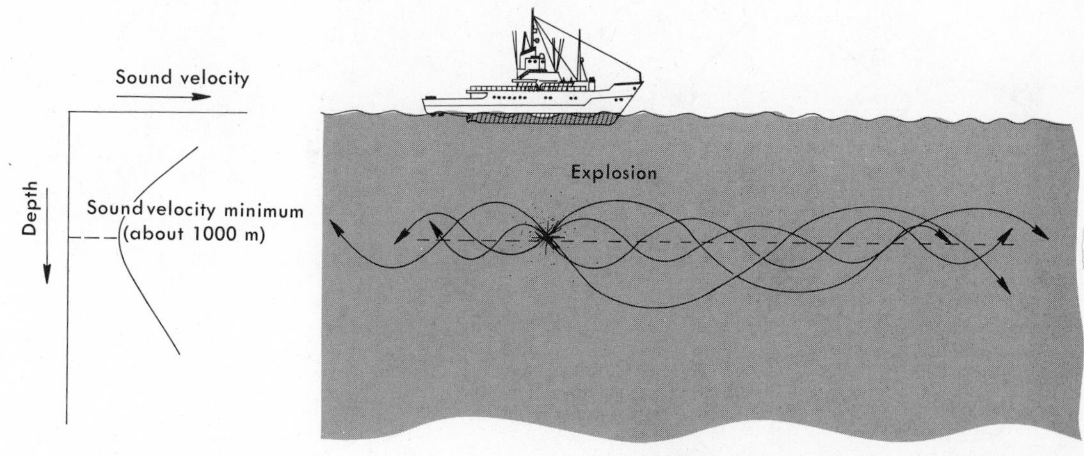

FIGURE 4.30 The sound velocity minimum (*A*), usually located at a depth of about 1000 meters, results in the Sound (SOFAR) Channel (*B*) in the oceans. Sound waves produced near the sound velocity minimum could travel for thousands of kilometers. Sound energy is refracted toward the sound velocity minimum.

municate with them and perhaps train them to respond to orders transmitted through sound. Certain sounds made by animals in the sea may be played back underwater and might serve as a lure. The study of marine biological sounds should, at least, help us to better understand the ecology and behavior of the animals that produce the sounds.

OTHER PROPERTIES OF SEAWATER

Viscosity, or the resistance of a liquid to flow, decreases with increasing temperature (similar to the increased runniness of syrup when heated). Salinity, however, increases the viscosity of water, but only slightly. The relationship of temperature and viscosity causes tropical water to be less viscous than temperate or polar water, thus providing less resistance to the sinking of tropical plankton. It is interesting that many tropical species of minute organisms have much longer spines and hairs than similar or even the same species found in colder, more viscous water.

Surface tension (the tendency of the liquid surface to resist penetration) decreases with increasing temperature and increases with increasing salinity. Surface tension of seawater is very important in supporting the weight of organisms that rest on the surface of the sea, such as the marine water-strider *Halobates* (Fig. 4.31), which is perhaps the only truly oceanic insect. Changes in surface tension due to temperature and salinity probably are not great enough to affect most organisms, however.

Materials that dissolve in water tend to spread out, or diffuse, until they are evenly distributed in the water. However, biological membranes (and other structures) serve

FIGURE 4.31 The oceanic water-strider *Halobates,* which "skates" over the surface of the sea, supported by the surface tension of the seawater. (After Marshall and Marshall, 1971.)

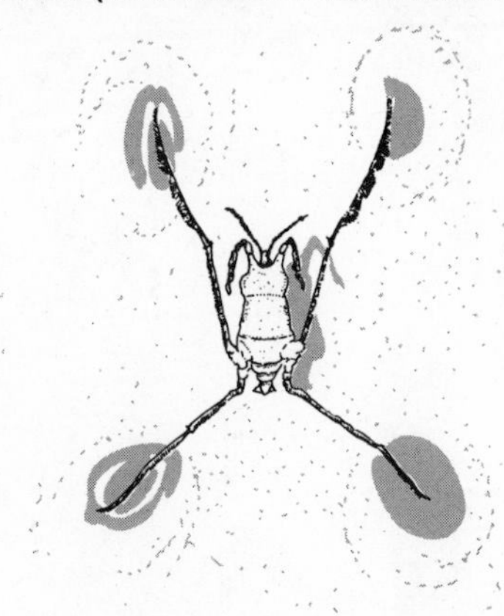

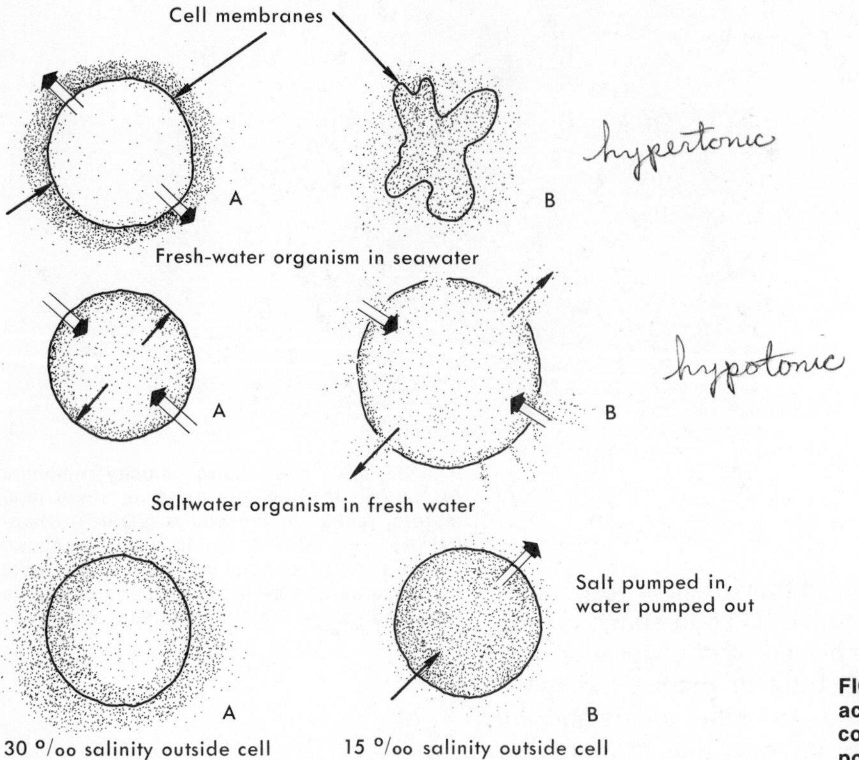

FIGURE 4.32 The flow of water across cell membranes under various conditions of salinity. *A,* Before exposure; *B,* After exposure. Flow of materials: □ H_2O ⇒; ⬚ salt →.

as barriers to restrict diffusion. The membranes that form the boundaries of plant and animal cells are selectively permeable, that is, they allow some molecules, such as water and oxygen, to pass through freely, while others such as the salt ions are restricted in their movement across the membrane (Fig. 4.32). This selective flow of materials is called ***osmosis***. For example, if a selectively permeable membrane separates salt water from fresh water, there will be a net flow of fresh water into the salt water. The pressure on the membrane, caused by the flow of water, is the ***osmotic pressure***.

A freshwater organism placed in the sea tends to shrink as water diffuses through its membranes into the sea, leaving less water in the cell than before. If a strictly marine organism is placed in fresh water, the cells may rupture owing to the diffusion of water into the cells.

Animals such as the green crab (*Carcinus maenas*) that inhabit regions of fluctuating salinity can live in water with a lower salt concentration than is found in their cells. They survive by actively pumping water out of the cell. Their internal salt concentration varies somewhat, but not as much as that of the external water.

SUGGESTED READINGS

Bowditch, Nathaniel. 1965. *American Practical Navigator.* Washington, D.C., Government Printing Office.

Donn, William L. 1965. *Meteorology,* 3rd ed. New York, McGraw-Hill.

Neumann, Gerhard, and Willard J. Pierson, Jr. 1966. *Principles of Oceanography.* Englewood Cliffs, New Jersey, Prentice-Hall.

Othmer, Donald, and Oswald Roels. 1973. Power, Fresh Water and Food from Cold Deep Seawater. *Science,* ***182***:121–125.

Pickard, George L. 1963. *Descriptive Physical Oceanography.* New York, Pergamon Press.

Sverdrup, H. U., M. W. Johnson, and R. H. Fleming. 1942. *The Oceans.* Englewood Cliffs, New Jersey, Prentice-Hall.

von Arx, William S. 1962. *An Introduction to Physical Oceanography.* Reading, Mass., Addison-Wesley.

On April 4, 1975, the 170 meter (557 ft) oil tanker *Spartan Lady* broke in two and sank in heavy seas about 200 km southeast of New York. The dark streaks in the photos are oil spilling onto the sea. (Courtesy of US Coast Guard.)

CHAPTER 5

Ocean Waves

The most readily observable phenomena in the sea are the waves. They range from small ripples to gigantic and highly destructive waves such as tsunamis, which may rise over 30 meters above normal sea level by the time they reach the shore. Tides, which are caused by the gravitational attraction of the moon and the sun on the waters of the earth, also qualify as waves and will be discussed in the next chapter. Waves are of tremendous importance to us, not only because they make us seasick and enable surfers to surf but also because of their sometimes harmful effects on beaches and artificial structures. Someday, electricity may be generated from ocean waves. In addition, waves have many other far-reaching influences on the oceans. For instance, waves breaking on the sea surface make it easier for atmospheric oxygen and carbon dioxide to be dissolved in seawater. Wave action is also responsible for the mixing of the surface water layer. Both of these aspects are of great significance in biological oceanography. Ocean waves are important even in space exploration. The ability to predict wave conditions in the oceans is vital to the safe return of astronauts, because recovery operations would be almost impossible on a rough sea. For example, one Gemini capsule flipped over and sank in somewhat choppy seas. Luckily, the astronauts were rescued.

Most ocean waves are caused by wind. However, they can also be caused by submarine earthquakes, submarine landslides, submarine volcanic eruptions, landslides into the sea, ships, and the gravitational attraction of the moon and the sun on the earth.

NATURE OF WAVES

Figure 5.1 shows several uniform ideal ocean waves generated by any of the forces mentioned earlier. The distance between two adjacent crests (or troughs) is called the ***wavelength (L).*** The vertical distance between the crest and the trough is the ***wave height (H).*** The time required for two successive crests (or troughs) to pass by a fixed point is the ***period (T)*** of the wave. The ***speed (C)*** of the wave is the distance the wave travels in a unit of time such as meters per second, or kilometers per hour. A

how many in a certain period of time

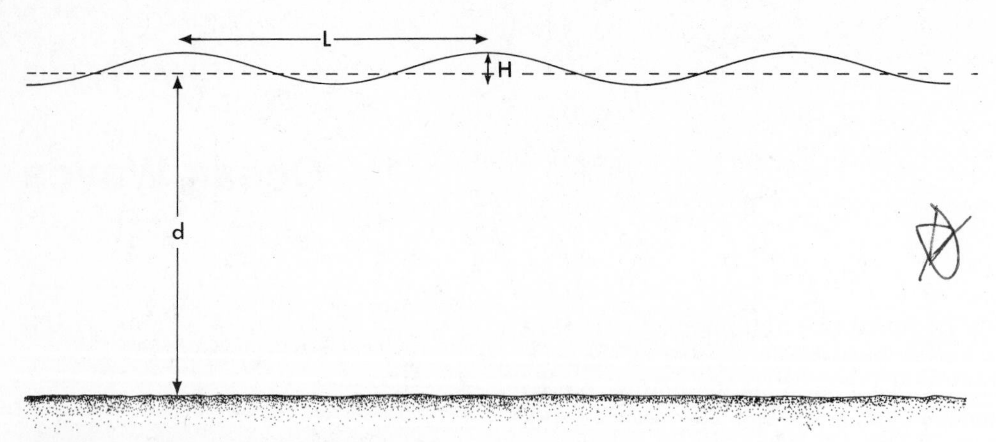

FIGURE 5.1 Ideal ocean waves. L = wavelength, H = wave height, d = water depth.

very useful but simple relationship using the preceding factors for any wave is L = CT. For example, if the period of a wave is 10 seconds, and the wave speed is 15 meters per second, the wavelength is 150 meters.

In reality, ocean waves are never as simple as those shown in Figure 5.1; the real ocean surface is much more irregular. Waves moving in different directions, which may or may not have the same periods, lengths, and heights can interfere with each other, producing a complex sea surface pattern (Fig. 5.2).

DEEP AND SHALLOW WATER WAVES

Ocean waves are classified into two major categories: ***deep*** and ***shallow water waves.*** A wave is called a deep water wave if the ratio of water depth to wavelength (d/L) is greater than 1/2, and it is called a shallow water wave if this ratio is less than 1/25.[1] If the ratio d/L is between 1/2 and 1/25, it is called an intermediate wave. It is important to note here that the terms "deep" and "shallow" water do not necessarily imply deep and shallow in the usual sense. For example, a tsunami having a wavelength of 150 kilometers will travel as a shallow water wave in waters less than 6 kilometers in depth (d/L = 6/150 = 1/25); ripples, however, are classified as deep water waves even near the beach or in large puddles. Deep water waves become shallow water waves as they move into shallow areas in which d/L is less than 1/25.

Many differences exist between deep and shallow water waves. Figure 5.3 shows the motions of water particles in these two types of waves. The water particles as-

[1]Note that "d" refers to the ocean depth from the surface to the ocean bottom at that location and should not be confused with "H," which refers to wave height as defined earlier.

FIGURE 5.2 Complex wave pattern on the open sea. (Courtesy of US Coast Guard.)

sociated with deep water waves move in circles, while those associated with shallow water waves move in ellipses. In deep water waves, the circular orbital speeds as well as the size of the circles decrease with depth and become almost negligible at a depth equal to half the wavelength (L/2). For shallow water waves, the horizontal particle speeds and the horizontal dimensions of the ellipses remain unchanged from the surface to the bottom of the ocean. The ellipses, however, get flatter with depth, and at the bottom, the particles move back and forth, almost horizontally. Thus in the case of tsunamis and other large waves which fit the definition of shallow water waves, water movement can be expected even in the deepest portions of the oceans. These waves can have

FIGURE 5.3 Particle motions of (A) deep and (B) shallow water waves. The straight arrows indicate the direction of wave advancement.

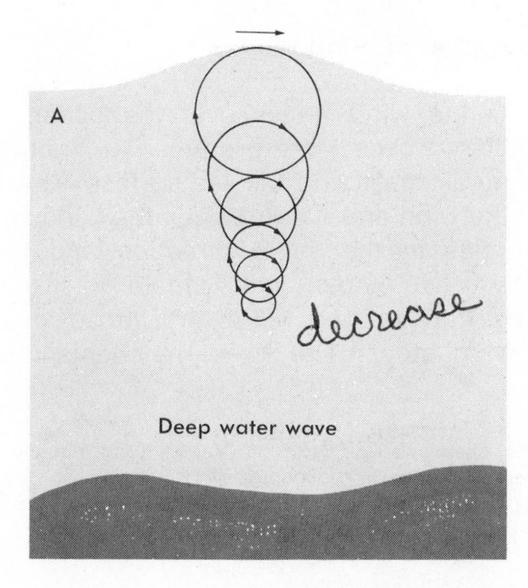

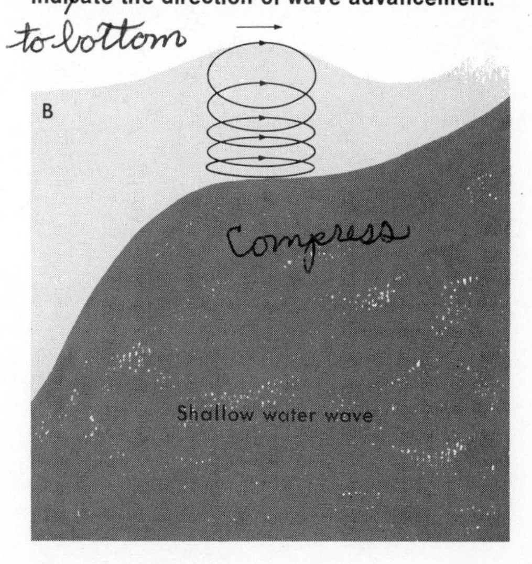

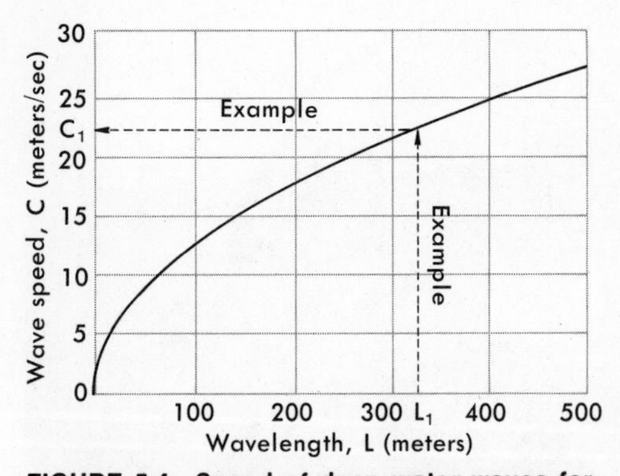

FIGURE 5.4 Speed of deep water waves for various wave lengths. This graph can be used to determine the speed of the waves (C_1) by knowing the length of the wave (L_1). In this case, a deep water wave with a length of 325 meters moves at a speed of about 22.4 meters per second.

great influence on shallow bottom areas such as continental shelves and portions of the continental slopes, although their effects on the deep sea bottom are not considered to be significant.

Surprisingly, there is almost no forward motion of water in waves until they break near the shore. This effect can be demonstrated by observing a piece of driftwood floating in the ocean. The wood does not move forward with the waves but remains nearly stationary as the waves pass. Or watch the movement of a spot of dye or food coloring in a tank of water. The waves will pass by, leaving the dye almost undisturbed.

Another important difference between deep and shallow water waves is their respective speeds. The speed of a deep water wave depends only on the wavelength, while that of a shallow water wave depends only on the depth of water.

Figure 5.4 shows the speed of deep water waves for various values of wavelengths. Figure 5.5 shows the speed of shallow water waves for various values of depths. These graphs can be used to determine the speed of the waves by knowing the wavelength in deep water or the depth in shallow water. The speeds of waves are given by the following equations:[2]

Deep water wave ($d/L > 1/2$): $C_{deep} = \sqrt{gL/2\pi}$

Shallow water wave ($d/L < 1/25$): $C_{shallow} = \sqrt{gd}$

where C = wave speed in meters per second; g = gravitational acceleration of the earth (9.8 m/sec²); L = wavelength in meters; d = depth of water in meters; and $\pi = 3.14$.

WIND WAVES

Generation of Wind Waves

As the wind blows over the ocean, ripples (called "capillary waves") are produced by friction between the moving air mass and the sea surface. These ripples move with the wind and die out when the wind stops. If the wind continues in the same direction and the wind speed exceeds two knots, more stable waves (called "gravity waves") are formed, which will move forward even after the wind stops. The exact mechanism by which wind

[2]For intermediate waves (d/L between 1/2 and 1/25) the wave speed in meters per second is given by the formula: $C_{int} = \sqrt{(gL/2\pi)\tanh(2\pi d/L)}$. Values of tanh (hyperbolic tangent) can be found in standard mathematical tables.

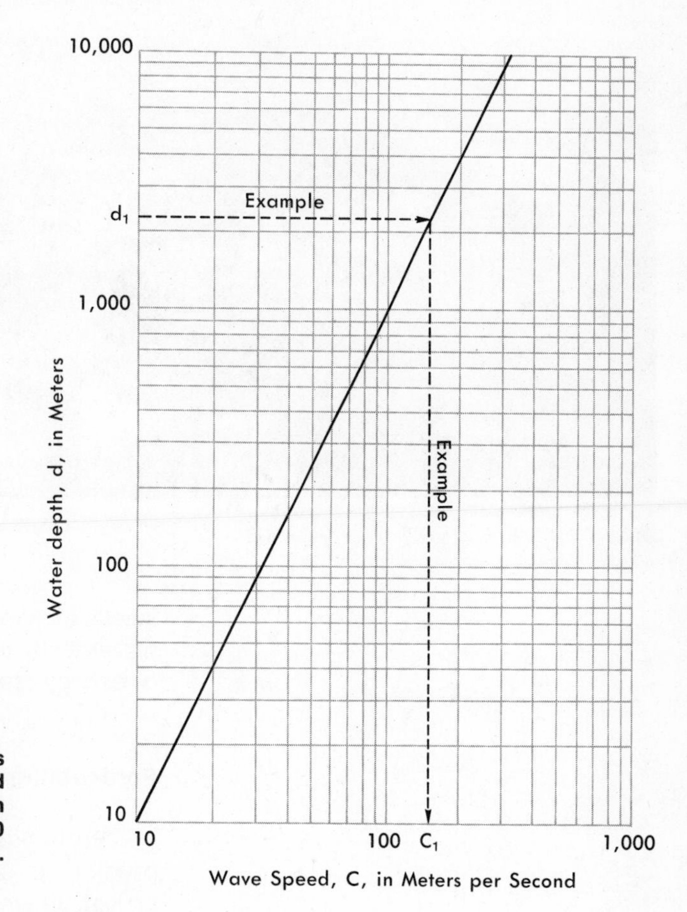

FIGURE 5.5 Speed of shallow water waves for various depths. This graph can be used to determine the speed of the waves (C_1) by knowing the water depth (d_1). In this case, a shallow water wave in water depth of 2,250 meters moves with a speed of 148 meters per second. Note that both scales are logarithmic.

energy is transformed into wave energy is not well understood. The ridges of the ripples (crests) certainly make it easier for the wind to pile up the water and push it forward. As long as the wind continues to blow, these waves will grow in size up to a maximum that is dependent on wind speed. Waves 12 meters high are about the maximum ordinarily present in the open ocean, although waves in excess of 15 meters have occasionally been observed from ships on the high seas. In 1933, the Navy tanker USS *Ramapo* encountered record high waves. One wave of about 35 meters was observed, with the ship steaming with the waves.

Waves in a storm area (generating area) are called ***sea*** (Fig. 5.6). The nature of these waves depends on wind speed, duration of the wind, and the distance over which the wind blows (called the ***fetch***). The sea waves are choppy and have sharp, short crests which are almost randomly oriented. Away from the storm, the small short-period waves decay and die out, leaving only the more uniform, smooth, long-period waves, called ***swells.*** However, irregularities in swells may be produced by local winds existing outside the generating area or by interference from

FIGURE 5.6 Waves in the storm area are called sea. (Courtesy of Woods Hole Oceanographic Institution.)

swells arriving from different generating areas. Swells can travel for thousands of miles in deep water with little loss of energy (see Fig. 5.8*b*).

Forecasting of Wind Waves

Perhaps most important to mankind is our ability to predict the size of waves, their speed, path, and time of arrival. Wave forecasting is important to navigation and shore installations as well as to sailing and surfing. Even oceanographical cruises require predictions so that scientific observations can be made safely in relatively quiet seas.

Wave forecasting is made empirically, on the basis of numerous past observations at sea and on statistical analysis of data, rather than by pure theory. The sea condition at any place depends on the following factors:

(1) wind speed
(2) duration of the wind
(3) fetch, the distance over which the wind blows
(4) distance away from the storm area
(5) local bathymetric conditions (bottom contours)

The deep water waves in the generating area (the sea), however, depend only on wind speed, duration, and fetch, information about which can be obtained from meteorological sources. There are many techniques of forecasting deep water waves. One of these, developed by C. L. Bretschneider, is shown in Figure 5.7. Knowing the wind speed, the duration, and the fetch, Figure 5.7 can be used to predict the ***significant wave height*** which is the average height of the highest one third of all the waves that will be observed in the area (generally the

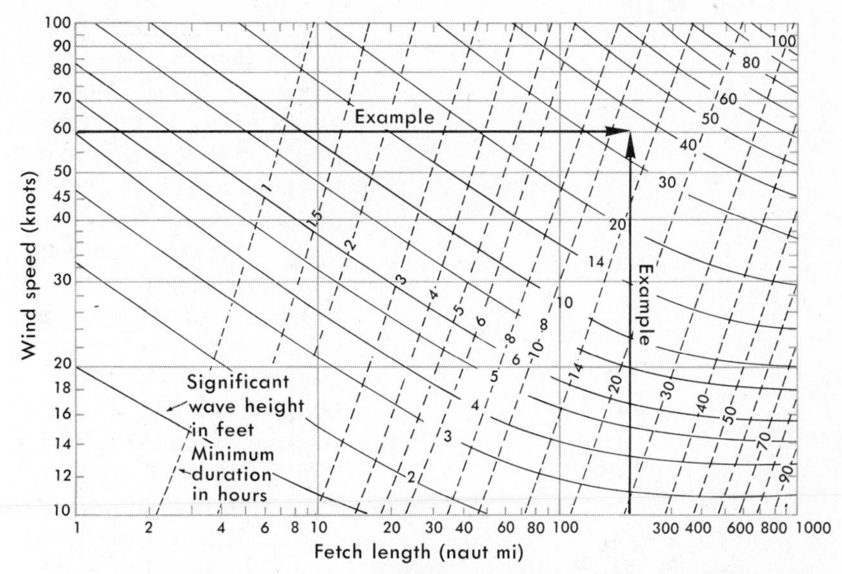

FIGURE 5.7 The Bretschneider deep water wave forecasting curves. Example: For a wind speed of 60 knots, a fetch length of 200 nautical miles, and a minimum wind duration of 12 hours, the significant wave height is predicted to be 35 feet. (Modified from US Army Coastal Engineering Research Center, 1966.)

height most responsible for damage). Knowing the significant wave height, one can easily compute the average height of all the waves, the average of the highest 10 per cent of all waves, and the most frequent wave height, using the following relationships:

average wave height = 0.63 × significant wave height

average height of the 1/10 highest waves = 1.27 × significant wave height

most frequent wave height = 0.50 × significant wave height

In many instances, information about wave conditions near coastal areas (swells) or in areas far away from the generating area is desired. The nature of these waves will depend, however, on the distance away from the storm area, the interference of waves from other generating areas, local wind conditions, and bathymetry. Forecasting of swells is somewhat more complex and involved than forecasting of sea, and the reader is referred to the references at the end of the chapter for information on this subject.

WAVE REFRACTION

A casual trip to the beach raises many interesting questions about waves. Why do waves always move toward the beach, usually with crests parallel to it? Why are they more closely spaced (piled up) near the beach than offshore? Why is there more wave action on some parts of the beach than others? To answer these questions, consider a deep water wave. It will travel unchanged in direction at a constant speed which depends on its wavelength. When water depth equals half the wavelength, the wave starts "feeling" the bottom and progressively slows down. In addition, it will bend or undergo ***refraction,*** depending on the angle of approach with respect to the shallower bottom contours (Fig. 5.8). Actually, the part of the wave to first feel bottom (in shallow water) slows down, allowing the parts of the wave that are still in deeper water to "catch up" until they, too, feel bottom. Soon the wave becomes nearly parallel to shore. The principle involved in the refraction of water waves (Snell's Law) is similar to that for the refraction of seismic waves, light waves, and sound waves discussed earlier. (The refraction of waves is discussed in detail in Appendix B.) In Figure 5.8, the arrows show the direction of wave advancement (the ray path) from deep to shallower water. Refraction occurs only when the ray paths are not perpendicular to the bottom contours.

As waves pass through progressively shallower waters, the ray paths become more nearly perpendicular to the bottom contours, as shown in Figure 5.8; that is, the wave crests will be more nearly parallel to the beach. Thus it can be said that waves are "drawn" toward the beach by refraction. (See also Figure 5.9.)

When waves reach shallow water, whether they are refracted or not, their speed and wavelength decrease and wave height increases, but their periods do not change. In order to keep the same period, the wave crests must get closer together (pile up) as they approach the beach.

An important consequence of wave refraction is that waves concentrate their energy at headlands (points) and disseminate energy into bays (Fig. 5.10). The coastline in the illustration has two points or headlands and a bay between them. The bathymetric contours in the shallow water area, as in many parts of the world, are roughly par-

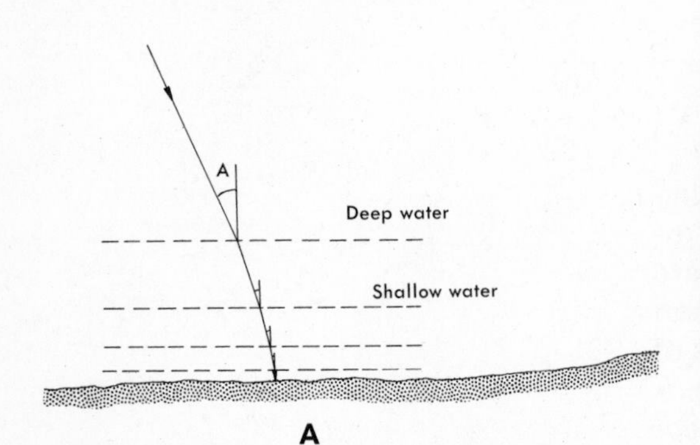

FIGURE 5.8 *A,* **Refraction of waves as they pass through regions of decreasing water depth. Dotted lines represent bottom contours.** *B,* **When swells reach shallow water, sometimes having traveled thousands of kilometers, they may refract, becoming nearly parallel to the shore. (From US Naval Oceanographic Office H.O. Pub. No. 606-e[1950].)**

allel to the coastline. As the waves enter shallow water, they are refracted. Most of the wave energy is concentrated at the headlands; therefore, less energy reaches the bay, which will be relatively calm. The headlands will undergo greater erosion because of this concentrated wave action, the bay will receive sediments, and the eventual result will be a smooth coast. Even if the waves approach the shallow water area from a different direction than that shown in the figure, as they near the shore, refraction will eventually change them to approximately the orientation shown.

Wave refraction studies are of great importance in choosing locations for new harbors, jetties, and breakwaters. It is essential to know in advance whether the harbor will be filled with sand or will receive too much wave energy, and whether the jetties and breakwaters will be strong enough to withstand the impact of wave action. Another application of wave refraction data is in prediction of the paths as well as the time of arrival of tsunamis (discussed later in this chapter) in various areas so that damage can be minimized.

BREAKERS AND SURF

As waves travel from deep water into shallow water areas ($d/L < 1/2$), they begin to "feel" the bottom and to slow down, as the speed is controlled by the water depth. All waves—whether they are seas or swells, deep or shallow water waves—will break before the ***wave steepness*** [ratio of wave height to wavelength (H/L)] reaches 1/7. These waves break because the water particles tend to move faster than the speed of the waves. In shallow water, as the wave slows down, the wavelength shortens and the wave height increases. Shallow water waves break when water depth equals 1.3 times the wave height ($d = 1.3H$). Thus, one can determine the depth of coastal oceans merely by knowing the location and height of various breaking waves. This simple but useful technique was used during World War II landings on many coasts where information on the bottom topography was lacking.

Breakers are classified as spilling, plunging, or surging breakers (Fig. 5.11).

Spilling breaker: breaks over a large distance and over relatively flat bottom; the crest breaks and "slides" down the face of the wave.

Plunging breaker: front of crest curls and breaks once, over bottom of intermediate steepness, forming a tunnel along the face of the wave.

Surging breaker: does not really break but surges up the slope of very steep beaches or sea walls.

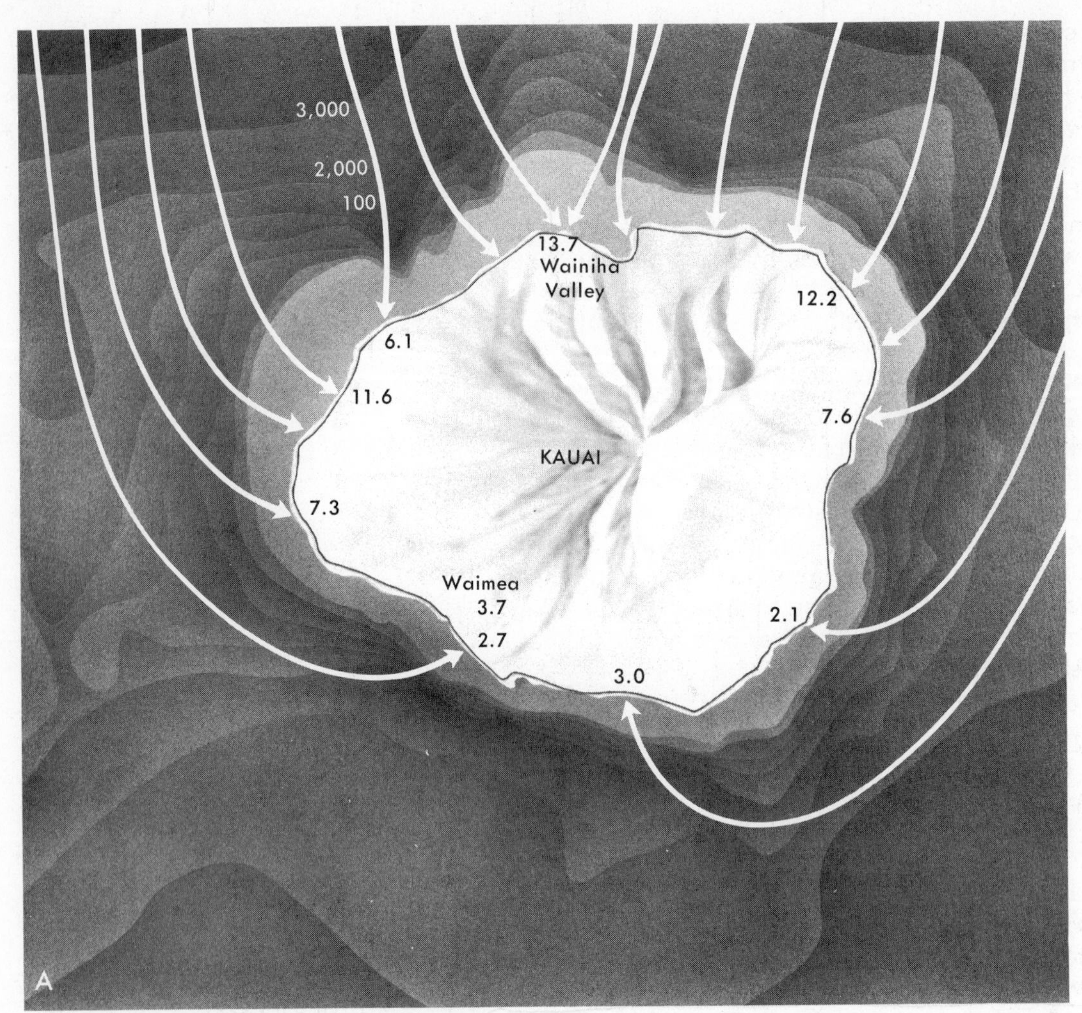

FIGURE 5.9 *A,* **Refraction of tsunamis around the island of Kauai, in the Hawaiian Islands. The tsunamis originated in the Aleutian area following the 1946 earthquake. Solid lines indicate ray paths, and the arrows show their direction. Notice that these tsunamis also arrived on the south side of the island, even though they originated about 4,000 km to the north. Tsunami heights along the shore and ocean depths are given in meters.**

Waves, especially those in the surf zone, may provide a free ride for porpoises and people. A person may ride a surfboard on the forward surface of the wave and slide downward. Because the wave crest is always moving forward, the surfer may ride the wave onto the beach. By moving at angles to the crest less than 90°, one can adjust the forward motion of the surfboard to the speed of the wave and ride the wave at higher speeds than the wave itself travels. The surfer maintains a relative position on the wave by traversing a path nearly parallel to the wave crest, attaining maximum speed. One might even surf under the crest of a plunging breaker in a sort of aqueous tunnel or cave, called a "tube" by surfers.

The forces involved in surfing are shown in Figure 5.12. The force of buoyancy (B) is always perpendicular to the surface of the water, and the force of gravity (G) is always downward. Thus, the resulting force (F) is directed

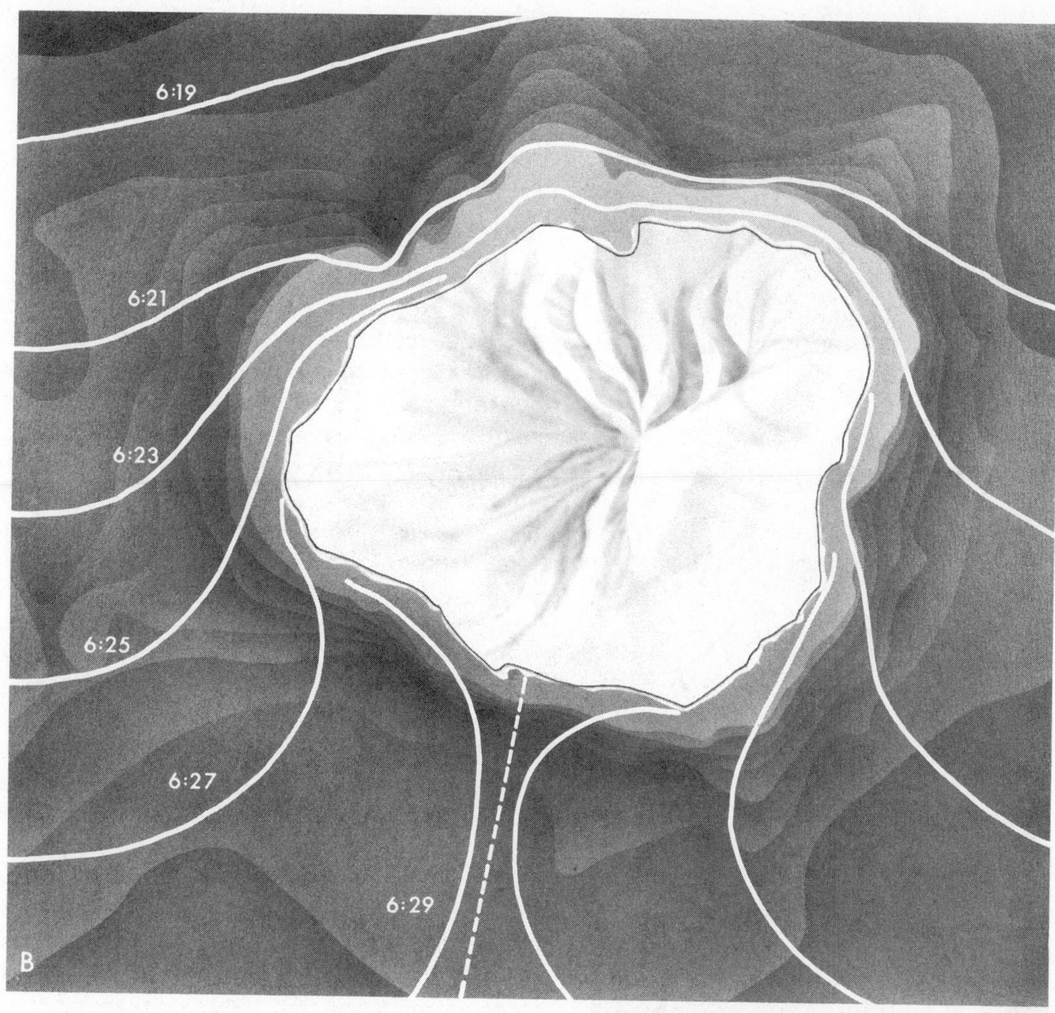

FIGURE 5.9 (*Continued*) *B*, Position of wave crest at different times around the island. Note that the waves traveled around the island in less than 10 minutes. (Modified after Shepard et al., 1950.)

forward and downward along the wave face, producing a forward motion of the surfboard. The principles depicted in the diagram apply also to surfing without a board ("body surfing") and are used by porpoises below the water

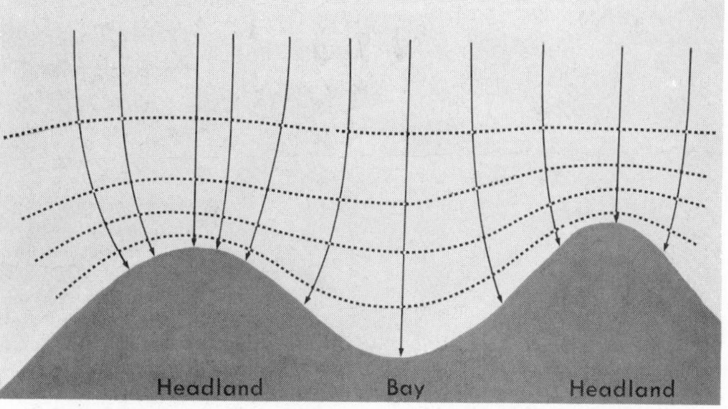

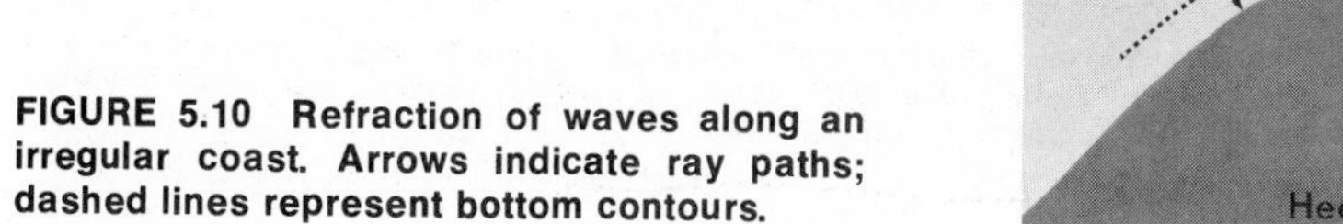

FIGURE 5.10 Refraction of waves along an irregular coast. Arrows indicate ray paths; dashed lines represent bottom contours.

A

B

C

FIGURE 5.11 *A*, Spilling breaker; *B*, Plunging breaker; *C*, Surging breaker. (From US Army Coastal Engineering Research Center, 1973.)

surface. The porpoises can ride on a "surface" of equal pressure in the wave. These small whales have been known to ride the wakes of ships with little effort.

TSUNAMIS

Tsunamis or ***seismic sea waves*** are giant ocean waves usually caused by large submarine earthquakes. They are sometimes called "tidal waves" although they have nothing to do with tides. In the open ocean, tsunamis usually have lengths of about 150 kilometers but may have lengths of up to 1,000 kilometers, speeds sometimes exceeding 800 kilometers per hour, and heights of about one meter. The period of a tsunami ranges from about 10 minutes to an hour. Thus, they are unnoticed in the open sea and cause no damage to ships. But when they approach shallow water, their wave heights increase drastically, sometimes running up the shore more than 30 meters above normal sea level. Tsunamis may advance like a ***bore*** (an abruptly advancing ridge of water) in very shallow water, sometimes exceeding 15 meters per second (about 34 miles per hour). Tsunami damage (Figs. 5.13 to 5.15) is usually great near coasts where the offshore slope is very steep. Because of their long wavelengths, they travel as shallow water waves ($C = \sqrt{gd}$) over most portions of the oceans. For an ocean depth of 4,000 meters, $C = \sqrt{9.8 \text{ m/sec}^2 \times 4{,}000 \text{ m}} = 715$ km/hr (445 mph). Tsunamis can travel thousands of miles with little loss of energy and are able to cause tremendous damage to coastal areas far from their point of origin. For example, the tsunami from the 1964 Alaskan earthquake caused many deaths and much property damage along the Pacific coasts of Canada and the United States. In Crescent City, California, 12 lives were lost and property damage totaled $11 million. The tsunami of May 23, 1960, at Hilo, Hawaii, caused by an earthquake in Chile, resulted in the death of 61 people and about $20 million damage to property.

Most tsunamis are caused by sudden vertical displacement of the sea floor, as by vertical faulting (Fig. 5.16), submarine landslides, or landslides into water, all of which are caused by earthquakes located near coastal areas. Most tsunami-producing earthquakes have magnitudes greater than 6.5 on the Richter scale[3] and originate

---

[3]Magnitude refers to the energy released by the earthquake at its source. Because the magnitude scale is logarithmic, an increase of one "magnitude" actually involves about a hundredfold increase in energy release.

Some examples of magnitudes of well-known earthquakes include the following: San Francisco, 1906 (8.3); Chile, 1960 (8.0); Alaska, 1964 (8.4); and San Fernando (near Los Angeles), 1971 (6.5).

A

B

FIGURE 5.12 ***A,*** **Forces involved in surfing. B = buoyancy; G = gravity; F = resultant force.**
B, **Surfing on (and under) a large wave. (Photos by Dan Merkel, courtesy of *Surfer Magazine.*)**

at depths usually less than 50 kilometers. It is important to note that not all such earthquakes result in tsunamis. For example, the California earthquakes do not produce tsunamis because the faults associated with them move hori-

FIGURE 5.13 *A*, Crescent City, California, before the Alaska earthquake of March 27, 1964. *B*, Enlarged view of the outlined area two days after the tsunami hit the city. Note the destruction of buildings. (Photos courtesy of NOAA.)

zontally rather than vertically. It is generally believed that in order for a tsunami to be produced vertical displacement must be at least several meters in a large area and the fault length at least 100 km. Earthquakes may trigger submarine landslides (Fig. 5.16) in regions where large quantities of sediment rest on the continental slope. This is the mechanism which is responsible for some turbidity currents and the resulting breakage of submarine telegraph cables (see Chapter 2). Tsunamis can also be produced by landslides into water and submarine landslides that are unrelated to earthquakes, and by submarine volcanic eruptions or explosions. The explosion of the volcano Krakatoa (1833) produced large tsunamis in the Pacific, which killed more than 35,000 people. However, landslides into water and submarine landslides are generally observed to cause only local tsunamis.

Tsunamis are most common in the Pacific, because earthquakes are more common there, but they have occurred in all oceans. In fact, one of the largest tsunamis in history occurred in the Atlantic Ocean after the 1755 submarine earthquake near Lisbon, Portugal. Waves rose up to about 15 meters along the coast of Morocco, and at Tangier the water rushed inland 2.4 kilometers. This tsunami radiated in all directions and was observed as far away as the West Indies.

A tsunami is not a single wave but rather a series of waves which may persist for hours. The first motion may be either an advance or a withdrawal of water. Many tide records indicate an initial rise in sea level about a third as large as the succeeding withdrawal. The initial lowering of water may be like a very low tide, exposing animals and seaweed that are normally covered. People have been known to be lured toward the sea by this lowering of water, only to be killed by the advancing wall of water in

FIGURE 5.14 *A*, The port area at Seward, Alaska, before the March 27, 1964, earthquake. *B*, The same location after the earthquake, showing tsunami damage along the waterfront. (Photos courtesy of US Geological Survey.)

the next large wave. The most destructive waves, however, arrive within the first four hours. Figure 5.17 shows the records made at two locations of the tsunami that resulted from the Alaskan earthquake of 1964.

Although tsunamis cannot be prevented, they can be predicted, and death and damage to property can be minimized. The National Oceanic and Atmospheric Ad-

FIGURE 5.15 Tsunami damage at Kodiak, Alaska, following the 1964 earthquake. (US Navy photo.)

ministration of the Department of Commerce operates the Seismic Sea-Wave Warning System with headquarters in Honolulu, Hawaii, serving the entire Pacific coast. The principles involved in the warning system are as follows: (1) knowledge of the exact location of earthquakes in coastal areas, using critically placed seismic stations around the Pacific; (2) detection of waves having tsunamilike characteristics generated after the earthquake at intermediate locations; (3) a communications network to transmit the information; and (4) construction of probable paths and determination of expected arrival time of the tsunami at various coastal areas, using knowledge of the principle of wave refraction. Since we know the major tsunami-producing areas of the Pacific, refraction diagrams and travel times for numerous hypothetical tsunamis can be worked out ahead of time. Figure 5.18 is a wave refraction diagram for the 1960 Chilean tsunami. The success of any warning system depends mainly on the willingness of people to obey evacuation instructions. Many deaths have been prevented since the inception of the system in 1948. However, the system is ineffective in areas close to the source of the tsunami.

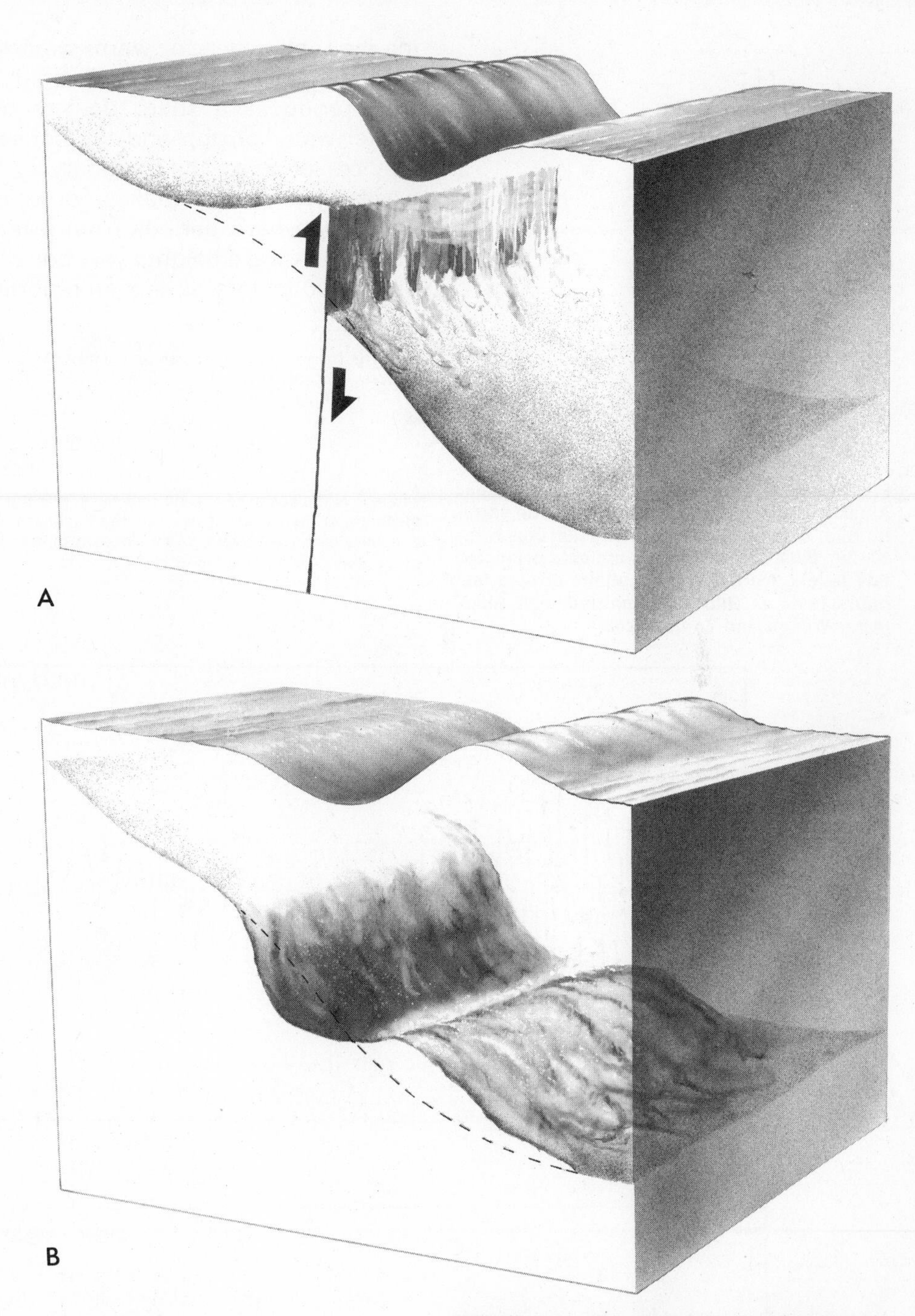

FIGURE 5.16 Mechanics of tsunami generation. *A*, Fault; *B*, Submarine landslide. The dashed line in both cases indicates the previous profile of the sea floor.

INTERNAL WAVES

Just as wind waves are produced at the air-sea boundary by pressure differences, waves known as ***internal waves*** (waves at an internal boundary) may be formed between layers of water of different densities. It takes less energy to produce an internal wave between layers of water than to produce a comparable surface wave between the sea and the air. Fresh water from river run-off, ice melt, or a layer of warm water at the surface may be much lighter than the salty or cold water below. This contrast produces a sharp density boundary between the lighter water on top and the denser water below. Waves may be set up on the boundary by forces such as surface waves, tides, earthquakes, or ships' propellers. These waves can have periods from a few minutes to hours or days. Their wave heights vary considerably, and heights of about 100 meters have been reported. Their speeds,[4] how-

[4]The following equation can be used to compute the speed of internal waves in a two-layered ocean:

$$C = \sqrt{\frac{gd''d'}{d}\left(\frac{\rho'' - \rho'}{\rho''}\right)}$$

where C is the speed of the internal wave, g is the acceleration of gravity, d′ is the thickness of the surface layer, d″ is the thickness of the deeper layer, d is the total thickness of the two layers, ρ' is the density of the surface layer, and ρ'' is the density of the deeper layer (see Fig. 5.19).

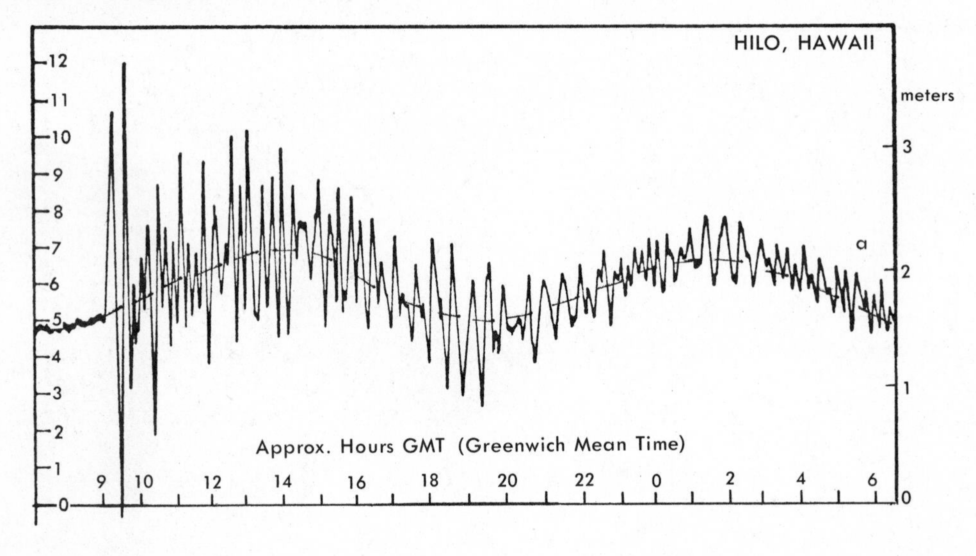

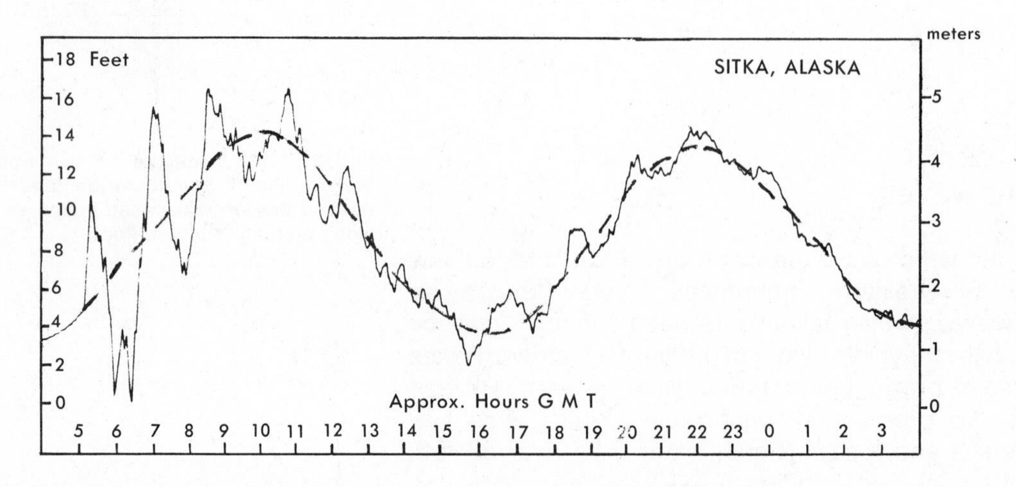

FIGURE 5.17 Tsunami resulting from the 1964 Alaskan earthquake as recorded by tide gages at Hilo, Hawaii, and Sitka, Alaska on March 28–29, 1964. Dotted curves indicate projected tide levels. Note that the tsunami arrived four hours later at Hilo as compared with Sitka. (After Wilson and Torum, 1968.)

ever, are very low compared to ordinary surface waves. Internal waves may also develop in areas where there is a continuous increase of density with depth.

The most familiar internal waves are the short period waves produced by slow-moving ships in a stratified ocean, where a thin layer of low-density water overlies dense water. Ships are known to lose speed when turbulence from the propellers churns into the density boundary (Fig. 5.19) and sets up internal waves. The ship "sticks" and makes very little progress. This is known as "dead water." At speeds of about two knots or less, much of the ship's energy may be spent in making internal waves rather than in propelling the ship forward. The severity of this problem depends on the propeller design and may be overcome by increasing the ship's speed to about five knots.

Although internal waves do not result in any noticeable waves at the surface, they are often accompanied by parallel slicks at the sea surface, which can be used to track the progress of the internal waves. They may also be observed by noting temperature and/or salinity fluctuations with time at various depths. Internal waves sometimes cause submarines to rise and fall, and they were suspected of causing the loss of two nuclear submarines, *Thresher* and *Scorpion.* They also may interfere with the transmission of sound in the sea. The phenomenon of internal waves may be important for biological processes in the oceans in that they contribute to mixing of deeper nutrient-rich water with surface water, especially if the internal waves break. Nutrients from deeper waters are thus brought to the surface, where sufficient light is present for primary production by plants.

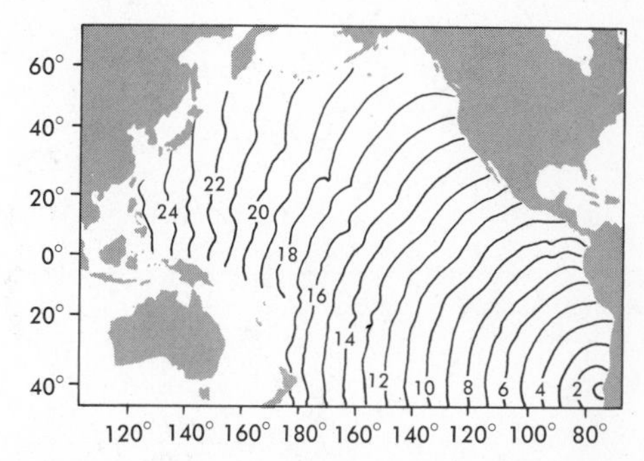

FIGURE 5.18 Wave refraction diagram of a tsunami that originated on the coast of Chile in 1960. Numbers indicate the hours of travel time from the source. (After The Committee for the Field Investigation of Chilean Tsunami of 1960, Report on the Chilean Tsunami of May 24, 1960, as observed along the Coast of Japan, December, 1961.)

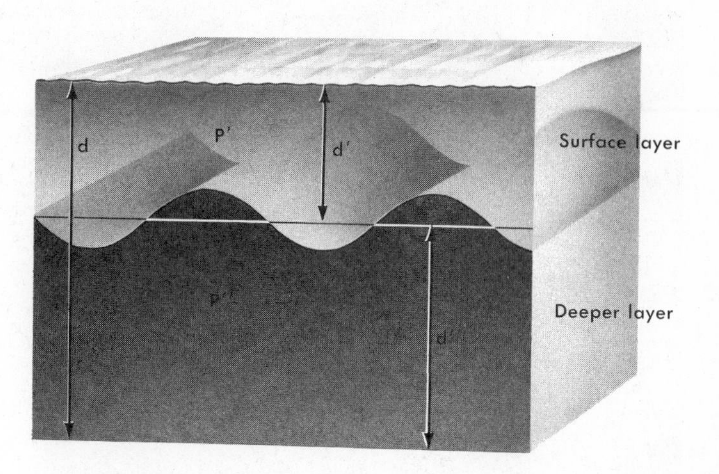

FIGURE 5.19 Internal waves in the ocean. d′ = thickness of the surface layer of density, p′; d″ = thickness of the deeper layer of density p″; and d = total water depth.

STORM SURGES

Storm surges, also known as ***storm waves*** or ***storm tides,*** are caused by the strong and steady winds associated with hurricanes (tropical cyclones) or other severe storms. Unfortunately, storm surges and tsunamis are both popularly referred to as "tidal waves" even though they have different origins. In storm surges, the winds simply pile up the water against the coast. The rise in water level is greatest in partially enclosed bodies of water such as the Gulf of Mexico and is negligible around islands. The rise in water level is augmented if it coincides with high tides, sometimes reaching heights of more than six meters. Damage from storm surges can be tremendous, especially in low-lying coastal areas (Fig. 5.20). Hundreds of thousands of people have died as a result of destructive storm surges in places surrounding the Gulf of Mexico, the Bay of Bengal, and the North Sea. The storm surge that hit Galveston, Texas, in 1900 caused thousands of deaths. In 1938, a hurricane and its storm surge hit the New England coast, resulting in the death of almost 500 people and about $300 million in damage. The disastrous storm surge that inundated Bangladesh in 1970 is estimated to have resulted in one half million deaths.

In addition to flooding of coastal areas, very strong winds and heavy rain usually accompany storm surges, making evacuation efforts very difficult.

FIGURE 5.20 Storm surge damage at Sea Isle City, New Jersey, after the March, 1962, hurricane. (From US Army Coastal Engineering Research Center, 1973.)

SUGGESTED READINGS

Barnstein, J. 1954 (August). Tsunamis. *Scientific American.*

Bascom, Willard. 1964. *Waves and Beaches.* Garden City, N.Y., Doubleday.

———. 1959 (August). Ocean Waves. *Scientific American.*

Pierson, W.J., G. Neumann, and R.W. Jones. 1955. *Practical Methods for Observing and Forecasting Ocean Waves by Means of Wave Spectra and Statistics.* U.S. Navy, H.O. Pub. 603.

U.S. Navy. 1944. *Breakers and Surf: Principles in Forecasting.* H.O. Pub. 234.

U.S. Navy. 1944. *Wind Waves at Sea, Breakers and Surf.* H.O. Pub. 602.

The tidal power station on the Rance estuary in France. (Courtesy of French Embassy Press and Information Division.)

CHAPTER 6

Ocean Tides

Tides, the periodic rise and fall of the oceans, are the most predictable of all the motions of the oceans. Yet, paradoxically, many aspects of the tides are not fully understood. Knowledge of tidal fluctuations as well as tidal currents is important not only in navigation but also in the ecology of the coastal oceans. A lack of understanding of tides caused Julius Caesar's fleet to meet disaster on the beaches of England. His army invaded Kent just before a high tide, which subsequently caused his ships to be wrecked upon the shores.

Although the enormous energy of the ocean tides may never be fully harnessed, man has already begun to produce power from the tides, as in the Rance estuary in Brittany, France. Similar facilities may someday be set up in other areas of the world.

TIDES ON AN IDEAL EARTH

Let us consider tides from an ideal point of view, that is, tides on a spherical earth covered by a uniform ocean, the bottom of which is assumed to be frictionless. Ocean tides are caused by the gravitational attraction of the moon and the sun on the waters of the rotating earth, and the centrifugal forces resulting from the mutual revolutions of the earth-moon-sun system. (Contrary to popular belief, the moon does not really revolve about the earth; both the earth and the moon revolve about a common center of gravity located between their centers.) The earth's own gravity and the centrifugal forces resulting from the earth's rotation on its axis in no way cause oceanic tides, as these forces do not vary with time. If we ignore the effects of the sun for the time being, we can state that every water particle in the oceans is subjected to two forces, the gravitational force of the moon and the centrifugal force due to the earth-moon revolution about each other. The tidal effects of the sun are similar to those of the moon, but only about half as great (46 per cent) as those produced by the moon.

The gravitational attraction due to the moon is greatest for a water particle directly beneath the moon and least for a water particle on the other side of the earth (Fig. 6.1*a*). The earth and the moon revolve about each

FIGURE 6.1 ***A,*** **Gravitational attraction between the moon and various water particles.** ***B,*** **Centrifugal forces due to the effect of the earth-moon revolution on selected water particles. (Modified after Turekian, 1968.)**

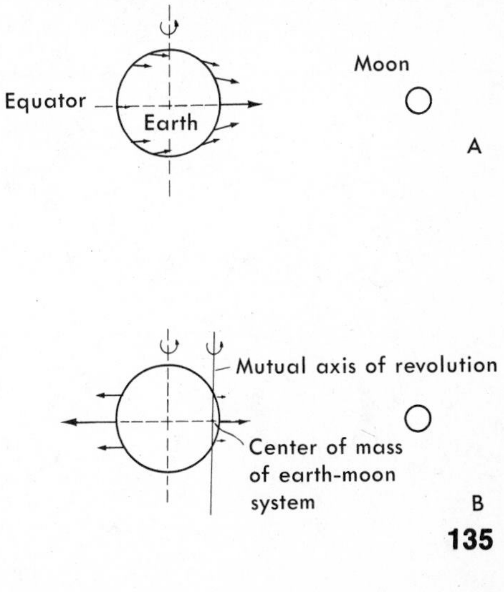

other (that is, the earth goes around the moon and the moon goes around the earth) approximately every 27 days. This mutual revolution takes place about an imaginary axis passing through the center of mass of the earth-moon system, which at any instant is located about 1700 km below the earth's surface. Because the earth rotates on its own axis, the position of this center of mass is continually changing. The mutual revolution of the earth-moon system results in centrifugal forces which are always perpendicular to the axis of revolution (Fig. 6.1*b*) and are directly proportional to the perpendicular distances between the axis and the water particles.

The gravitational and centrifugal forces acting on a water particle augment or oppose each other, resulting in the net forces shown in Figure 6.2*a*. These forces are symmetrical with respect to the earth's axis of rotation when the moon is above the equator and are distorted when it is not over the equator. The ***declination*** of the moon may be as much as 28½° north or south of the equator. The interval between northern and southern declinations is 13.7 days. Figure 6.2*b* shows the horizontal components of the tide-producing forces. These are the forces acting parallel to the sea surface, causing two equal bulges of water (Fig. 6.2*c*). The vertical components simply cause a minute decrease in the earth's gravity, since they are oppositely directed. Because of the

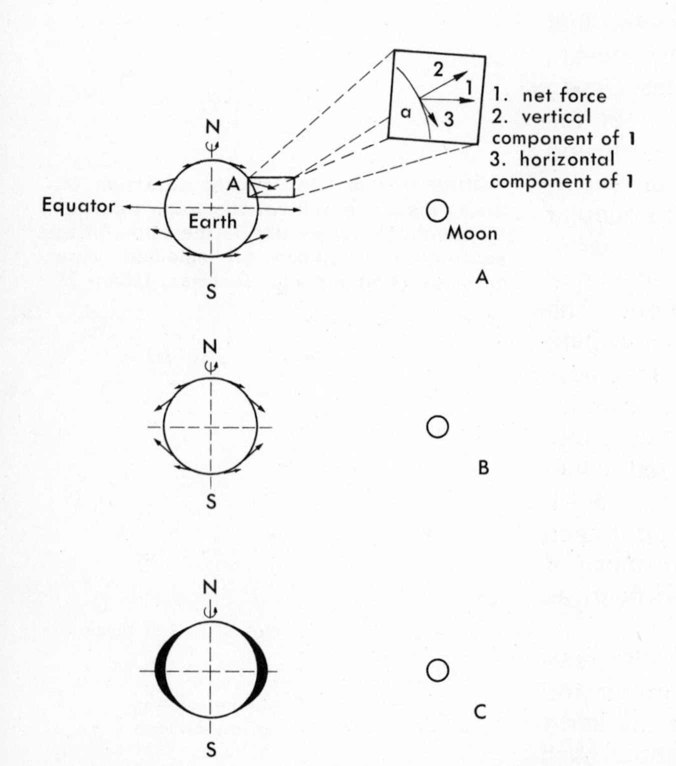

FIGURE 6.2 ***A***, **Net gravitational and centrifugal forces on water particles;** ***B***, **Horizontal components of the tide-producing forces;** ***C***, **Bulges produced by the forces shown in** ***B***.

earth's rotation and the moon's revolution around the earth, most places on an ideal earth would experience two equal high tides (bulges) and two equal low tides during a period of 24^h 50^m or one ***lunar day,*** the time between two successive appearances of the moon over the same longitude. If the earth did not rotate, these tides would occur only twice each lunar month.

So far, we have ignored the tidal effects of the sun, which are similar to those produced by the moon but only about half as great. When the sun, the moon, and our ideal earth are on a line (new moon or full moon), the resulting two tidal bulges will be larger than those produced by the moon alone (Fig. 6.3*a*). These tides, called ***spring tides,*** would be repeated every two weeks during full moons or new moons. Spring tides produce the highest high tides and the lowest low tides. The ***tidal range*** (the difference between two successive high and low tide levels) reaches a maximum during spring tides.

When the sun, the moon, and the earth are at right angles to each other, ***neap tides*** develop (Fig. 6.3*b*). This occurs during the first or third quarter of the moon (about one week after spring tide). Because the sun's weaker tidal effects are directed perpendicular to those of the moon, the net effect is a reduction in the size of the bulges produced by the moon alone. There would still be

THE MYSTERIOUS GRUNION

A mysterious relationship exists between the tide and the grunion *(Leuresthes tenuis)*, **a small fish (about 15 cm in length) which swims up the beaches of Southern California to spawn during the peaks of the high spring tides of spring and summer. The grunion spawn for three to four nights following the full or new moon. This ensures that the eggs will remain undisturbed for at least 10 days, the time necessary for development before hatching. The female burrows tail first into the sand just after the high tide and lays her eggs. The male arches himself around the head of the female and releases the sperm into the moist sand. After fertilization, the eggs develop for almost two weeks until the next high spring tide washes away the sand and releases the young fish, which then swim toward the sea. Many of the young grunion are eaten by other fish before they mature. Some of the adults are captured by birds, eager Californians, and tourists who flock down to the beach as the grunion stream up the beach. However, enough of the young grunion survive to continue the species, and some of the adults scramble back to the sea and eventually may return to spawn again.**

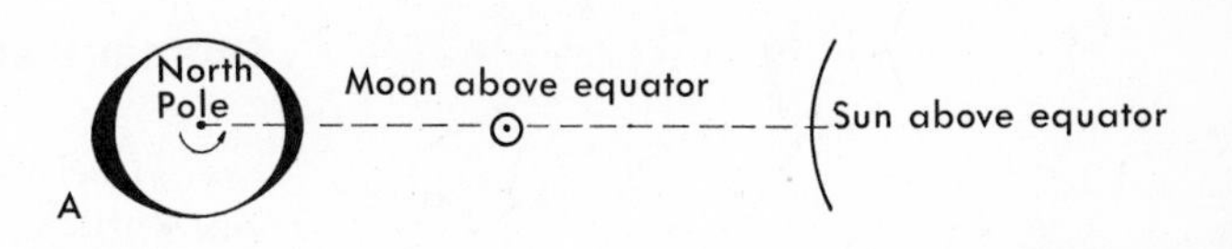

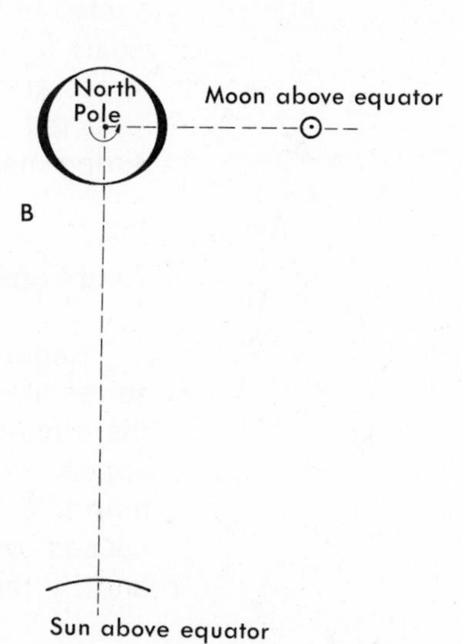

FIGURE 6.3 Tidal bulges produced by various orientations of the earth, moon, and sun. *A*, Spring tide; *B*, Neap tide. Note that the moon could be on the opposite side of the earth for *A* and *B*, yet the bulges will remain the same.

two high tides and two low tides during a lunar day, but the tidal range would be at a minimum. Between spring and neap tides, a more complex situation should develop.

TIDES ON THE REAL EARTH

Tides on the real earth are more complex than those expected on an "ideal" earth. Some of these differences and their causes are discussed on the following pages.

Types of Tides

Figure 6.4 shows the tides observed at various locations throughout the world. At some locations (Fig. 6.4*a* and *b*) there are two approximately equal high tides and two low tides every day. These are ***semidiurnal*** (semidaily) ***tides***, and their periods are about 12^h 25^m (one half lunar day). Tides of this nature are found along the Atlantic coast of the United States. In some areas (Fig. 6.4*c*, *d*, and *e*), however, two unequal high tides and/or two unequal low tides may occur each lunar day. These are called ***mixed tides***, and they are common along the Pacific coast of the United States. ***Diurnal*** (daily) ***tides,*** with only one high and one low tide each lunar day, predominate in areas such as the Gulf of Tonkin (Fig. 6.4*f*). Notice that no tide is purely semi diurnal or purely diurnal.

Diurnal Inequality

The twice-daily high and low waters for the semidiurnal and mixed tides are rarely equal in height. Only when the moon is above the equator can these tides be expected to be equal. At all other times, a ***diurnal inequality*** exists (Fig. 6.5). During these periods, the tide-producing forces are more diurnal than semidiurnal. This situation is reflected on many of the tidal curves shown in Figure 6.4 during maximum declinations of the moon.

Time Lags

Regardless of type, tides more or less repeat themselves at each location on successive lunar days, although there may be a time lag of up to a few hours between the appearance of the moon overhead and the occurrence of a high tide. This time lag is constant for each locality and is caused by friction between the ocean bottom and the water and by the fact that continents block the free transfer of

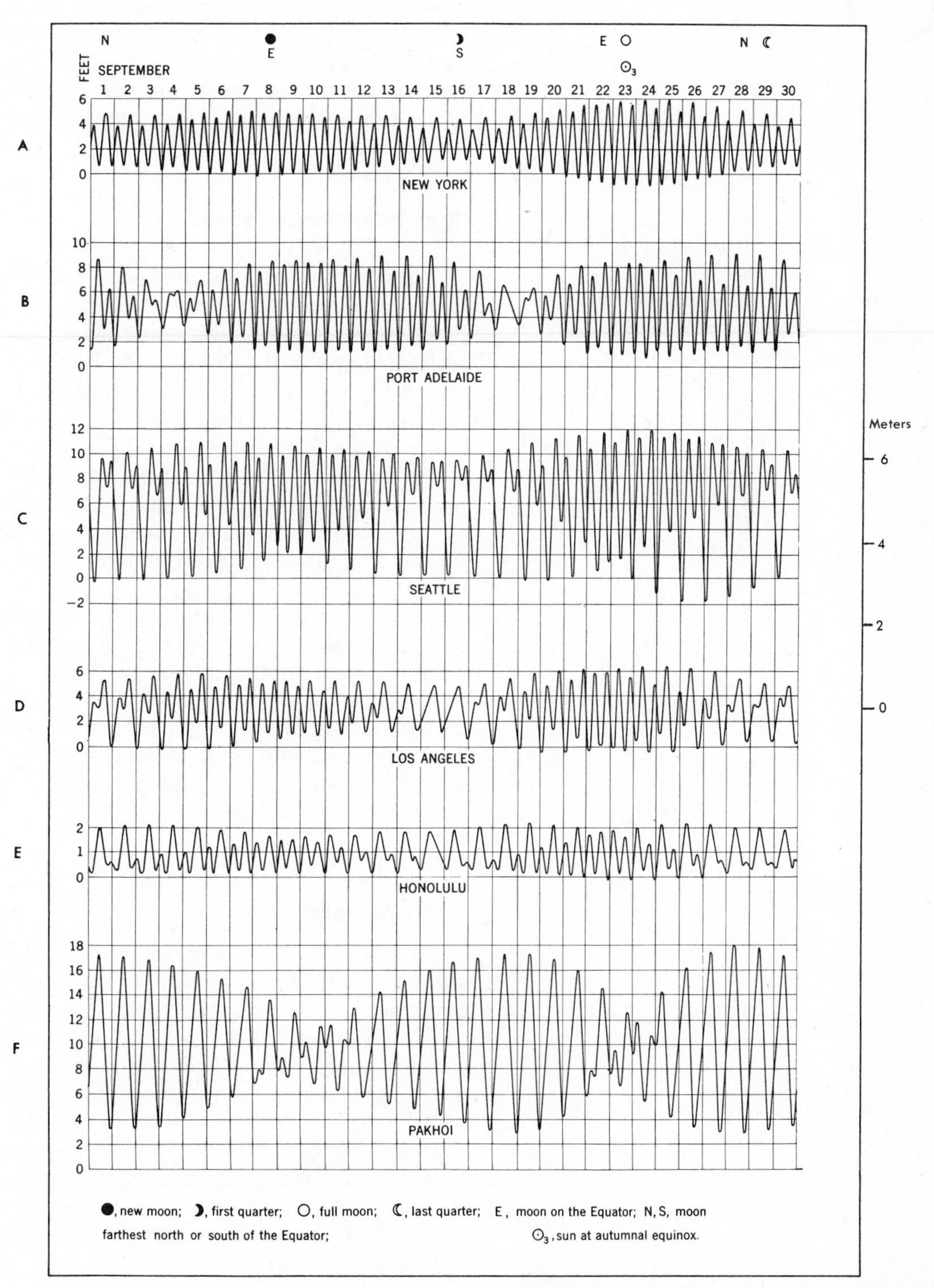

FIGURE 6.4 Tidal curves for various localities. Pakhoi is on the Gulf of Tonkin. Note that the meter scale on the right does not apply to Honolulu, as its vertical scale is exaggerated. (From Bowditch, 1962.)

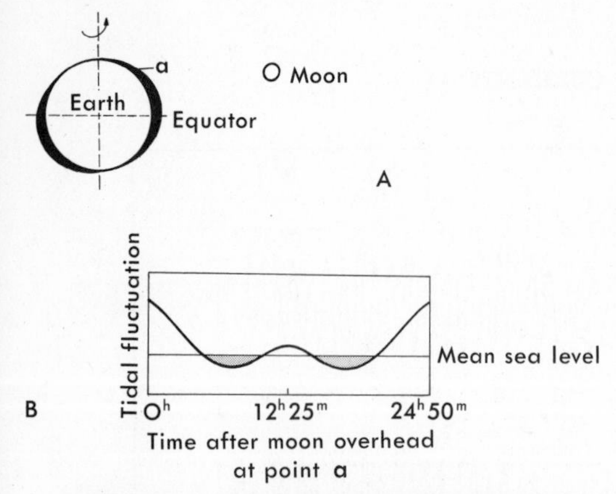

FIGURE 6.5 *A,* **Tidal bulges when moon is not over the equator.** *B,* **Resulting tidal curve for point a.**

water around the globe. It is also obvious from Figure 6.4 that spring and neap tides do not exactly coincide with the alignment of the earth, the moon, and the sun discussed under ideal tides. That is, spring tides, for example, do not always coincide with full or new moons. There is usually a few days' delay, which is also constant for each locality. The reasons for this delay are not well understood. Perhaps it too is related to friction and continental blockage.

Other Tidal Complications

Tidal ranges and periods vary not only from day to day, but drastic differences in tidal ranges and periods are sometimes observed over short distances. For example, the tidal range at the head of the Bay of Fundy (Fig. 6.6) may be as much as 18 m, whereas at the mouth of the bay the tides range only about 3 m. Tides along the northern shores of the Gulf of Mexico (Fig. 6.7) are predominantly diurnal in nature, and the tidal range is less than one meter, whereas along the east (Atlantic) coast of Florida, the tidal ranges are nearly two meters and are of the semidiurnal type. We will now investigate why tides vary in this manner.

In addition to the continental blockage of free transfer of water and the frictional forces between the ocean bottom and the water, another effect, the Coriolis force (see Appendix C) must be considered. Still another factor is the size and shape of the ocean basins. All of these elements actually control the tides locally, although their ultimate causes are the tide-producing forces described in our discussion of ideal tides.

Resonance

The water in a rectangular (enclosed) basin such as a bathtub can be set in motion by application of some external force. The water will rock back and forth (Fig. 6.8) with a period which depends on the length of the basin and the water depth. This is the ***natural period*** of the basin. There will be no vertical motion in the middle of the basin, the ***nodal line***. The waves in the basin are ***standing waves*** produced by reflection from opposite sides of the basin. In standing waves, the water surface simply moves up and down, and there is no forward motion of the waves. The waves discussed in Chapter 5 are called ***progressive waves,*** in which the wave form advances in some direction. If the period of the external force and the natural period of the basin coincide, ***resonance*** will be set up, causing amplification of the rocking

A

B

FIGURE 6.6 Tides in the Bay of Fundy. *A,* High tide; *B,* Low tide. (Photos courtesy of National Film Board of Canada.)

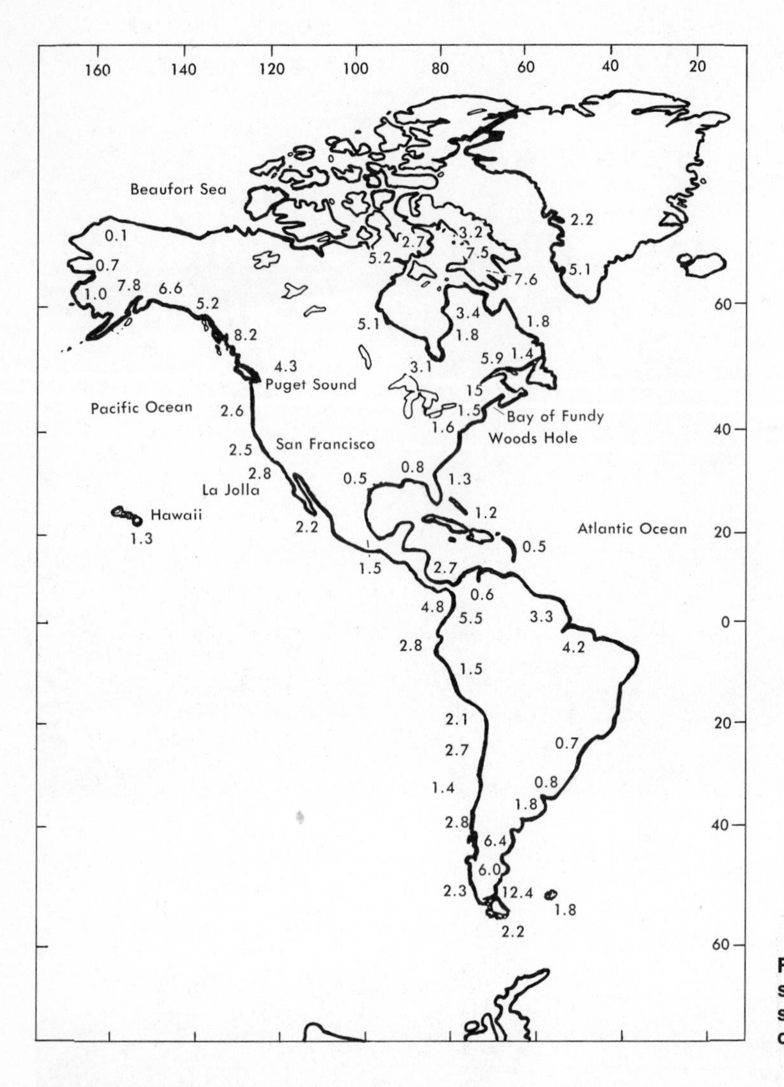

FIGURE 6.7 Map of North and South America, showing maximum tidal ranges in meters at selected locations. (After Doty, 1957, and others.)

motion. A good example of resonance is that of pushing a swing; each push is good for only a small amount of motion, but if the pushes are properly timed, the swing goes higher and higher. For a basin that is open on one end, such as the Bay of Fundy, the situation is very similar to that of the enclosed basin, except that the nodal line is located at the mouth of the bay. The period of such a basin is twice that of an enclosed basin of the same length and depth, and is given by the following equation:

$$T = \frac{4L}{\sqrt{gd}}$$

where T is the natural period of the open basin; L, its length; d, its depth; and g, the gravitational acceleration of the earth. Thus, the period increases with increasing length and decreasing depth of the basin. The natural period of the Bay of Fundy (about 12 hours) is very close to the period of the ocean tide entering the bay. Thus resonance is set up in the bay. This explains why the tide is a maximum at the head of the Bay of Fundy and a minimum at its mouth (the nodal line), where the tides are those of the nearby Atlantic.

Except during equinoxes, the sun is either north or south of the equator, and hence the sun's tide-producing forces for most places on earth are more diurnal than semidiurnal, as is true during the northern and southern declinations of the moon (see Fig. 6.5). Thus, if a basin has a natural period of about 24 hours, it will respond preferentially to the diurnal forces rather than to the smaller semidiurnal forces. This is exactly what happens to certain large basins such as the Gulf of Mexico and the Gulf of Tonkin, which have predominantly diurnal tides.

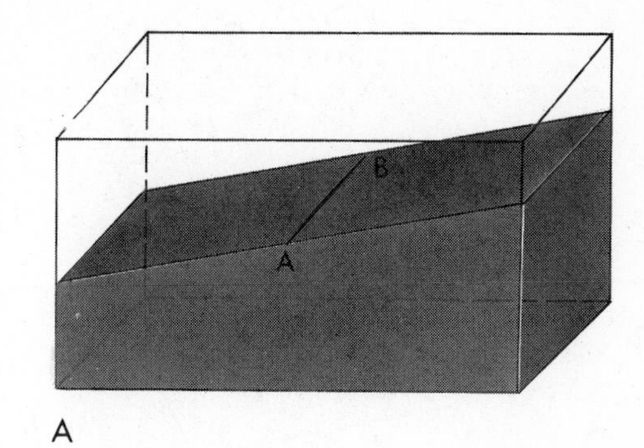

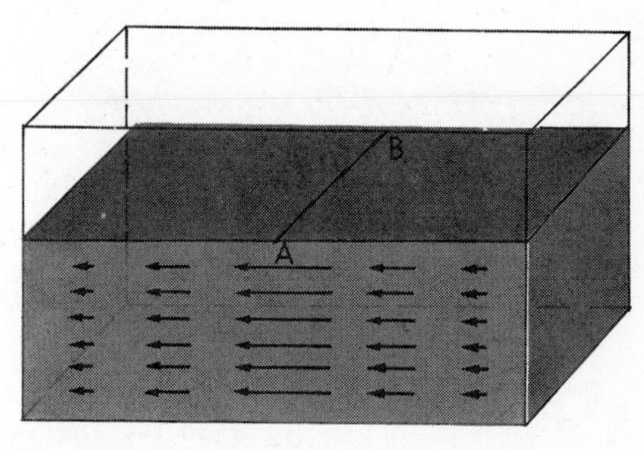

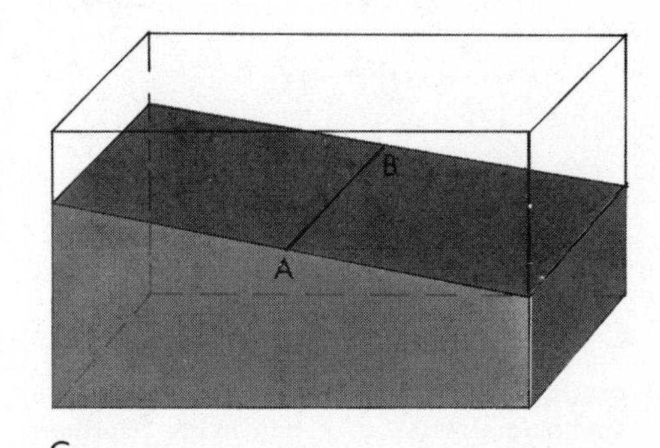

FIGURE 6.8 Resonance set up in a rectangular (closed) basin for half a tidal cycle. Arrows indicate resulting current. Note that no tidal fluctuations occur along AB, the nodal line.

EFFECT OF THE EARTH'S ROTATION

In addition to the resonance mentioned above, tides in wide bays and estuaries are modified by the earth's rotation, namely, by the effect of ***Coriolis force*** (Appendix C). Coriolis force deflects all moving objects near the surface of the earth, including moving water particles, which are deflected to the right in the northern hemisphere and to the left in the southern hemisphere. Coriolis force increases from zero near the equator to a maximum at the poles. Figure 6.9 illustrates the effect of Coriolis force on semidiurnal tides in a bay in the northern hemisphere. Starting with high tide at the head of the bay, as the tide begins to retreat, Coriolis force deflects the water toward the right. Thus, high tide in the bay will move in a counterclockwise direction around the bay. Note that there is no tidal fluctuation at point A, in the middle of the bay. This is similar to the nodal line of Figure 6.8, except that the line has been reduced to a point—the **amphidromic point**—by the effect of Coriolis force. It is evident from this illustration that the tides do not always occur at the same time at different points even at the head of the bay.

TIDAL BORES

In many shallow, steep, funnel-shaped rivers, the tide advances as a single roaring wall of water, called a ***tidal bore*** (Fig. 6.10). Bores may move at speeds of up to 25 km/hr and may have heights up to 8 m. The generation of bores is essentially the same as that of breakers in shallow water. Spectacular bores occur in the Tsientang River in the People's Republic of China, in the Amazon, the Bay of Fundy, and in many English and French rivers.

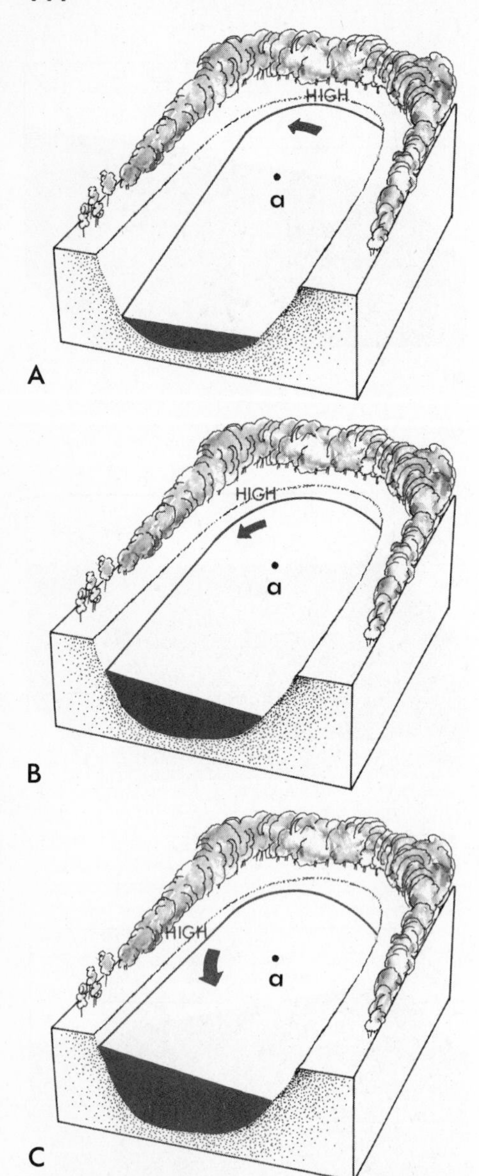

FIGURE 6.9 The effect of Coriolis force on semidiurnal tides in a bay in the Northern Hemisphere. The tides move in a counterclockwise direction about point a, the amphidromic point. Only a quarter of the tidal cycle is shown. The tidal conditions shown will be repeated 12 hours later.

FIGURE 6.10 Tidal bore in the Bay of Fundy. (Courtesy of NOAA.)

Tides in the Open Ocean

Only in the Antarctic Ocean can tides be expected to move completely around the globe from east to west, as seen in ideal tides. The other oceans and seas can be thought of as gigantic basins, and the tides present in them are determined by their natural periods, modified by Coriolis and frictional forces. The natural period of the Atlantic Ocean "basin" is about 12 hours and hence it responds to semidiurnal tide-generating forces. In addition, the semidiurnal tides from the Antarctic Ocean travel northward into the Atlantic. The resulting tides in and around the Atlantic Ocean are semidiurnal in nature. In contrast, mixed tides are found in the Pacific and Indian Oceans, because the dimensions of these oceans are such that they respond to both diurnal and semidiurnal forces.

Because there are no fixed points that can be used as reference points, there are no reliable means of measuring tides in the open ocean. They can only be computed, using calculations based on numerous tidal observations along the coasts and considerations of the size and shape of the ocean basins. It is estimated that the maximum tidal range in the open ocean is about one meter. Tides may be considered as a type of wave with a very long period, which progress as shallow water waves because the ratio of the wavelength to ocean depth is very large (see Chapter 5 for definition and discussion of shallow water waves). Therefore, tidal ranges are generally amplified near the coasts, similar to amplification of height of other ocean waves. Figure 6.11 shows the computed tidal map of the oceans for an assumed position of the moon directly above the equator on the Greenwich meridian (0° longitude). The effects of the sun are not considered. The lines shown are called ***cotidal lines*** and represent points connecting simultaneous high tides for each hour after the moon has crossed the Greenwich meridian. Low tides would occur about six hours before or after high tides. Tides move around the amphidromic points, counterclockwise in the northern hemisphere and clockwise in the southern hemisphere because of Coriolis force. No lunar tide will occur at the amphidromic points shown. For this reason, at islands such as the Solomon Islands, which are close to an amphidromic point of lunar tides, the tides follow the sun rather than the moon. High tides, for example, would occur each day at the same time in this case.

Tidal Currents

Tides may involve large quantities of water. The water piled up along the coast during high tide must return to

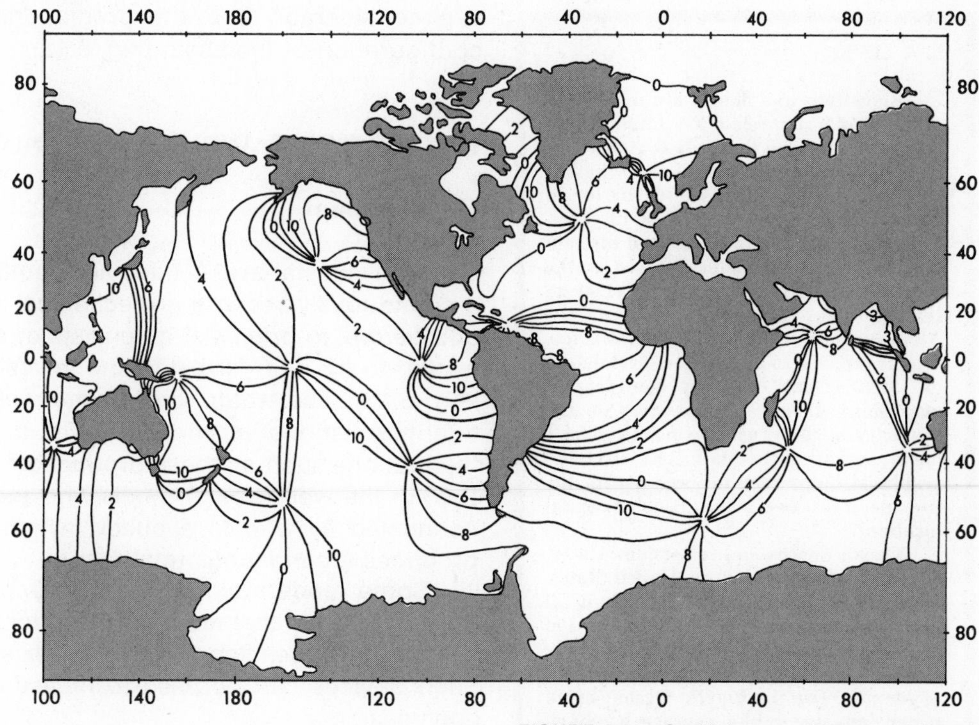

FIGURE 6.11 Cotidal map of the world for an assumed position of the moon over the equator, at 0° longitude (the Greenwich meridian). Numbers indicate times of high tide (in hours) after the moon has crossed the Greenwich meridian. Note the many amphidromic points, where no tide will be produced by the moon. Tides rotate around these points. (After Cartwright, 1969.)

the sea, thus resulting in a current. Knowledge of tidal currents is important in navigation, especially in bays and estuaries. Tidal currents are capable of cutting channels in such bodies of water, which further affect navigation. Tidal currents are influenced by the same forces that govern tides, and their patterns can be complicated. In the open ocean, tides can travel as progressive shallow water waves, and hence the water particles travel in elliptical paths. Although the theoretical horizontal velocities of these currents should be the same from top to bottom, frictional forces reduce them considerably at depths. Maximum currents (usually a fraction of a knot) will be found associated with the crests and troughs of these tidal waves. Tidal currents in the open ocean, however, are influenced by Coriolis force. As a result, clockwise-rotating currents will be formed in the northern hemisphere (one complete rotation during one tidal cycle).

In many estuaries and bays, as well as in areas where standing tidal waves are present, maximum currents occur below the nodal line (see Fig. 6.8) when there is no tidal fluctuation in the bay. The current velocities at other locations are maximal between high and low tides. Theoretically, no currents should exist during high and low waters, as they should begin to form just after high and low tides. However, the actual tidal currents observed may vary considerably from this prediction, depending on the configuration of the basin (Fig. 6.12).

SEA LEVEL

Elevations and depths are referred to some datum or reference level. Topographical maps in the United States give elevations above mean sea level, which is the average height of the sea surface for all stages of tides over a period of 19 years. The concept of sea level as used in mapping in the United States is based on the determinations of mean sea level at 26 tidal stations along the Pacific Ocean, the Atlantic Ocean, and the Gulf of Mexico. This is referred to as the "sea level datum of 1929." Tides can vary considerably over short distances, depending on the geometry of the ocean basins. Therefore, mean sea level as indicated on topographical maps may not agree with the true mean sea level at a particular coastal location.

Ocean depths and tide tables are referred to other data. In the United States, mean low water (the average height of all low tides at a place each day over a period of 19 years) is used for charts and tide tables of the Atlantic Ocean and the Gulf of Mexico. For the Pacific Ocean, mean lower low water (the average height of lower of two low tides each day over a period of 19 years) is the standard. Land features on nautical charts, however, are usually referred to mean high water (the average height of all high tides at a place each day for a period of 19 years).

MEASUREMENT AND PREDICTION OF TIDES

Measurement

Near coastal areas, tides are commonly measured by a continuously recording mechanical device known as a ***tide gauge***. In principle, it consists of a well or shaft connected to the sea, far below the lowest possible tide level (Fig. 6.13). The water level in the well responds only to tidal fluctuations and not to waves or other short-period fluctuations such as those produced by a passing ship. A float in the well rises and falls with the tides. The float is connected by wire to a pulley, which moves a pen and produces a continuous tidal record.

Some modern electronic tide recording devices have been developed that measure the differences in pressure at a point on the sea floor caused by sea surface fluctuations. Such devices can be used to measure tides even in the open sea.

Measurements of tidal current are considerably more difficult than measurements of tidal height. In addition, these measurements are complicated by nontidal currents existing in the area. Numerous current-measuring devices are available, the most common ones being current meters. Nontidal currents and various current-measuring devices are discussed in the next chapter.

Prediction

Tides at any place cannot be accurately predicted by knowing only the positions of the moon and the sun relative to that place, since they also depend on other factors such as the size and configuration of the basin. Accurate tidal predictions, therefore, require actual tidal observations for at least a year and, preferably much longer to establish these factors. Such predictions may be made using a mechanical instrument called a ***tide predicting machine*** (Fig. 6.14), originally invented by Lord Kelvin in 1872, although in the United States electronic computers are now being used to predict tides. Tide prediction tables, including data for tidal currents, are published annually for many representative coastal locations throughout the world. In the United States, they are published by the National Oceanic and Atmospheric Administration (NOAA) of the Department of Commerce.

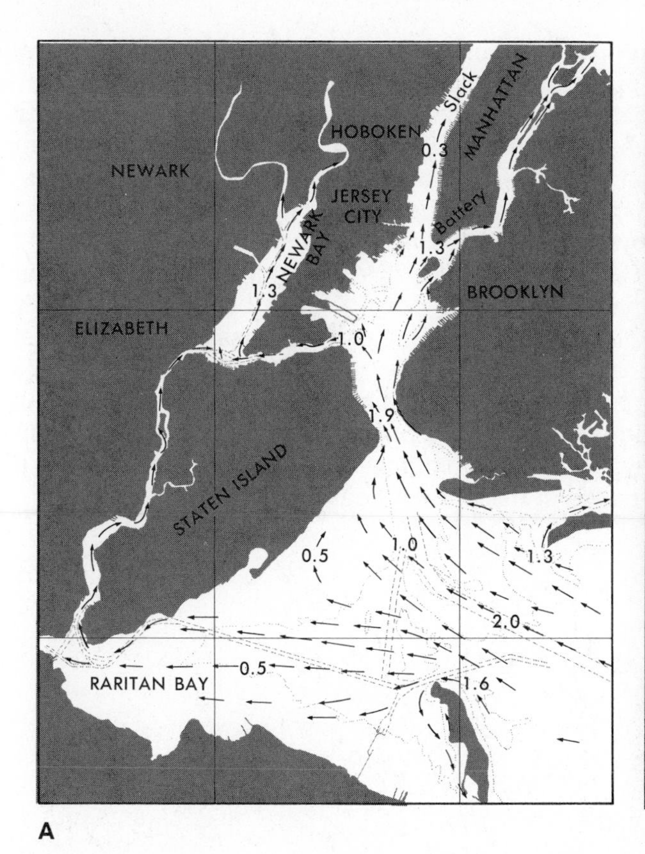

A

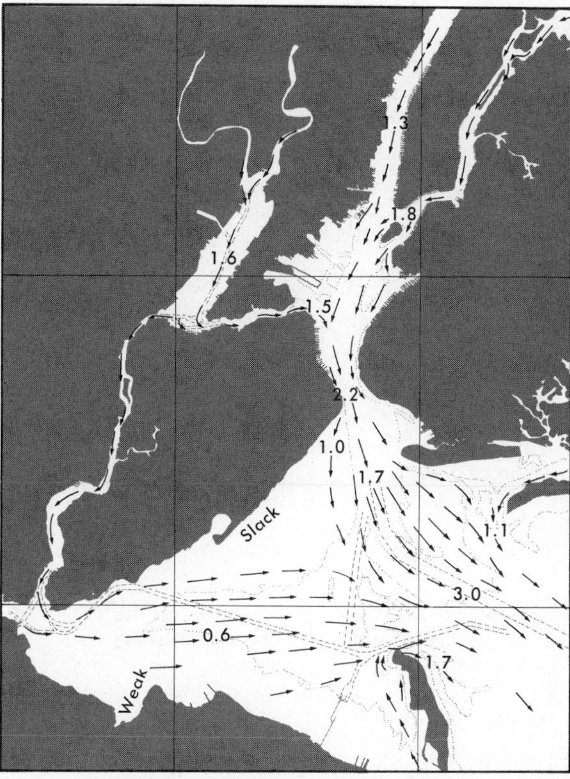

B

FIGURE 6.12 Tidal current charts for New York. *A,* Four hours after low water, *B,* Four hours after high water, at the Battery. Tidal current speeds are shown in knots. (After NOAA: Tidal Current Charts, New York Harbor, 7th ed. Washington, D.C., 1956.)

Tides do not repeat themselves from one year to the next, because the relative positions of the earth, the moon, and the sun are approximately repeated only once every 18.6 years. Thus, if tides were to be predicted statistically, at least 18.6 years of actual tidal observation would be required.

POWER FROM TIDES

The tidal ranges in many bays and estuaries throughout the world are very large. For example, the maximum tidal range in the Rance estuary in France is about 13 m, and in the Bay of Fundy it is about 18 m. Approximately 170 million cubic meters of water enter and leave the Rance estuary during a semidiurnal tidal cycle. In 1960, an electric power plant was built in a dam near the mouth of the Rance (Fig. 6.15). The blades of the turbines are turned by the tidal flow in and out of the estuary. When in full operation, this power plant is capable of producing about 670 million kilowatt-hours of electricity annually. This is roughly half the amount produced by a typical hy-

FIGURE 6.13 A tide gauge station.

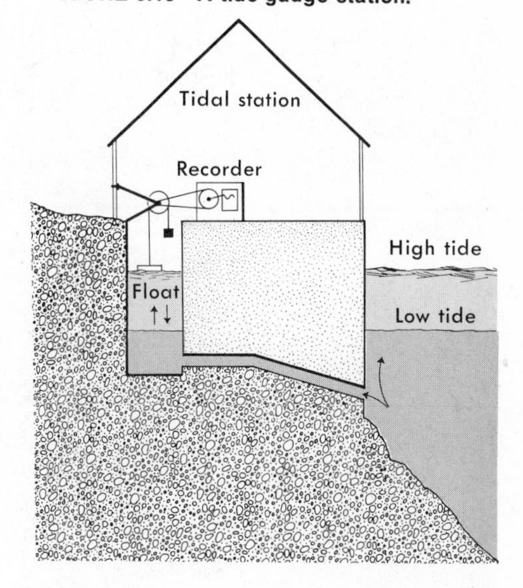

FIGURE 6.14 A tide predicting machine. This method of predicting tides has been replaced by computers. (Courtesy of NOAA.)

droelectric power plant in France. The only other tidal power plant is in the Soviet Union, near the White Sea.

Tidal power plants may be built in many parts of the world. A joint United States—Canada tidal power plant has been proposed for many years to be built in the Passamaquoddy Bay near the mouth of the Bay of Fundy. The tidal flow in this area is more than 10 times that involved in the Rance estuary. No firm decision, however, has yet been made. Far more power could be generated by a power plant built in the Bay of Fundy itself. Although tidal power can account for only a small portion of the world's energy needs, it is pollution-free and dependable on a long-term basis. The major obstacles are economic feasibility and possible ecological damage from the changes it would cause in the normal tidal flow.

FIGURE 6.15 The tidal power station on the Rance estuary. (Courtesy of French Embassy Press and Information Division.)

SUGGESTED READINGS

Clancy, Edward P. 1968. *The Tides.* Garden City, N.Y., Doubleday.
Defant, Albert. 1960. *Ebb and Flow.* Ann Arbor, University of Michigan Press.
Goldreich, Peter. 1972 (May). Tides and the Earth-Moon System. *Scientific American.*
McDonald, James E. 1952 (April). The Coriolis Effect. *Scientific American.*
Turekian, Karl K. 1968. *The Oceans.* Englewood Cliffs, N.J., Prentice-Hall.

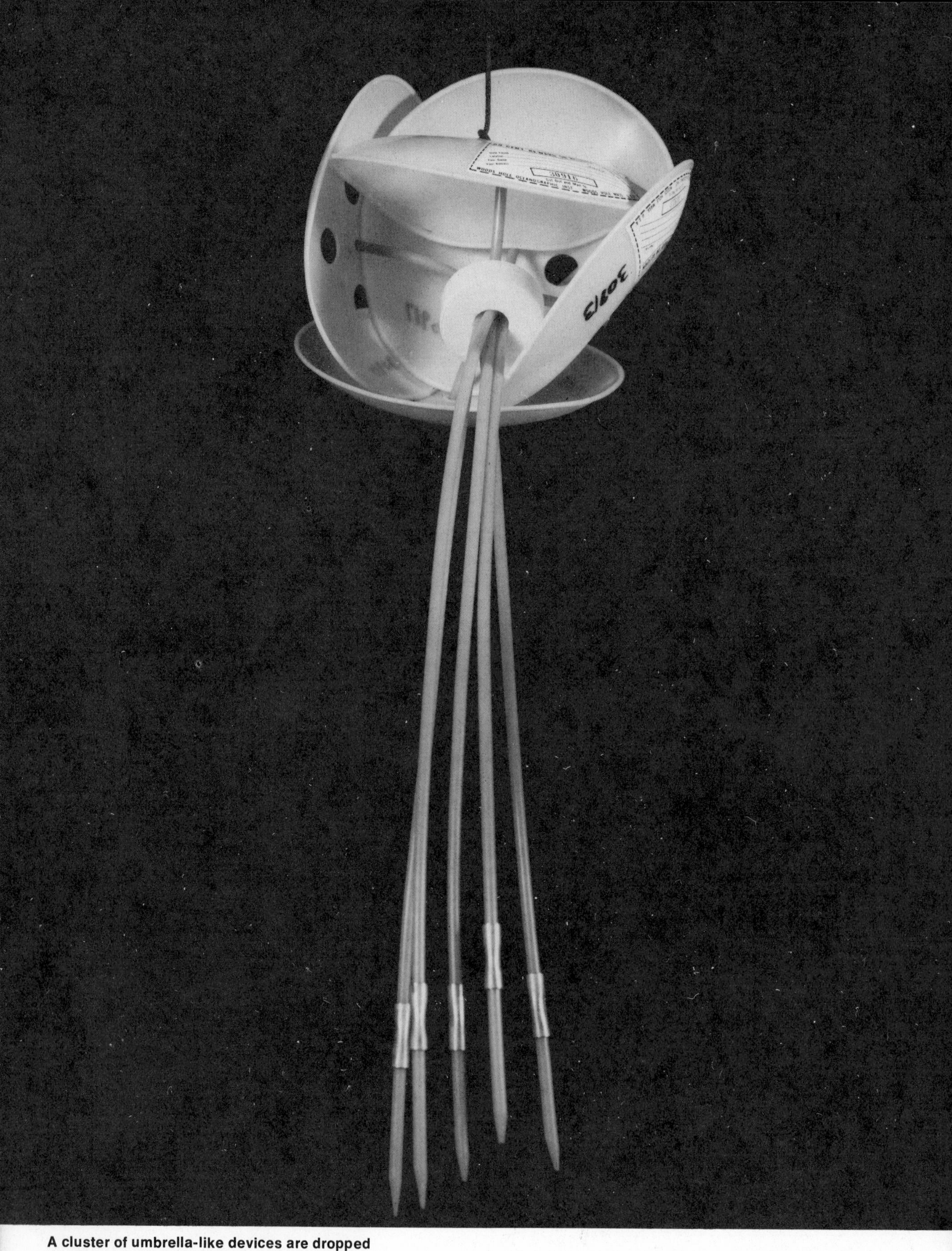

A cluster of umbrella-like devices are dropped into the sea at a predetermined location. They sink, separate, and drift with bottom currents. Finders are requested to return them to the address shown on the label, and to state the locations at which they were found, providing the researcher with important data regarding current movements. (Courtesy of Woods Hole Oceanographic Institution.)

CHAPTER 7

Ocean Currents

Although not as perceptible as the surface waves, water from the surface to the bottom is constantly in motion as a result of ocean currents. Surface currents are caused primarily by the force of wind against water, whereas deep ocean circulation is related to density differences of the water within the ocean. Currents are of great significance to all of us and to ocean life. Knowledge of ocean circulation is important to sailors who wish to take advantage of surface currents in their voyages. It is also useful in prediction of weather, and it affects the movement of nutrients and pollutants. Ocean circulation is responsible for bringing nutrients to surface waters in the biologically productive regions of the oceans. In fact, the oceans would be nearly lifeless if it were not for their circulation.

SURFACE CURRENTS

Atmospheric Circulation

The primary cause of surface ocean currents is wind. The ultimate driving force of the winds over the earth is solar energy. As seen earlier (Chap. 4), the equatorial regions of the earth receive more solar energy per unit area than does the rest of the earth. Figure 7.1 shows an idealized atmospheric circulation pattern of the earth. The equatorial air masses rise as a result of increased heating and eventually move poleward at higher altitudes. Consequently, low pressure belts known as the ***doldrums*** are created along the equator. In order to compensate for the rising air, air masses move toward the equator near the surface. The poleward moving equatorial air gradually cools and sinks at about 30° north and south latitudes (the subtropical high pressure belts known as the ***horse latitudes***), and upon reaching the earth's surface, part of it spreads poleward and some moves toward the equator. The poleward moving air meets the cool, dense polar air moving toward the equator at about 60° north and south latitudes, forming the ***polar front***.

Because we are dealing with a rotating earth, ***Coriolis force*** (see Appendix C) must be considered. This force

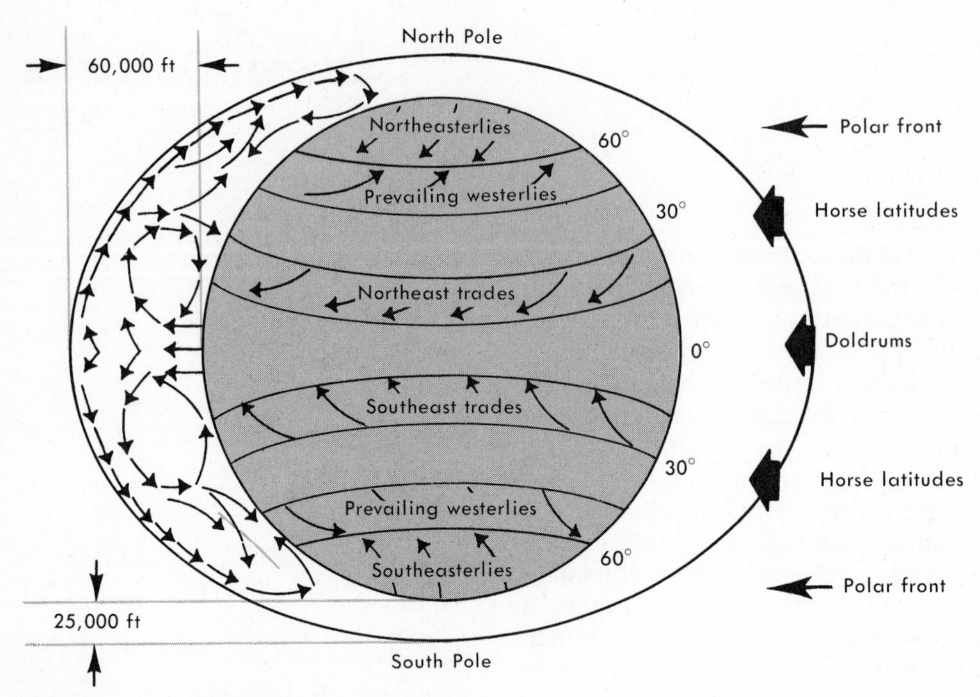

FIGURE 7.1 Idealized atmospheric circulation. (After Bowditch, 1962.)

deflects all moving objects toward the right in the Northern Hemisphere and to the left in the Southern Hemisphere. The equatorward flow of wind near the earth's surface, for example, is deflected by Coriolis force toward the west, but because of the effects of friction, the deflection of air masses is not complete. Thus, in the Northern Hemisphere, surface winds tend to blow to the southwest and in the Southern Hemisphere toward the northwest. Figure 7.2 shows average wind patterns over the oceans during winter and summer. Except for distortions caused by the presence of land masses, the wind patterns of both the Northern and Southern hemispheres are roughly symmetrical with respect to the equator. The most prominent of all features shown are the ***trade winds*** (between about 25° north and south latitudes), which generally blow from the northeast in the Northern Hemisphere and from the southeast in the Southern Hemisphere. Between about 30° and 60° latitudes are the ***prevailing westerlies.*** They blow to the northeast and southeast in the Northern and Southern hemispheres, respectively. Poleward of the prevailing westerlies are the ***polar easterlies.*** Before attempting to

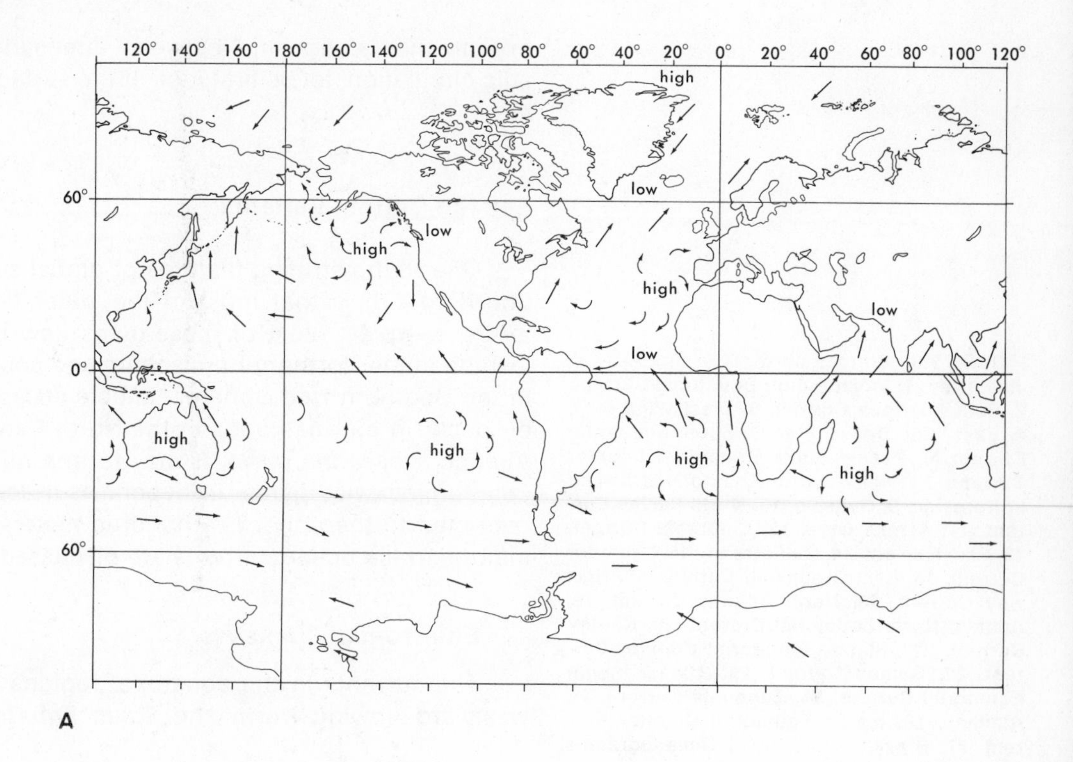

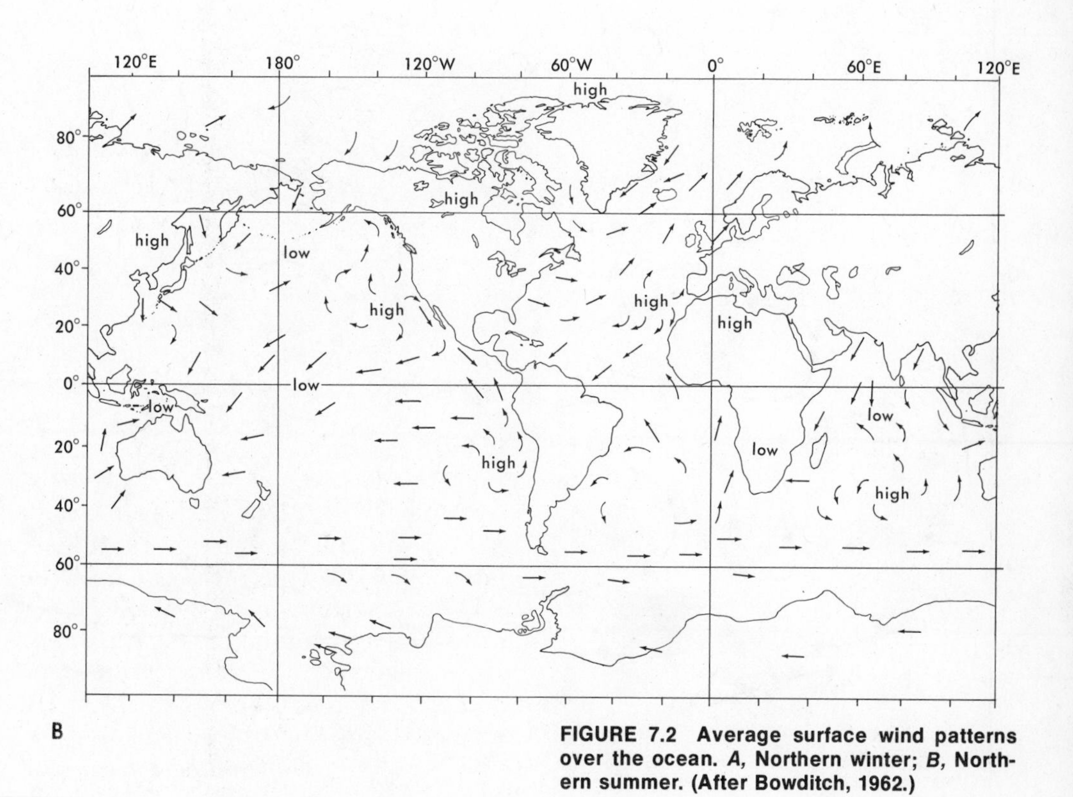

FIGURE 7.2 Average surface wind patterns over the ocean. *A*, Northern winter; *B*, Northern summer. (After Bowditch, 1962.)

explain the relationship between atmospheric and oceanic circulation, let us first look at the surface current patterns.

Surface Current Patterns

One of the striking features of global surface circulation (Fig. 7.3) is that most of the water flows in closed loops, or ***gyres***. Most of these gyres flow in a clockwise fashion in the Northern Hemisphere and counterclockwise in the Southern Hemisphere. Notable exceptions, however, occur in high latitudes in the North Pacific and North Atlantic, where the currents may form a number of small counterclockwise gyres. In general, surface currents are restricted to the upper few hundred meters. A few of the major surface ocean currents are discussed below.

Equatorial Currents

The currents in the equatorial regions consist of the westward flowing ***North*** and ***South Equatorial Currents***,

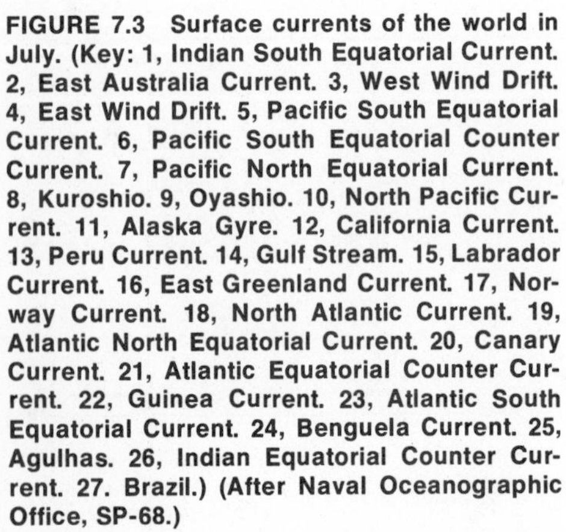

FIGURE 7.3 Surface currents of the world in July. (Key: 1, Indian South Equatorial Current. 2, East Australia Current. 3, West Wind Drift. 4, East Wind Drift. 5, Pacific South Equatorial Current. 6, Pacific South Equatorial Counter Current. 7, Pacific North Equatorial Current. 8, Kuroshio. 9, Oyashio. 10, North Pacific Current. 11, Alaska Gyre. 12, California Current. 13, Peru Current. 14, Gulf Stream. 15, Labrador Current. 16, East Greenland Current. 17, Norway Current. 18, North Atlantic Current. 19, Atlantic North Equatorial Current. 20, Canary Current. 21, Atlantic Equatorial Counter Current. 22, Guinea Current. 23, Atlantic South Equatorial Current. 24, Benguela Current. 25, Agulhas. 26, Indian Equatorial Counter Current. 27. Brazil.) (After Naval Oceanographic Office, SP-68.)

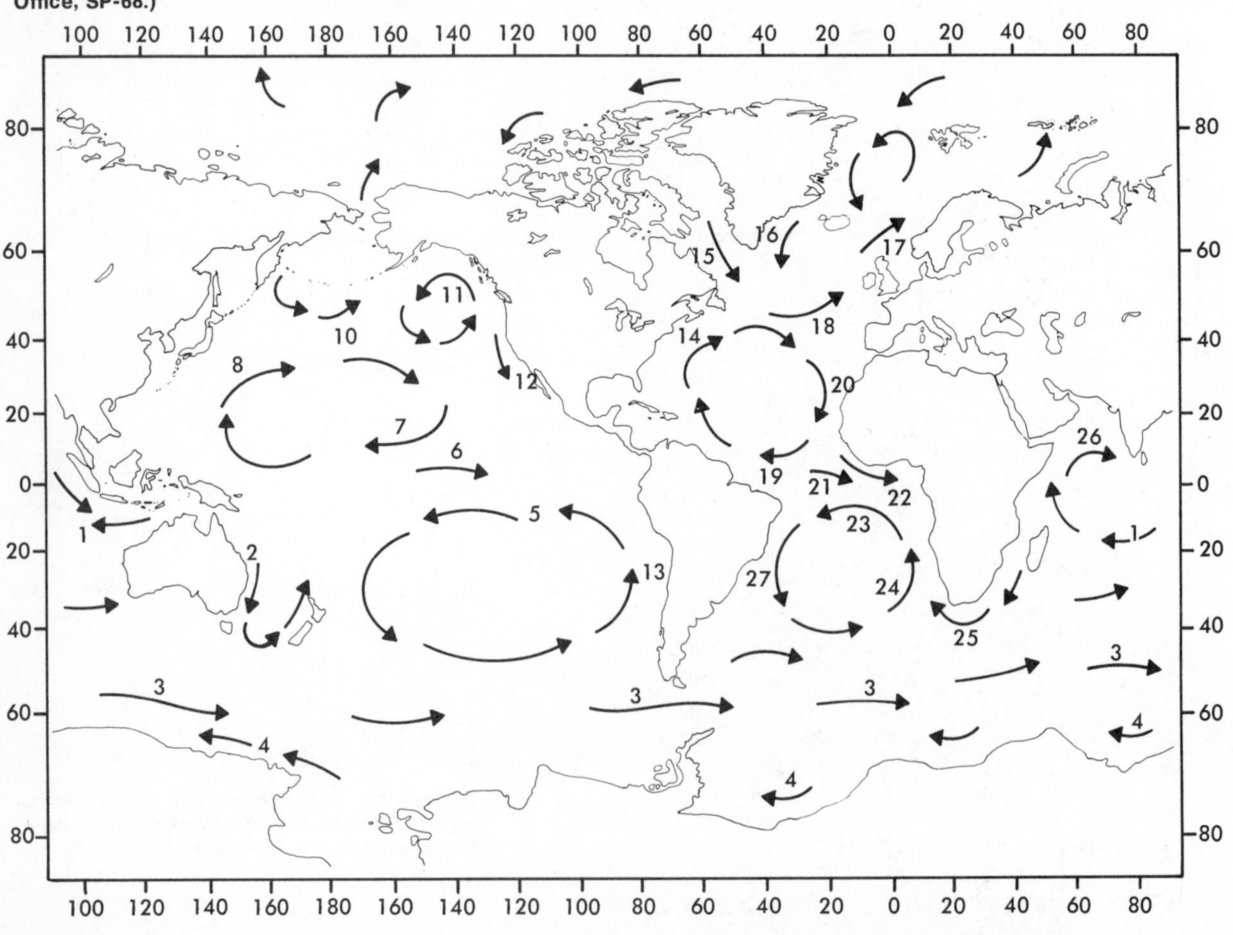

which are separated by the eastward flowing ***Equatorial Counter Currents.*** In the Atlantic, the Equatorial Counter Current is most highly developed in the east, near the coast of Africa. These currents generally have speeds of about 25 cm/sec (0.5 knot) and may flow at speeds of up to 125 cm/sec (Table 7.1). The direction of the North Equatorial Current in the Indian Ocean, however, is reversed during the Northern Hemisphere summer (when the monsoon winds blow from the southwest) and flows eastward, as does the Equatorial Counter Current.

Western Boundary Currents

The warm westward flowing equatorial currents are deflected poleward near the western boundaries of the oceans (the east coasts of continents). These currents are known as ***western boundary currents*** and include some of the most intense ocean currents (sometimes exceeding 200 cm/sec). These boundary currents are the ***Kuroshio*** and the ***East Australian currents*** in the Pacific Ocean, the ***Florida***, ***Gulf Stream***, and ***Brazil currents*** in the Atlantic Ocean, and the ***Agulhas Current*** in the Indian Ocean. Some western boundary currents are quite extensive vertically, flowing at depths as great as 1500 m or more, considerably deeper than most surface currents.

The northward flowing Gulf Stream is not easily distinguished by temperature from the warm ***Sargasso Sea*** along which it flows. However, the current flows along a boundary between the warm Sargasso Sea water on its right and cold water on its left and acts as a barrier, preventing the mixing of the two waters. Its flow is not straight but is deflected eastward toward Europe and is further characterized by a number of meandering loops along its margins. In addition, the current pattern varies

Table 7.1 SPEED AND TRANSPORT OF SELECTED MAJOR SURFACE CURRENTS*

CURRENT	*MAXIMUM SPEED (cm/sec)*	*AVERAGE SPEED (cm/sec)*	*TRANSPORT (million m^3/sec)*
Gulf Stream	300	100	100
Kuroshio	300	90	50
Florida	250	35	25
California	–	40	–
Brazil	–	–	10
Peru	100	–	20
West Wind Drift	50	–	100–200
East Wind Drift	150	–	–
North and South Equatorial	125	20–35	45
Pacific Undercurrent (Cromwell)	150	–	40

*51.5 cm/sec = 1 knot (nautical mi/hr). Data compiled from various sources, including Bowditch (1965), Defant (1961), Warren (1966).

with time. The pattern of the Gulf Stream is similar to that of the Kuroshio and other western boundary currents.

OTHER MAJOR SURFACE CURRENTS

In the Pacific Ocean, the North Pacific Current carries most of the water from the Kuroshio toward the east, until it eventually splits into northward and southward flowing branches. The southward moving current, the ***California Current***, completes the major clockwise gyre in the North Pacific Ocean when it joins the North Equatorial Current. The northward flowing water contributes to a smaller counterclockwise ***Alaskan Gyre.*** Similarly, in the Atlantic Ocean, the ***North Atlantic Current*** carries water eastward. The southward flowing branch of this current, the ***Canary Current***, completes the major clockwise gyre in the North Atlantic. At the same time, water is deflected northward from the North Atlantic Current to form several smaller counterclockwise subpolar gyres.

The surface circulation in the Southern Hemisphere is virtually a mirror image of that in the Northern Hemisphere, except for the absence of the minor subpolar gyres. In the South Pacific, the counterclockwise gyre is completed by the globe-circling ***West Wind Drift*** and the northward flowing ***Peru Current***. Similar current patterns exist in the South Atlantic and South Indian oceans. Between the West Wind Drift and the Antarctic continent is a narrow westward flowing current, the ***East Wind Drift***.

Causes of Surface Currents

EKMAN TRANSPORT

In 1902, V. W. Ekman, a Swedish oceanographer, provided the mathematical basis for understanding wind-driven ocean currents. When a wind blows over the ocean, the water particles near the surface tend to move in the same direction as the wind. Once this motion is initiated, Coriolis force begins to act, deflecting the particles to the right in the Northern Hemisphere and to the left in the Southern Hemisphere. The motion of a thin layer of surface water will be about 45° to the right of the wind in the Northern Hemisphere. The wind energy, however, is transmitted even below the surface layer. Because of friction, the water at the surface tends to pull the underlying water with it but at a lower speed. Coriolis force further deflects this deeper water. This reaction continues to some depth at which the frictional forces and the water movement due to this process become negligible. The spirally moving water, from the surface down to this depth is called the ***Ekman spiral*** (Fig. 7.4). The water at some

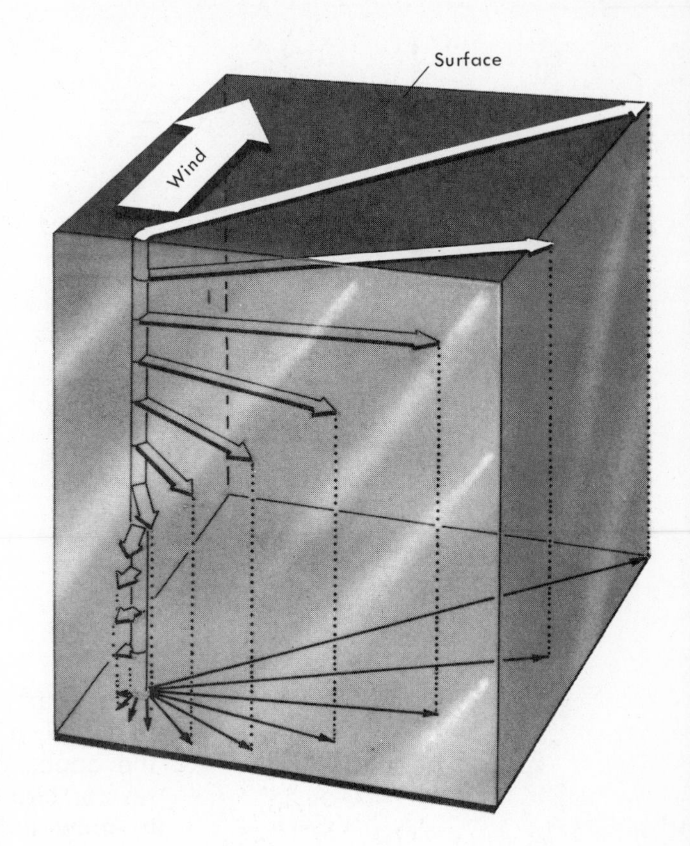

FIGURE 7.4 The Ekman spiral (the effect of wind on water near the surface) in the Northern Hemisphere. (After Baker et al., 1966.)

depth may even be moving in an opposite direction to that of the wind at the surface. The overall effect, however, is that a surface layer of water may have a net motion 90° to the wind. This is known as the ***Ekman transport***. The depth of this layer may extend downward to several tens of meters.

UPWELLING

Because of Ekman transport, an interesting phenomenon occurs along some coasts. Assume that a wind is blowing parallel to a coast in the Northern Hemisphere, as shown in Figure 7.5. The net transport, 90° to the right of the wind, is away from the coast. Deeper water must move toward the surface along the coast to replace the water that has moved offshore. This process is known as ***upwelling.*** Upwelling may transport water from as deep as 400 meters, although depths of 100 to 200 meters are more common. It is actually a rather slow process, involving rates of flow of a few tens of meters per day. The cold, deeper water is generally rich in nutrients, and when it is upwelled, surface productivity is enhanced. Many of the world's most productive fisheries are in upwelling areas, for example portions of the west coasts of North and South America (see also Chap. 11). Because wind condi-

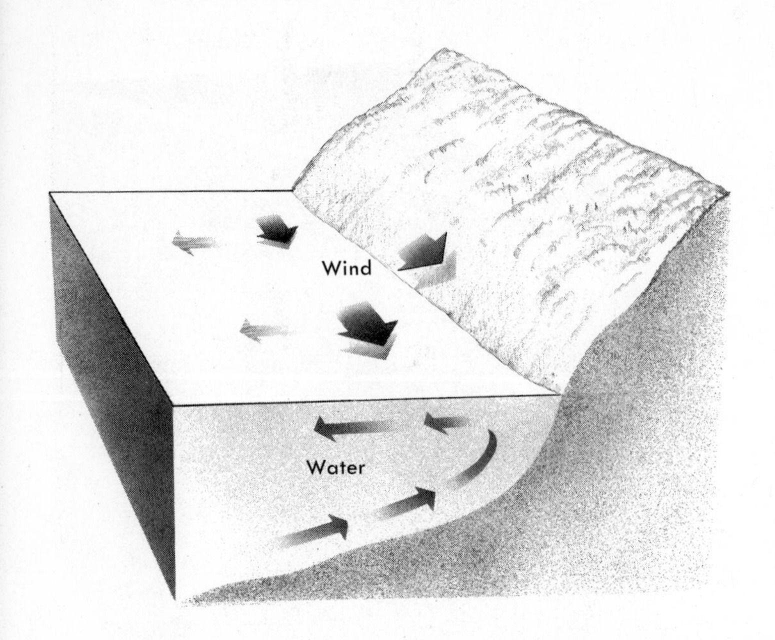

FIGURE 7.5 Upwelling in the Northern Hemisphere, with wind blowing parallel to the coast. The deeper water replaces surface water transported offshore.

tions near the coasts are usually variable, upwelling is frequently a seasonal process. If the wind were to blow in the opposite direction to that shown in Figure 7.5, a reverse circulation would be established. This would tend to draw nutrient-poor water from offshore toward the coast and might eventually cause a decline in productivity.

Along the Peruvian coast, upwelling has produced one of the world's greatest anchovy fisheries. Occasionally, the normally strong trade winds are weakened in this area, causing a cessation of upwelling and a movement of warm, nutrient-poor, tropical water southward over the northward flowing Peru Current. The resulting decrease in productivity causes the death of many animals such as anchovies and guano birds. This results in anaerobic conditions and the production of hydrogen sulfide gas by anaerobic bacteria. Hydrogen sulfide discolors the hulls of ships and the exteriors of homes and is sometimes referred to as the ***Callao Painter,*** after a Peruvian port city. The phenomenon usually occurs around Christmas, so it is also known as ***El Nino*** ("the Child"). Recent occurrences of El Niño have been in 1972, 1965–66, 1957–58, 1953, and 1941.

Geostrophic Balance

In the preceding discussion, we have considered the effects of wind, friction, and the earth's rotation on the ocean surface. Although there is a good correlation between surface wind patterns (Fig. 7.2) and the observed currents (Fig. 7.3), a total agreement is lacking, especially when Ekman transport is considered. For example, the trade winds in the North Atlantic blow from the northeast, yet the surface current (North Equatorial Current) flows almost due west, except when it reaches the South American coast and is deflected toward the northwest by the land mass. In fact, the presence of land masses plays an important role in determining the configuration of many surface currents. According to the idea of Ekman transport, a net movement toward the northwest would be expected in the open ocean. There must be some other forces acting on the water to modify Ekman transport.

Figure 7.6 shows a sea surface profile across the equator in the Atlantic Ocean. It shows a topographic high or "hill" centered about 5° north latitude and topographic lows or "valleys" on both sides of it. Because the heights involved are only a few tens of centimeters, the topography is deduced by indirect means from the distribution of the density of seawater rather than from direct measurements. The distribution of seawater density is primarily due to the effects of evaporation, precipitation, and the transport of the lighter water by wind. The North Equatorial Current, for example, is flowing along a sloping surface which is directed upward toward the north. The slope restricts the Ekman transport to the northwest and causes the current to flow westward. This situation can occur only if the Coriolis force acting on the current is balanced by an equal and opposite force. This opposing force is the ***pressure gradient force,*** the horizontal force due to the decrease in pressure between two points, caused by the sloping sea surface. Any water that is transported up the

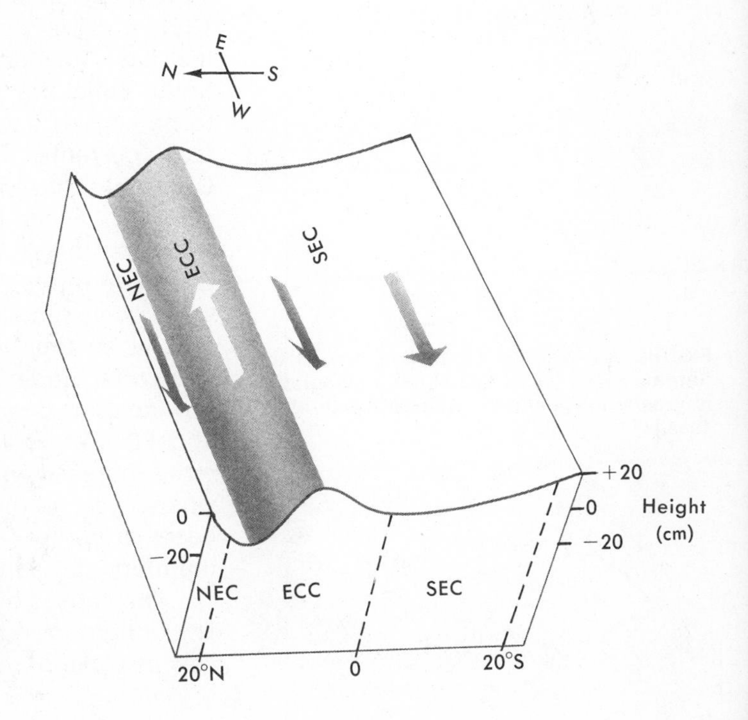

FIGURE 7.6 Sea surface topography across the equator in the Atlantic Ocean. NEC, North Equatorial Current: ECC, Equatorial Counter Current (shaded); SEC, South Equatorial Current. (After Defant, 1961.)

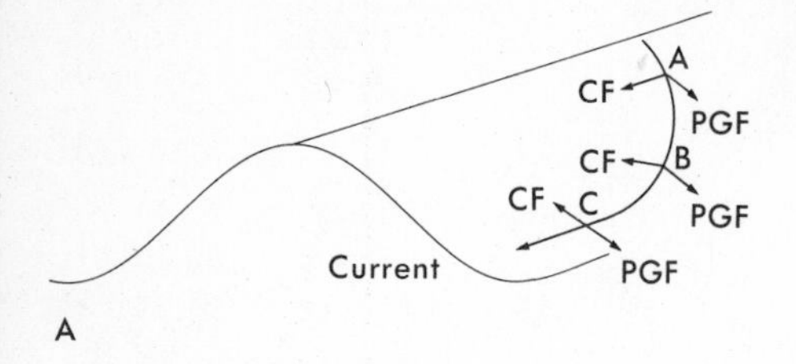

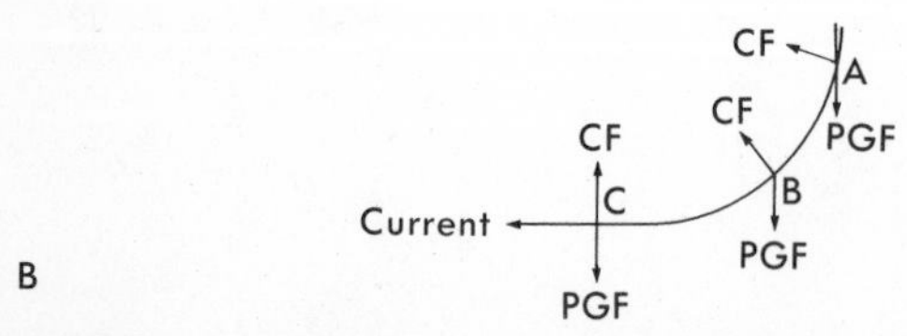

FIGURE 7.7 Deflection of a current due to Coriolis force and the pressure gradient force on a sloping sea surface in the Northern Hemisphere. Note that the Coriolis force and pressure gradient force are balanced at point C. (CF, Coriolis force; PGF, Pressure gradient force.)

slope would flow back down the slope owing to gravity and would be eventually deflected (Fig. 7.7). This type of flow, in which the Coriolis force is exactly balanced by the pressure gradient force, is called a ***geostrophic current.*** A purely geostrophic current is assumed to be frictionless and unaccelerated, because no forces acting in the direction of the current are considered. Of course, wind and friction are quite significant factors, as mentioned earlier. The west-flowing South Equatorial Current and the east-flowing Equatorial Counter Current can also be accounted for by geostrophic flow (Fig. 7.6). Furthermore, water piles up in the western parts of the tropical oceans owing to the strong equatorial currents flowing toward the west. The resultant eastward sloping sea surface may contribute to the eastward-flowing Equatorial Counter Current.

Although wind is the primary driving force, most ocean currents, especially surface currents, tend to flow around hills of light water, as in the clockwise-flowing gyre around the Sargasso Sea "hill" (Fig. 7.8), or around valleys of denser water, as in the counterclockwise Alaskan gyre.

EQUATORIAL UNDERCURRENTS

In addition to the surface currents mentioned earlier, another type of current, although not at the surface, is important. These currents are the ***equatorial undercurrents***, one of which was initially reported in the Atlantic in 1876 by Buchanan, one of the investigators on the *Challenger* Expedition. This phenomenon was largely ignored until 1951, when an undercurrent was accidentally observed in the East Pacific, moving under the westward flowing South Equatorial Current. Tuna-fishing gear was observed to be carried eastward below the westward flowing surface currents. This undercurrent is known as the ***Cromwell Current*** after Townsend Cromwell, who first studied it. Since that time, equatorial undercurrents have been confirmed for the Indian as well as for the Atlantic Ocean. In the Indian Ocean, however, the undercurrent is well developed only during the northeast monsoon season, when the North Equatorial Current is present. The equatorial undercurrents are thin ribbons of eastward flowing water centered about the equator. They average about 250 km in width and 250 m in thickness. These currents increase in speed toward the core of the ribbon, where they may reach 150 cm/sec (about 3 knots). The cores are also characterized by high-salinity water and therefore can be traced easily.

The causes of undercurrents are not fully understood. The undercurrents probably represent a subsurface return flow of some of the water that has been piled up along the

FIGURE 7.8 Surface circulation around the Sargasso Sea "hill." Sea surface topography is greatly exaggerated. (After Williams et al., 1968.)

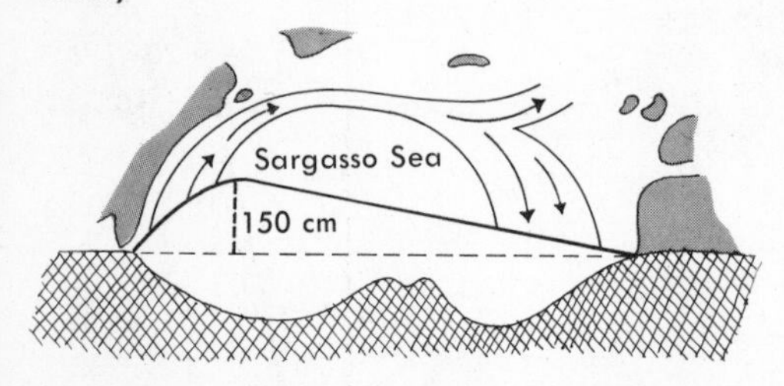

western margins of the oceans. Perhaps as a result of the westward moving surface currents, the sea surface (and underwater pressure surfaces) slopes downward toward the east, resulting in a subsurface flow of water.

Climate and Surface Currents

The oceans play an important role in determining the climates of the earth (see also the discussion on Heat Budget in Chapter 4). Surface currents of the oceans determine to a certain extent the distribution of sea surface temperature. If it were not for the ocean currents, the surface temperatures would be nearly zonal in an east-west direction. However, the surface temperatures (see Fig. 4.10) deviate from a zonal pattern in a manner which can be easily accounted for by the current patterns shown in Figure 7.3. For example, in the North Atlantic these deviations can be explained by the southward-flowing cold Labrador Current, and the northeastward flowing Gulf Stream, North Atlantic Current, and Norway Current, which transport warm water poleward. Therefore, the coast of Norway is generally somewhat warmer than the coast of Labrador.

The oceans serve as a vast heat reservoir, contributing heat to a cold atmosphere and drawing it from a warm atmosphere. Meanwhile, the sea itself remains rather constant with regard to temperature. Thus, the sea serves as a moderating influence on climate, especially when the wind blows over the sea and onto the land. This is especially true along the west coasts of continents that experience prevailing westerlies. Along the east coasts, where winds generally blow from the land, the ocean's effect on climate is much less striking. Figure 7.9 shows the seasonal temperature distribution for San Francisco, California, and Norfolk, Virginia, which are at about the same latitude. The greater effect of the cool California Current produces a more moderate climate at San Francisco compared to Norfolk, which is less affected by the ocean.

It is important to stress that wind patterns are not constant and that shifts in winds may cause relatively short-term weather conditions such as storms and fog. Although the development and movement of a storm depends on a number of factors, such as atmospheric pressure and wind, the strength of a storm may be modified by ocean currents. For example, the warm waters of the Kuroshio and the Gulf Stream evaporate rapidly when cold winter winds blow over them and are warmed. The warmed air can carry much more moisture than cold air. Therefore, a great amount of energy and water vapor is transferred to the developing storms in these regions, making the storms more severe than they would otherwise

FIGURE 7.9 Seasonal temperature distribution for San Francisco, California, and Norfolk, Virginia, at about the same latitude. Both cities have an average temperature of about 55° F (12.8° C). (After Trewartha, 1954.)

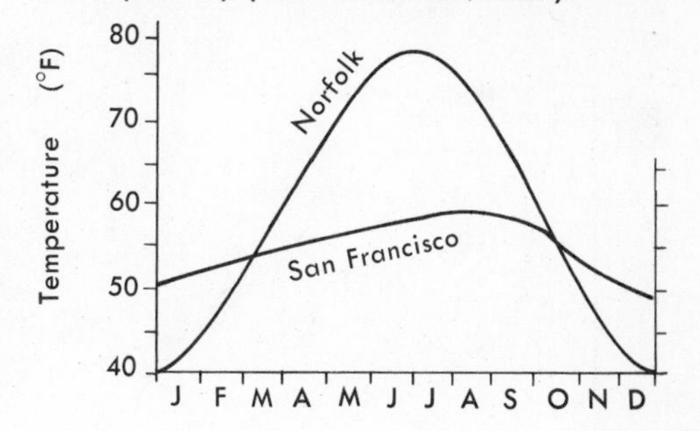

be. Conversely, when warm, moisture-laden winds blow over cold water, the air is cooled and its ability to hold moisture is decreased. Therefore, fog may occur as the water vapor condenses. This situation frequently occurs when winds that have blown over the Gulf Stream meet the cold Labrador Current flowing over the Grand Banks area off Newfoundland, posing a serious problem in the spring, when both fog and icebergs occur in this region. Fogs may also occur when very cold air flows over warm water, resulting in rapid evaporation and subsequent condensation of water vapor. This phenomenon is known as ***steam fog*** and occurs in the fall over semi-enclosed bodies of water such as lakes and Norwegian fiords.

Current Measurements

There are numerous methods of measuring surface currents. Basically, they fall into two broad catagories; the float method and the flow method. In the float method, an object such as a piece of wood may be observed as it drifts with the current; the direction and distance traveled in some unit of time (and therefore the speed) may be determined. In the flow method, a current may be observed flowing past a fixed point such as a ship at anchor or a

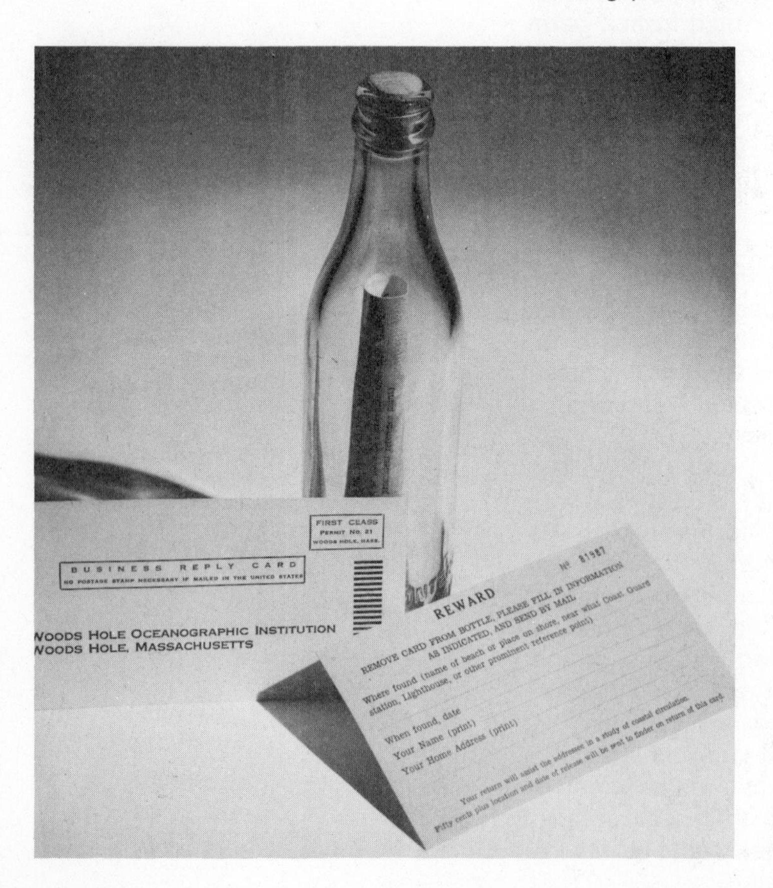

FIGURE 7.10 Drift bottle used to trace the movement of surface currents. (Courtesy of WHOI.)

moored buoy. The current, in this case, may be measured with a specially designed current meter which records the flow of water through the instrument.

Records of ships that have drifted off course or have traveled faster or slower than expected provide much general information concerning ocean currents. International cooperation in the compilation of ships' data has continued since 1853, and ships' records were analyzed even before this date. Based on this type of data, Benjamin Franklin published a chart of the Gulf Stream in 1786.

In addition, sealed bottles known as ***drift bottles*** (Fig. 7.10) containing post cards have been used to study surface drift between the points of release and capture. The person who picks up the bottle is requested to record the location and time of capture on the card and to mail it to the researcher (and to keep the bottle as a souvenir). This method has been simplified by replacing the bottle with a plastic covering for the post card. Of course, this method does not provide detailed information concerning the path but only the beginning and the end points of the drift, and it may have considerable error because of direct wind effects on the bottle. This technique is generally most useful in studying currents in coastal areas.

Although ship drift and drift bottle records have provided much of what is known about the general pattern of surface currents, other more sophisticated techniques are available for studying the details of ocean circulation. For example, a specially designed float called a ***drogue*** may be observed as it drifts with the current. A drogue (Fig. 7.11) is a weighted float with a subsurface vane or parachute that moves with the current at a desired depth, making the float less subject to drifting with the wind. It usually has a flag and a light attached to make it easier to follow and in addition may have a radio transmitter or a

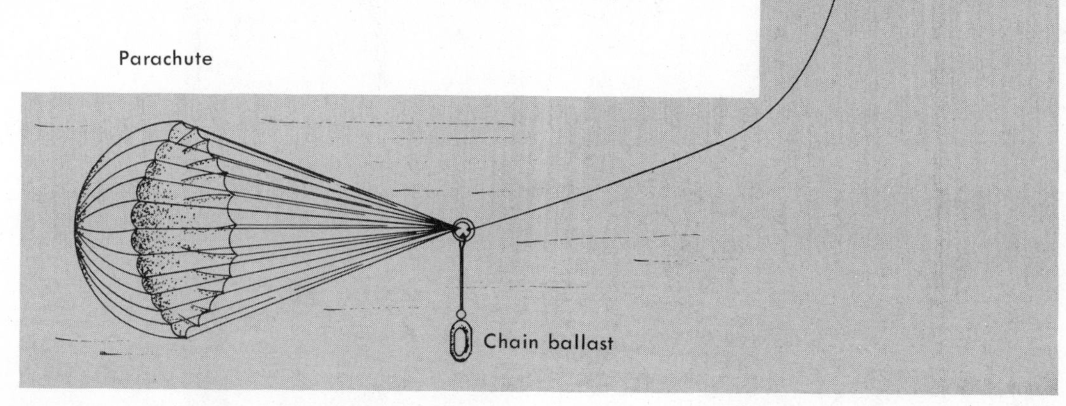

FIGURE 7.11 A parachute drogue, which drifts with the current (after Von Arx, 1962).

metallic radar reflector. Another float, the ***Swallow float*** (named after its developer, J. C. Swallow), may be adjusted to drift with the current at a particular density level (depth) in the sea. It produces a sound that may be picked up by a ship's sonar, and it is particularly applicable in the study of subsurface currents.

In addition to the above mentioned devices, currents may be studied by marking the water with a dye that can be detected even when greatly diluted. This method is especially useful in aerial surveying of currents, as the dye may be dropped from the air and photographed as it drifts.

Current meters, or flow meters, measure the speed and frequently the direction of the current. In some instruments, the current flows past a propeller, the rotation of which indicates the speed of the current. A compass mounted on the instrument may provide information about the current's direction. A great variety of current meters are available (Fig. 7.12). Unfortunately, the use of current meters is an expensive and time-consuming operation as compared to other methods and is therefore done only when the other methods are either too impractical or too imprecise. The flow of water past a point may also be estimated by measuring the deflection of a wire supporting a weight with a known drag (Fig. 7.13). This procedure may be done rather easily from even a small boat but is effective only near the surface, since many currents are quite complicated vertically.

FIGURE 7.12 *A,* Typical current meter about to be lowered into the sea. *B,* "Pop-up" current meter. The two spherical floats return the instrument to the surface after bottom current measurements have been recorded. (*A,* Courtesy of WHOI; *B,* courtesy of Lamont-Doherty Geological Observatory of Columbia University.)

A

DEEP OCEAN CURRENTS

The currents we have been discussing so far are produced primarily by wind. They are restricted to the surface water, although the tremendous Gulf Stream carries water even at depths of 1,500 meters or more. Water movement in surface currents is basically horizontal. Deep ocean currents, on the other hand, involve horizontal and vertical flow extending into the deepest parts of the oceans. They are the result of minute horizontal and vertical density variations in the sea. In the long run the deep circulation is inseparable from that at the surface. Even deep water is returned to the surface, becomes part of the surface currents, and eventually sinks to rejoin the deep circulation. Thus a balance exists in the flow of water within each ocean as well as among oceans.

The deep ocean currents are responsible for carrying oxygen into the deepest parts of the seas, enabling marine animals to exist at all depths. Deep currents are also responsible for returning nutrients to the surface. If it were not for these deep currents, the seas would be stagnant and anaerobic (lacking oxygen) in the depths, and surface productivity would be much less rich.

Compared to surface currents, deep ocean currents are very slow, frequently of the order of a few centimeters per second. However, speeds of more than 40 centimeters per second (about one knot) have been found at 4,000 meters' depth in the Western Atlantic. Surface currents having speeds of 50 to 100 cm/sec or more are common, and speeds of about 300 cm/sec have been observed in the Gulf Stream. In fact, radioactive carbon analysis of seawater has provided estimates that it has taken about 750 years for water to travel from the surface in the Antarctic to the bottom of the North Atlantic Ocean. On the

FIGURE 7.13 Currents may be measured by the deflection of a weighted line.

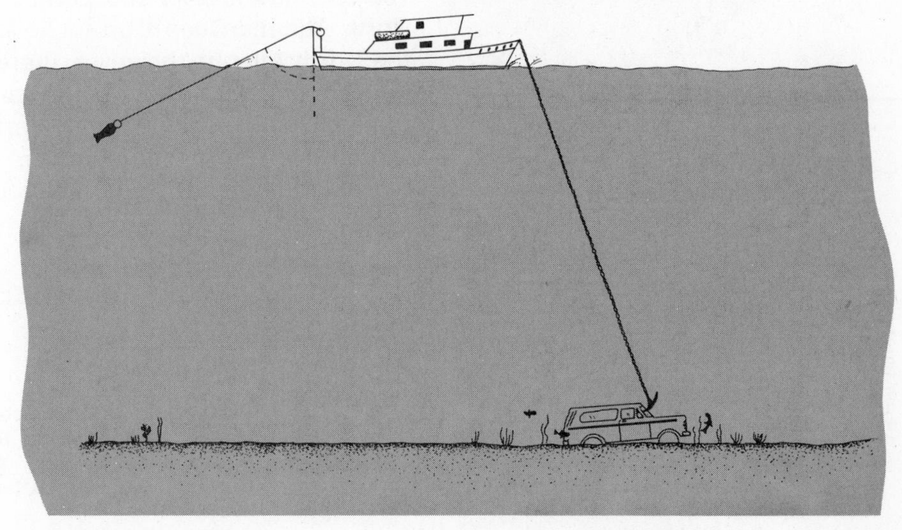

other hand, it takes about 1,500 years to travel from the Antarctic region to the bottom of the North Pacific Ocean. Because of their slow speeds, deep currents are much more difficult to study by direct means than are surface currents, although in recent times Swallow floats have been used to study them at a few thousand meters. Much of our present knowledge of these currents is based on indirect studies involving density distribution in the sea. Since oxygen enters or is produced in the sea only near the surface, oxygen depletion during the flow of deep water can also be used to trace these currents.

Patterns

Figure 7.14 shows the pattern of deep circulation in the Atlantic Ocean. This circulation is predominantly in a north-south direction. The ***Antarctic Bottom Current*** forms at the surface in the Weddell Sea near the Antarctic continent and flows along the ocean floor northward to about 40° N latitude. The ***North Atlantic Deep Current*** forms near Greenland as water in this region sinks toward the bottom and flows southward over the Antarctic Bottom Current. Beyond 40° S latitude, this water, while mixing with surrounding waters, rises from about 3,000 meters toward the surface until it joins the surface circulation in the ***West Wind Drift.*** Water from the North Atlantic is consequently carried into the Indian and Pacific oceans. This rising water, the ***Antarctic Circumpolar Current*** as it is sometimes called, is a global phenomenon, extending around the Antarctic continent. This "upwelling" provides nutrients to one of the world's most productive areas, the Antarctic Ocean. At about 50° S latitude in the Atlantic, water sinks to form the ***Antarctic Intermediate Current***, which flows downward and northward to a depth of about 700 to 1,000 meters and continues northward above the North Atlantic Deep Current to beyond 25° N latitude. In the North Atlantic, however, there is no comparable intermediate current flowing southward. All of these deep currents mix with surrounding waters and are thus weakened farther away from their sources.

In addition to the circulation shown in Figure 7.14,

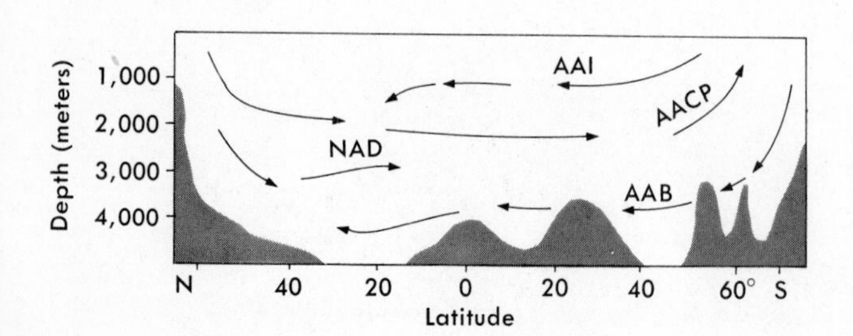

FIGURE 7.14 Deep ocean circulation in the Atlantic Ocean. AAB, Antarctic Bottom Current; AAI, Antarctic Intermediate Current: AACP, Antarctic Circumpolar Current; NAD, North Atlantic Deep Current. (After Wüst, 1936.)

dense water from the Mediterranean flows into the Atlantic (Fig. 7.15). This water travels over the Gibraltar Sill and sinks to a depth of about 1,000 meters as it fans out into the Atlantic and eventually contributes to the North Atlantic Deep Current.

The deep water in the Pacific Ocean is much more homogeneous than that of the Atlantic Ocean, as the Antarctic Bottom Current occupies most of the bottom of the Pacific (Fig. 7.16). This current actually originates in the Weddell Sea. The water flows first into the Antarctic Ocean, then part of it flows into the Atlantic, part into the Indian Ocean, and finally it reaches the Pacific, after mixing has modified its characteristics. Its indirect route results in sluggishness and greater depletion of oxygen in the Pacific as compared with the bottom water in other oceans. A current similar to the North Atlantic Deep Current is absent in the Pacific; instead, two minor currents are found flowing south toward the equator below the surface. These are the ***Subarctic Bottom Current***, which flows out of the Okhotsk Sea in the Northwest Pacific, and the ***North Pacific Intermediate Current***, which forms northeast of Japan in the region where the Oyashio Current converges with the North Pacific Current. The Antarctic Intermediate Current in the Pacific is somewhat weaker than the corresponding current in the Atlantic and extends northward only to the equator. Some of the water of the Antarctic Intermediate Current as well as of the Antarctic Bottom Current is returned to the surface in the Antarctic Circumpolar Current.

The deep circulation in the Indian Ocean is similar to that in the South Pacific. North of the equator, however, salty dense water from the Red Sea flows out and sinks to a depth of about 1,000 to 2,000 meters in the Northwest Indian Ocean. This source of water is similar to that found flowing out of the Mediterranean into the Atlantic Ocean.

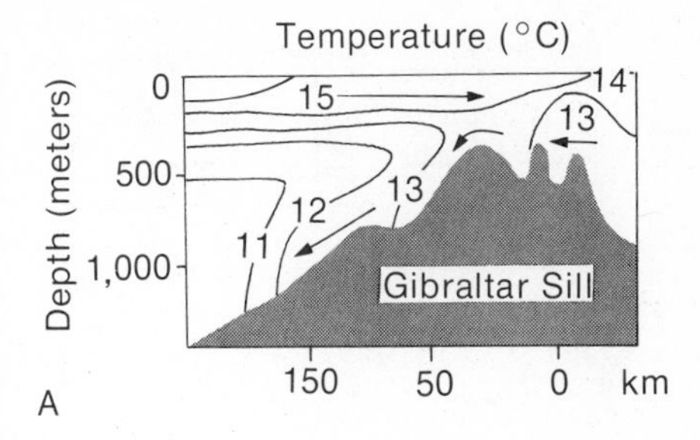

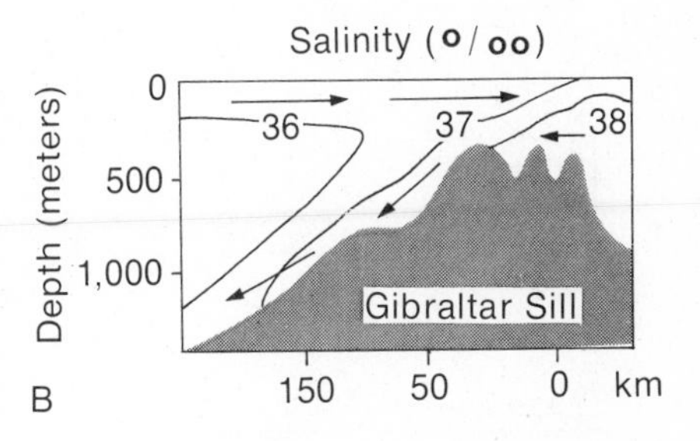

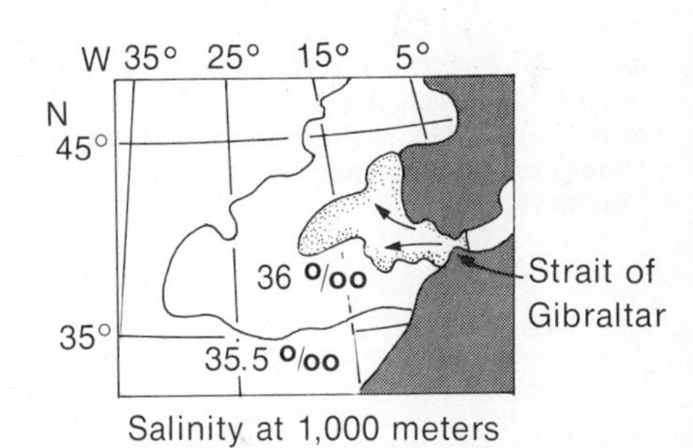

FIGURE 7.15 Flow of Mediterranean water into the Atlantic Ocean. *A*, Vertical temperature structure across the Gibraltar Sill; *B*, Vertical salinity structure across the Gibraltar Sill; *C*, Top view, indicating a westward movement of the plume of Mediterranean water. (*A* and *B* after Sverdrup et al., 1942; C after Groen, 1969.)

Causes of Deep Ocean Currents

In order for water to sink, it must be denser than the surrounding water. Let us consider a hypothetical ocean (Fig. 7.17*a*), in which the density increase with depth is the same everywhere. The surfaces of equal density would be horizontal, and there would be no deep-ocean currents, because no water in this ocean could sink to a lower level unless denser water were introduced from outside. Actually, real oceans have surfaces of equal density that are sloped (Fig. 7.17*b*). If denser water is introduced at the surface, it will sink until it reaches a level at which its density equals that of the surrounding water and will continue to flow laterally along the sloping equal-density surface.

FIGURE 7.16 Deep circulation in the Pacific Ocean. AAB, Antarctic Bottom Current; AAI, Antarctic Intermediate Current; AACP, Antarctic Circumpolar Current; NPI, North Pacific Intermediate Current; SAB, Subarctic Bottom Current. (After Pickard, 1963, and others.)

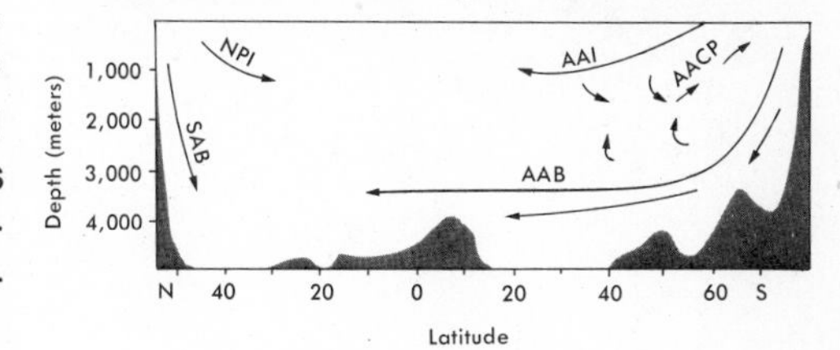

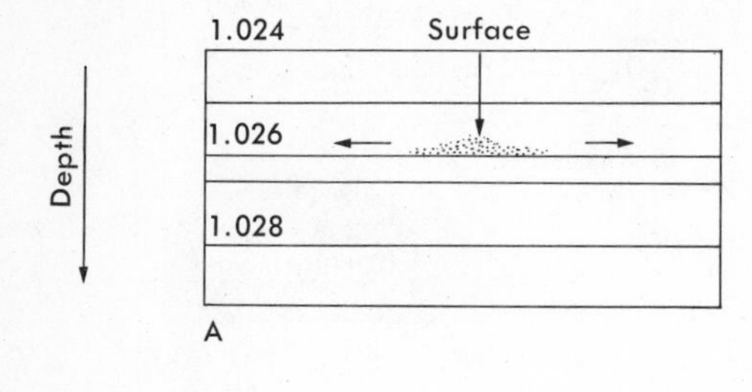

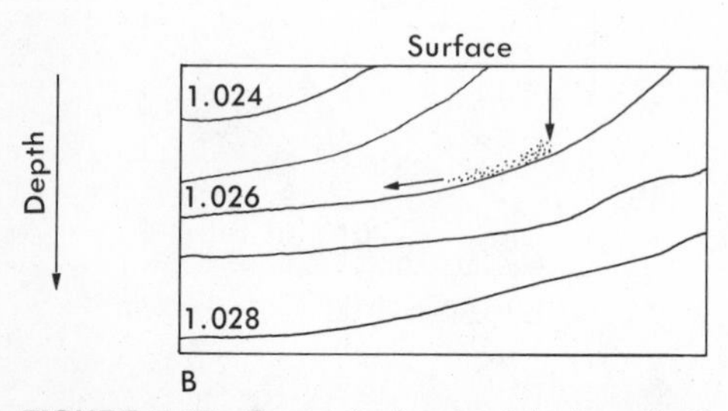

FIGURE 7.17 **Currents (arrows) in hypothetical oceans with (*A*) horizontal and (*B*) sloping density surfaces; assuming that water with a density of 1.026 is introduced at the surface. Numbers indicate density in grams per cubic centimeter.**

In order to maintain the deep ocean currents shown in Figures 7.14 and 7.16, a fairly constant source of dense waters must be produced at the surface, where they begin to sink. Evaporation (yielding increased salinity) and cooling both result in dense water. Also, the freezing of seawater into relatively salt-free ice causes an increase in salinity (and density) of the adjacent residual water. In addition, wind-driven surface circulation leads to transport of water toward regions where mixing and sinking are possible. For example, in the southern oceans, warmer water from the north and colder water from the south converge at about 50° S latitude ***(Antarctic Convergence)*** and mix. The resulting water is of nearly the same density as the downward-sloping density surface at the convergence. Therefore, this water is able to sink below the lighter water to the north and becomes the Antarctic Intermediate Current.

The Antarctic Bottom Current, which flows northward along the bottom in all oceans, is produced by the formation of very dense water by cooling and by the formation of ice near the Antarctic continent, especially in the Weddell Sea. The southward flowing North Atlantic Deep Current is formed by the sinking of water off Greenland. This water, although saltier, is warmer and has a lower density than that of the Antarctic Bottom Current and hence flows over it. The Antarctic Circumpolar Current is the result of water rising to compensate for the sinking of water associated with the Antarctic Intermediate Current to the north and the Antarctic Bottom Current to the south. In the Atlantic Ocean, the Antarctic Circumpolar Current is also fed by the North Atlantic Deep Current. Thus water from the North Atlantic eventually enters the other oceans via the West Wind Drift.

Indirect Study of Deep Ocean Currents

Unlike the surface currents, the extremely slow-moving deep ocean currents are difficult to measure directly with the present technology. The deep ocean circulation patterns shown in Figures 7.14 and 7.16 are derived mostly from indirect studies which involve measurement of temperature and salinity of seawater at various depths at different locations. We will now attempt in brief to see how these data are obtained.

Starting with the premises that dense water tends to sink and lighter water tends to rise and that density is a function of temperature and salinity, let us consider an idealized ocean. This ocean consists of two layers of water, each having a uniform temperature and uniform salinity (Fig. 7.18). Figure 7.18*c* shows a ***T-S diagram***, the plot of temperature versus salinity values of water sam-

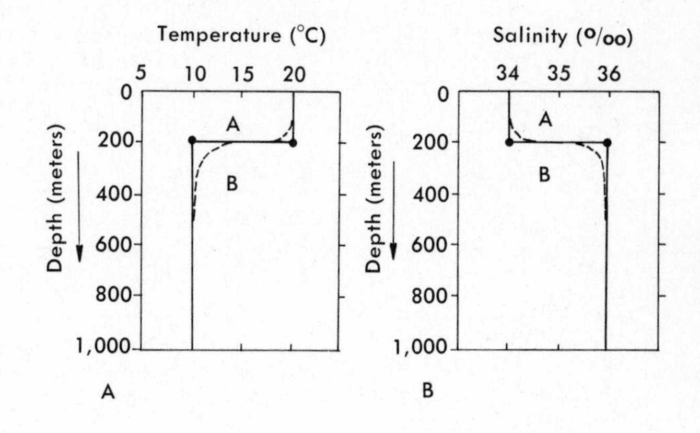

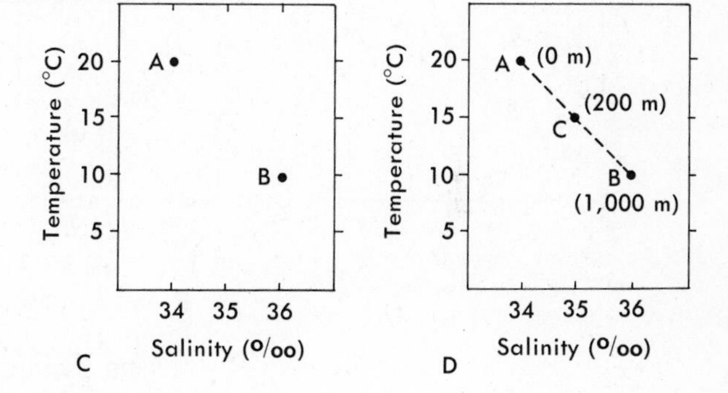

FIGURE 7.18 Temperature-salinity relationships in a hypothetical two-layer ocean (dashed lines represent results of mixing). *A,* Temperature; *B,* Salinity; *C,* T-S relationship before mixing; *D,* T-S relationship after mixing.

ples taken simultaneously from the top to the bottom of this ocean. Since only two temperature and salinity values are involved, the resulting T-S diagram is represented simply by two points corresponding to the two layers of water. If these two layers of water are slowly allowed to mix vertically across the boundary, the vertical temperature and salinity distribution will change to that shown in Figure 7.18*d*. At the original boundary, for example, water from both layers will mix in equal proportions and will have a temperature and salinity intermediate to the initial values (point C on Fig. 7.18*d*) of the two layers. Farther away from the boundary, there will be less mixing between the two layers. Mixing of two ***water types*** (each having a single temperature and salinity value) results in a ***water mass*** characterized by a range of temperatures and salinities and represented by a straight line on a T-S diagram.

Figure 7.19 shows the mixing of three layers of water (three water types) and the resulting T-S diagram. Note that in this case a curve (Fig. 7.19*d*) consisting essentially of two straight lines (two water masses) is obtained.

Figure 7.20 shows typical T-S diagrams for various oceanic regions. They are the most useful tools developed to date for studying deep ocean currents, especially if they are used in conjunction with other techniques, such

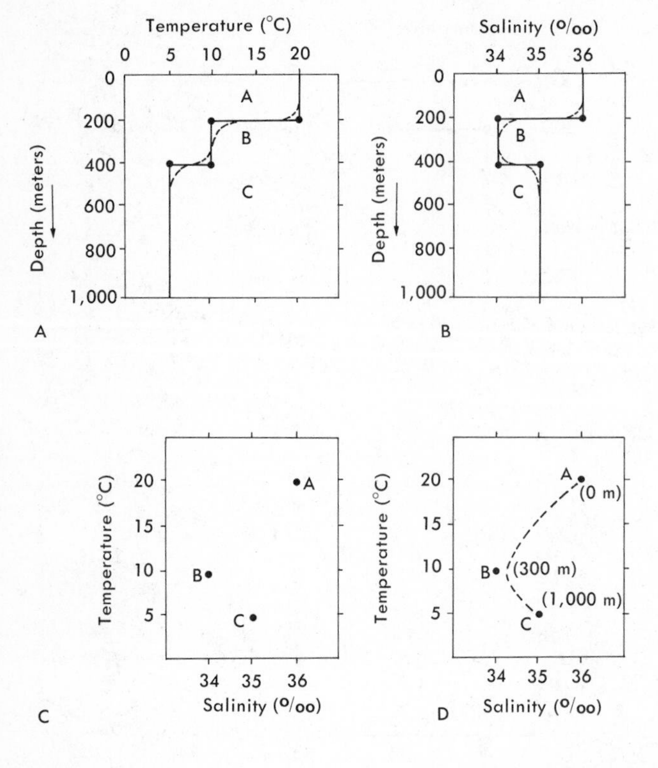

FIGURE 7.19 Temperature-salinity relationships in a three-layer ocean (dashed lines represent results of mixing). *A*, Temperature; *B*, Salinity; *C*, T-S distribution before mixing; *D*, T-S distribution after mixing.

as Swallow floats. T-S diagrams are unique for individual oceanic areas; those of the South Atlantic, for example, are characteristically different from those of the North Atlantic. Specific water masses are found in the various oceanic regions, produced and maintained by the mixing of different water types. Most of the water types are formed near the surface of the oceans as a result of physical processes such as evaporation, precipitation, and freezing of seawater. The resulting water, after mixing with other water types, may be carried to varying depths by currents such as the Antarctic Intermediate Current.

T-S diagrams can be used in tracing deep ocean currents. Figure 7.21 shows T-S diagrams from four oceanographic stations located along a north-south line in the Atlantic. The prominent feature of these diagrams is a "core" of low-salinity water of the Antarctic Intermediate

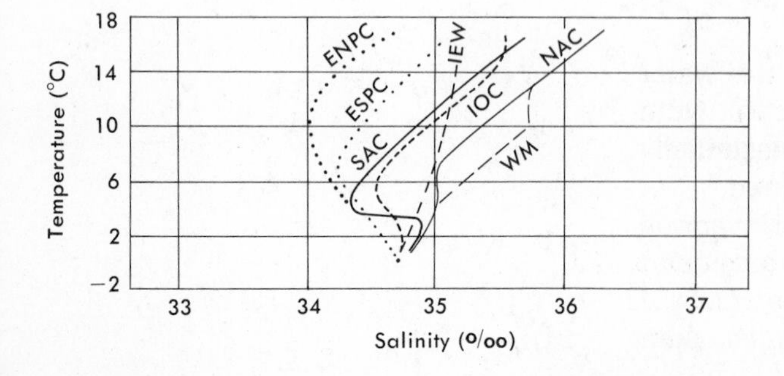

FIGURE 7.20 Typical T-S diagrams for various oceanic regions. Note that the lines all converge toward a point near the bottom of the graph. This point represents the water of the Antarctic Bottom Current. ENPC, Eastern North Pacific Central; ESPC, Eastern South Pacific Central; IEW, Indian Ocean Equatorial; IOC, Indian Ocean Central; SAC, South Atlantic Central; NAC, North Atlantic Central; WM, West of the Mediterranean Sea. (After Sverdrup et al., 1942.)

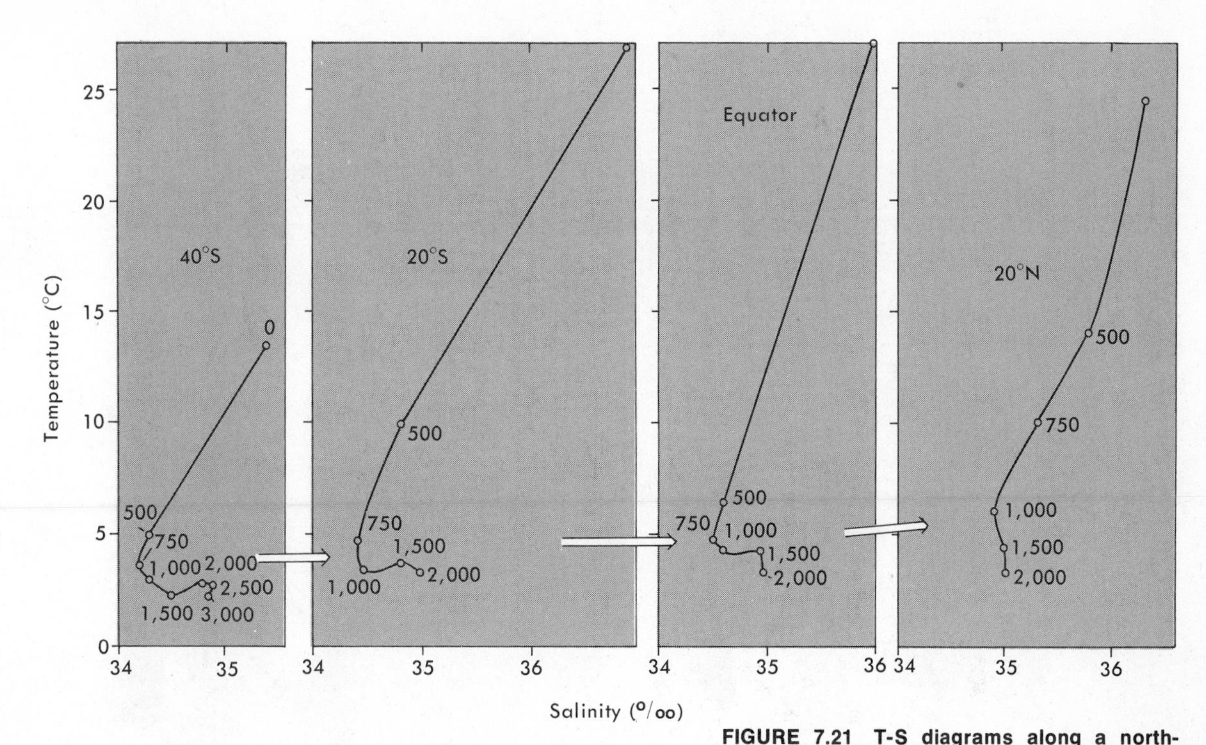

FIGURE 7.21 T-S diagrams along a north-south line in the Atlantic Ocean. Note that the low salinity core of the Antarctic Intermediate Current diminishes northward. Numbers along the curves indicate depths in meters. (After data of Wüst, 1936.)

Current, although cores are not necessarily characterized by a salinity low. They may also consist of a salinity high, or a temperature low or high, or a combination of them. Note that the core gradually disappears at northern stations. Thus, we can trace the movement of deep ocean currents simply by the use of T-S diagrams and by following the cores, if there are any. This method is known as the ***core method***.

Mathematical analysis of T-S curves can be used to estimate the speed of the currents involved as well as to determine the degree of mixing of the water types in question. Most of our information about deep ocean currents is obtained in this manner. In addition, the core method may be a highly useful tool in tracing the movement of pollutants, especially in coastal areas.

SUGGESTED READINGS

Knauss, John A. 1961 (April). The Cromwell Current. *Scientific American.*
McDonald, James E. 1952 (May). The Coriolis Effect. *Scientific American.*
Munk, Walter. 1955 (September). The Circulation of the Oceans. *Scientific American.*
Neumann, Gerhard. 1968. *Ocean Currents.* Amsterdam, Elsevier.
Pickard, George L. 1963. *Descriptive Physical Oceanography.* New York, Pergamon Press.
Stommel, Henry. 1958 (July). The Circulation of the Abyss. *Scientific American.*

Cannon Beach, Oregon. (Courtesy of Oregon State Highway Department.)

CHAPTER 8

Coastal Oceanography

The parts of the ocean most familiar to us are the coastal regions. They are especially important because most of the fisheries of the world are located here. They are also valuable as sources of petroleum and minerals. In addition, coastal regions include sites for harbors and ocean-oriented recreation. Although seemingly isolated, all of these uses interact and frequently conflict with one another. For example, port activities might result in coastal pollution, which in turn will have adverse effects on fisheries and recreation. The shore, which includes beaches and rocky coasts, undergoes natural changes owing to geological forces as well as to waves and coastal currents. When one interferes with these natural forces, detrimental effects may occur.

BEACHES

The ***beach*** includes the entire region of unconsolidated materials extending from the low-tide line to the uppermost region of wave action (the coastline), represented by cliffs, sand dunes, or permanent vegetation. Beaches, however, are transient features. They can grow or erode naturally on a daily, weekly, or seasonal basis, depending on the heights of the waves and the tidal fluctuations of the area. Many of the changes taking place along or near beaches are relatively slow, requiring tens or hundreds of years to have any large-scale effects. However, during severe storms and tsunamis, major coastal changes may occur within a few hours. It is still uncertain whether long-term gradual change or short-term catastrophic change is the most effective coastal shaper. We may further affect the nature of beaches by building shore structures such as jetties and breakwaters, which may protect or destroy the beaches.

Figure 8.1 shows various kinds of beaches. Some beaches are composed of nothing but sand, making them ideal for beachcombers and bathers. Others may be composed of pebbles or cobbles, and some coastal areas of the world may not have any beach at all but just a solid rocky bottom.

FIGURE 8.1 Various kinds of beaches. *A,* A sandy beach in Portugal. *B,* A pebble beach on the Olympic Peninsula, Washington. *C,* A steep cobble beach at Yaquina Head, Oregon. *D,* A rocky coast, with no beach, in Brittany, France. (*A* courtesy of Hayward Associates, Inc.; *B* courtesy of National Park Service, USDI; *C* photo by Lynn O'Connor, courtesy of School of Oceanography, Oregon State University; *D* courtesy of French Embassy Press and Information Division.)

Beach Materials

Beaches can be composed of different mineral grains of varying sizes. Much beach material is derived from the erosion of coastal rocks, but it may also be brought to the shore by rivers and wind. Once on the beach, these materials may be transported considerable distances along the beach.

As granite and granite-derived rocks are among the most common components of the continents, many beaches are composed of quartz and feldspar, the main constituents of granite. Depending on the amount of sorting (separation by size or weight) and the distance traveled (greater distance tends to destroy softer or more soluble minerals), some beaches may be almost wholly composed of any one mineral such as quartz (the most resistant material). Beaches located far from granite sources may be composed of bits of broken coral, as along the Florida coast. However, many Florida beaches are composed of about one half quartz material even though they are thousands of miles from obvious sources.

Beaches along volcanic coasts such as in Hawaii are generally composed of dark sands derived from the breakdown of volcanic rock. Some beaches in Brazil and India are composed of heavy black minerals that are the source of rare elements such as titanium and thorium. Wave and current activity along these shores has removed the lighter materials and concentrated the heavy minerals along the beach. In fact, many economic deposits of such minerals are found in areas that were once beaches.

Beach Profile

Figure 8.2 shows a generalized beach profile. Not every feature shown in this figure may be developed on any given beach. The beach profile can be divided into three major zones: ***offshore, foreshore,*** and ***backshore***.

The most seaward part of the profile, the ***offshore*** region, is the zone of breakers extending seaward from the low-tide line to the depth at which these waves first feel the bottom (see Chap. 5). In areas where the tidal range is relatively small, as well as in areas of gently sloping sea bottom, the offshore region is often characterized by several ***longshore bars*** and ***longshore troughs*** parallel to the coastline. A longshore bar is simply a submerged ridge produced by the deposition of sand seaward of the longshore trough. It may be exposed during low tides. Although the origin of longshore bars and troughs is not

FIGURE 8.2 Typical beach profile.

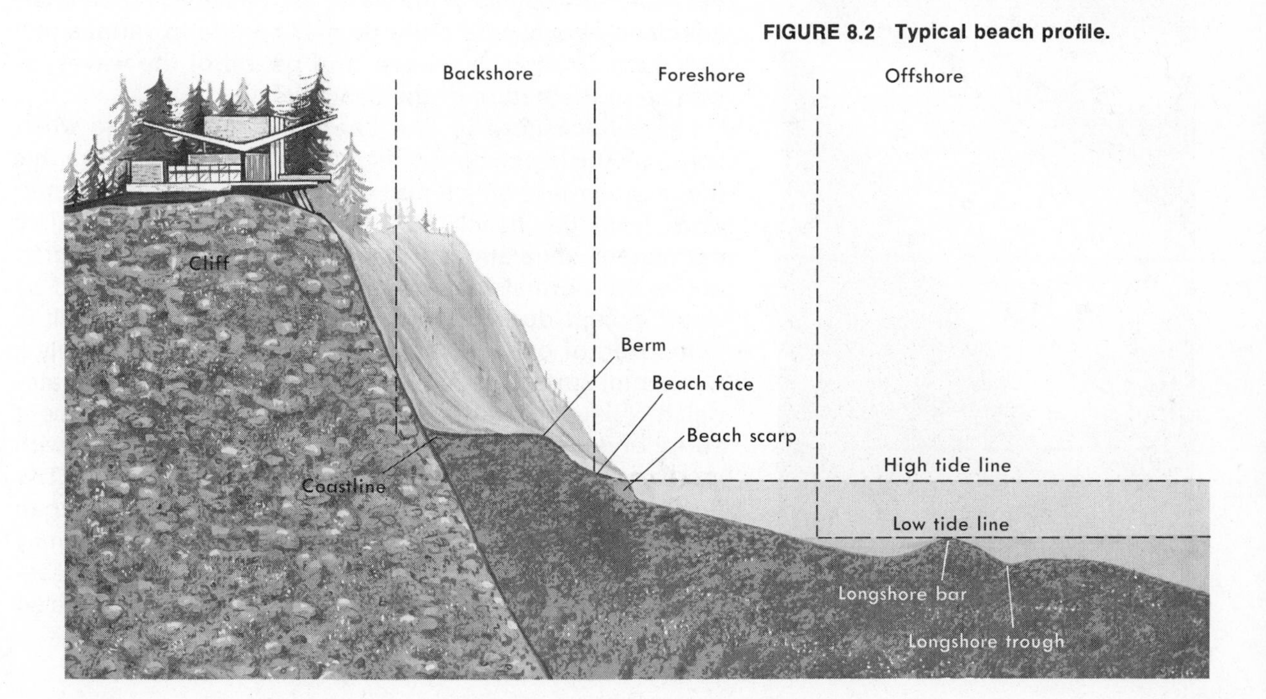

well understood, they are rarely developed in areas having large tidal ranges. This suggests that they develop only when surf and tidal conditions are fairly stable. Bars are often visible as light areas on aerial photographs owing to the shallowness of the water or waves breaking on the bars. Apparently, their size is a function of wave height, because higher winter waves usually result in larger, longer bars. In many cases, bars may be almost absent during the summer, when the waves are much smaller. In addition, bars and troughs usually migrate landward during winter or seaward during summer in response to changing wave heights.

The ***foreshore*** region consists mostly of the zone that is regularly exposed between low and high tides (the ***low tide terrace***) and the ***beach face*** (the sloping wave uprush region above high-tide line). The slope of the beach face depends on the grain size of the materials making up the beach face; the larger the grain size, the steeper the slope. For example, very fine sand (1/16 to 1/8 mm) produces a slope of approximately 1°, and very coarse sand (1 to 2 mm) results in a slope of about 9°. Cobbles (64 to 256 mm), on the other hand, produce a slope of about 24°. The lower limit of the beach face is sometimes characterized by a nearly vertical ***beach scarp*** produced by large waves and may have a relief of about one or two meters. The beach face is sometimes marked by a series of crescent-shaped depressions (small bays) separated by points called ***cusps*** (Fig. 8.3). Most are between one and 100 meters across. They tend to form under equilibrium conditions; that is, on beaches where the erosional and depositional aspects of the waves are balanced. Their characteristic shapes and patterns may be due to various factors such as the size, shape, and period of the waves, as well as to the nature of the beach itself.

FIGURE 8.3 Series of beach cusps, at El Segundo Beach, California. (From US Army Coastal Engineering Research Center, 1973.)

The ***backshore*** is "the beach" to most people when they visit the beach during high tides, but in oceanographic terminology it is only a part of the beach. It extends landward from the beach face to the cliffs, sand dunes, or permanent vegetation of the coastline. It is the region above the normal high-tide line and is rarely covered by water except during spring tides and severe storms. It is composed of one or more ***berms***. A berm is essentially a horizontal surface produced by deposition of beach material. It is usually built up (deposited) by long-period (about 10 seconds) spilling breakers. Berm height increases with increasing wave height and increasing tidal range. The berm materials are composed of larger particles than those making up the beach face. Even the berm itself may be composed of materials of varying particle size, increasing in size as one goes up the beach. Berm widths also vary considerably.

WINTER AND SUMMER BEACH PROFILES

Beaches are constantly undergoing changes, which are often periodic (Fig. 8.4). During the winter (as well as during severe storms and tsunamis), wave heights are considerably greater than at other times. These high waves are capable of eroding the beach (mostly the berm) and transporting the materials offshore to help form longshore bars. The berm width is reduced considerably, but its height increases as the uprush of higher waves transports some of the materials up the beach. During the summer, when the waves are much smaller, the process is one of wave deposition rather than erosion. The berm height decreases, and the longshore troughs are filled in. Eventually, a wider berm is produced.

BEACH DRIFTING, LONGSHORE CURRENTS, AND RIP CURRENTS

Waves approaching the shore obliquely cause movement of the materials along the beach by a process called ***beach drifting***. A sand grain, for example, is pushed up the beach face diagonally (parallel to the direction of wave advance) by the uprush, but during the backrush it may descend straight down the slope (Fig. 8.5). Beach materials can thus be transported great distances if the process is repeated.

When waves strike a coast obliquely, a current is formed in the surf zone which flows parallel to the shore

FIGURE 8.4 Typical summer and winter profiles of a sandy beach. (After Bascom, 1964.)

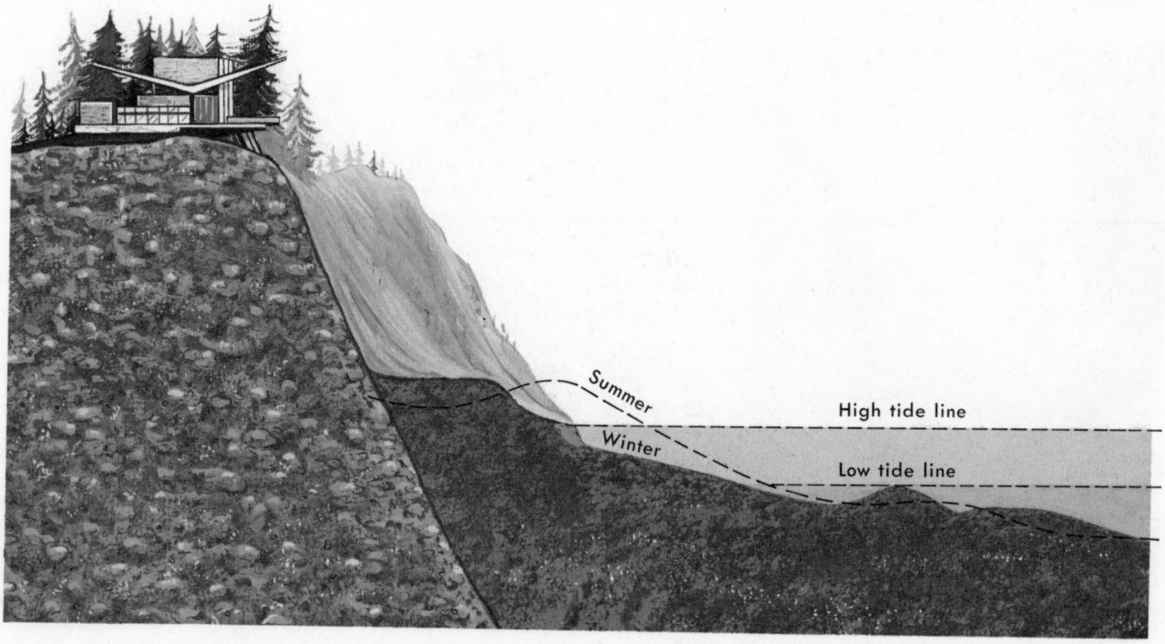

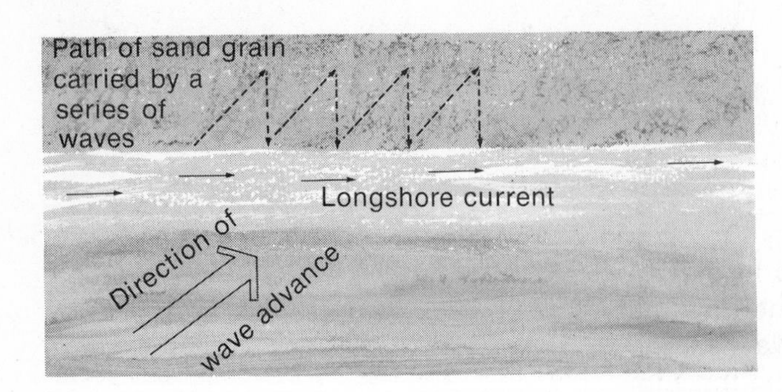

FIGURE 8.5 Beach drifting and longshore currents produced by waves striking the coast obliquely.

(Fig. 8.5). This flow of coastal waters is the ***longshore current***. It is best developed along straight coasts but may also be formed along irregular coasts. Longshore currents, which may pose a threat to landing crafts, usually have speeds of about one knot, although speeds of more than two knots have been reported. Like beach drifting, longshore currents are important in transporting sediments parallel to the coast. These sediments may be obtained locally by wave erosion or may be brought to the sea by rivers. Longshore currents are also responsible for the sediment deposition in many harbors and river mouths (Fig. 8.6), where the current is slowed because of greater water depth and the absence of an adjacent shoreline. Costly dredging operations are often required to counteract the effects of this type of deposition. The ***longshore*** (or ***littoral***) ***transport*** of material may be quite substantial. For example, along the South Atlantic coast of

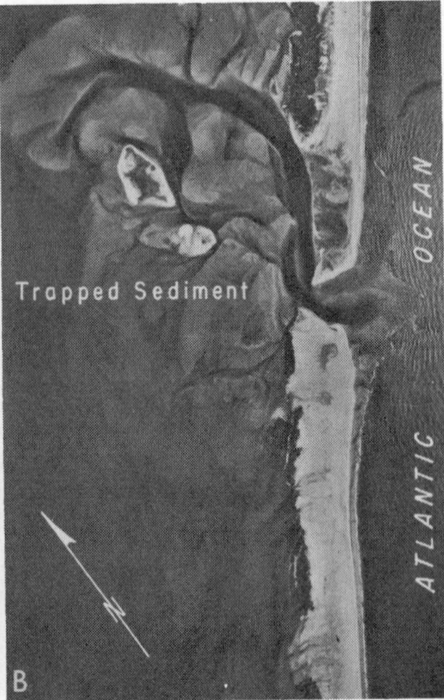

FIGURE 8.6 *A*, Sediment deposition at the mouth of the Alsea River in Oregon restricts its opening to the ocean. This in turn causes erosion on the southern bank of the river near its mouth. *B*, Sediment transported by longshore currents carried into Old Drum Inlet, North Caroline, by tides. (*A* courtesy of Oregon State Highway Department; *B* from US Army Coastal Engineering Research Center, 1973.)

the United States, the net southward transport past several points may be as much as half a million cubic meters per year.

The water brought to the near-shore region by breakers must eventually return to the sea. Some of the returning water may be localized to form ***rip currents***. Rip currents on a particular beach may vary depending on the nature of wave conditions present. Larger swells produce fewer but stronger rip currents than smaller swells. As seen in Figure 8.7*a*, a rip current consists of a ***feeder***, a ***neck***, and a ***head***. The feeder is usually the longshore current. The neck, or rip current proper, may be up to 30 meters wide, and the head can have ten times this width. The speed of rip currents may exceed four knots, but they are usually much slower. They may extend across the surf zone for about one kilometer.

Rip currents produced by waves reaching the shore head-on move out to sea straight across the surf zone, whereas those produced by waves striking obliquely usually move diagonally. Although rip currents can be found on straight coasts as well as on irregular coasts, they are usually more pronounced in small bays, such as on cusped beaches. The divergence of wave energy in bays and the resulting weak zone cause water to flow in from the convergent zones (points). Eventually, this water returns seaward, forming rip currents as shown in Figure 8.7*b*. Like the longshore currents, rip currents also transport sediments, but in this case sediments are moved to deeper water seaward of the surf zone. Rip currents are

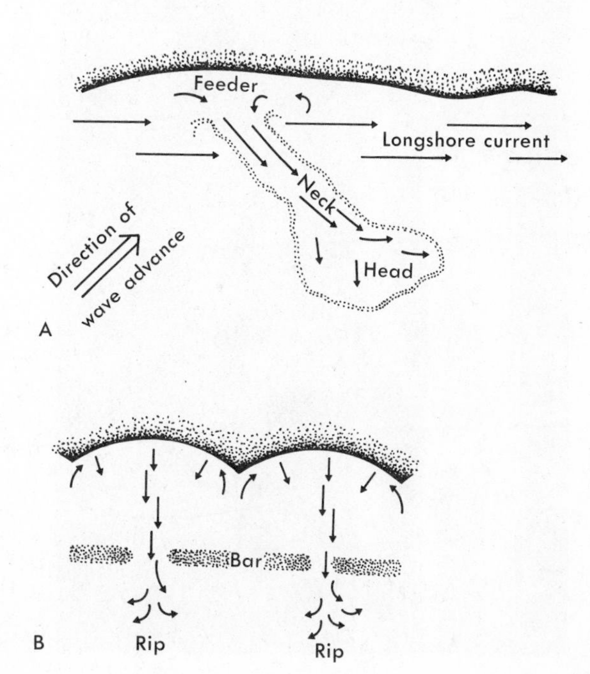

FIGURE 8.7 ***A*, Rip current produced by waves striking the coast obliquely. Note the longshore current feeding the rip. *B*, Rip currents produced by waves entering small bays separated by cusps.**

capable of cutting channels in the surf zone, even across the longshore bars.

Swimmers are sometimes caught in rip currents and swept out to sea. Even a powerful swimmer cannot swim against a strong rip. The best way to get back to land is to swim parallel to the shore until out of the rip and then swim back to shore. The presence of a rip current is usually indicated by the relative absence of breakers along the rip as well as by the higher turbidity of the water (Fig. 8.8).

OTHER COASTAL FEATURES

Barrier Islands and Barrier Spits

Other features of the coastal (offshore) region include ***barrier islands*** and ***barrier spits***, which are most common along the Gulf and Atlantic coasts of the United States. These features are usually formed nearly parallel to the shore by the sediment-laden longshore currents. Barrier

FIGURE 8.8 Rip currents along Ludlam Island, New Jersey. (From US Army Coastal Engineering Research Center, 1973.)

A

B

FIGURE 8.9 *A,* Apollo 9 photograph of the Cape Hatteras area, North Carolina, showing barrier islands. Cape Lookout is near the bottom of the photograph. Cape Hatteras juts farthest into the ocean. *B,* Skylab 3 photograph showing Sandy Hook, New Jersey, a barrier spit. (Courtesy of NASA.)

islands are essentially the same as longshore bars, except that they are permanently exposed (Fig. 8.9*A*). If one end is attached to the mainland, a barrier spit is formed. They are often curved toward the land like a hook (Figs. 8.6 and 8.9*B*). The curving of the barrier spits is believed to

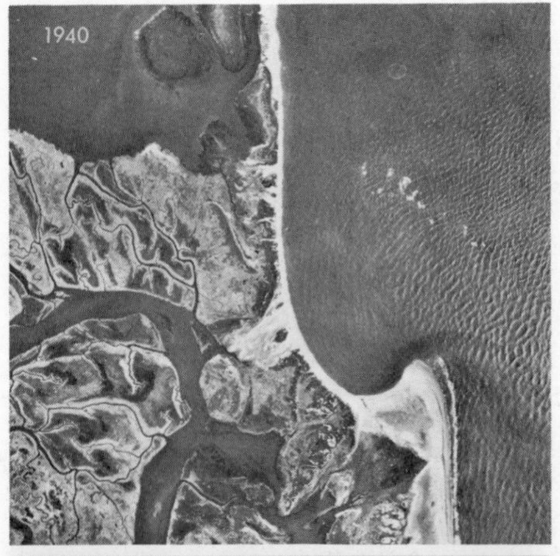

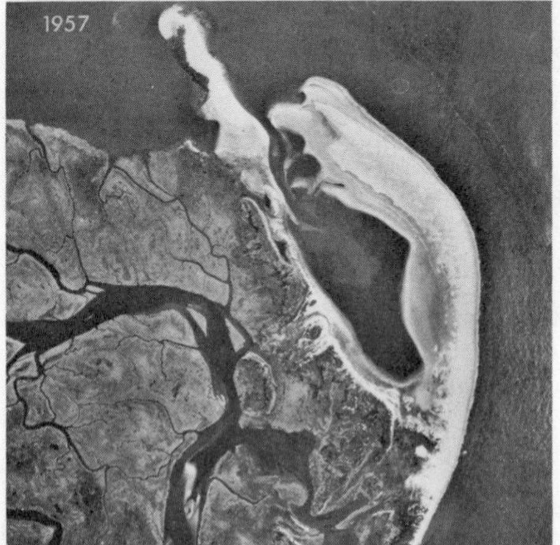

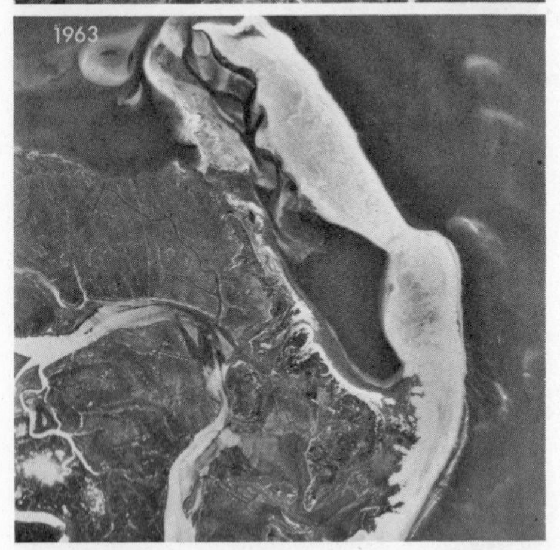

FIGURE 8.10 Changes in the coastline at Little Egg Inlet, north of Atlantic City, New Jersey. Note the growth of the spit in the lower two photos. The width of the area photographed is about four kilometers. (Courtesy of U.S. Geological Survey.)

be caused by the influence of waves reaching the shore from two or more directions or by wave refraction taking place in these regions. Lagoons and salt marshes may develop behind barrier islands and spits, as the sediments brought from land accumulate and are prevented from reaching the open sea.

All coastal features undergo constant change (Fig. 8.10). Barrier spits and islands may grow or may undergo rapid erosion during severe storms, earthquakes, and tsunamis.

Sand Dunes

Sand dunes characterize many coasts, especially in tropical and temperate areas. They are located above the backshore. Sand dunes are mounds of various sizes and shapes produced by wind deposition. Wind is capable of transporting dried beach sand and then depositing it when an obstacle such as a plant is encountered. Once initiated, the dunes begin to grow in size by blocking and trapping sand carried by the wind. Eventually, plants and even trees may be buried by the sand. The orientation of these coastal dunes varies, depending on the direction of the wind.

Dunes provide natural protection for many coasts against beach erosion by large waves. They also prevent the loss of coastal sand to the ocean by acting as retaining walls. The growth of sand dunes can be accelerated, as well as stabilized, by planting various grasses and even discarded Christmas trees or by construction of fences. In areas where there are no natural sand dunes present, artificial dunes may be produced in this manner.

Long-Term Coastal Changes

In addition to the action of waves and currents, geological forces can cause vertical movement of coasts. These are the same forces that are responsible for making mountains, volcanoes, and earthquakes. Except for volcanic activity and earthquakes, which cause rapid changes, their action is slow, and their effects may require hundreds or thousands of years to be perceived by us. Examples of long-term changes of coasts have been provided by archaeologists. Some coastal areas around the Mediterranean have either been uplifted or have sunk below the sea. Some ancient port cities built by the Romans, the Greeks, and the Phoenicians more than 2,000 years ago serve to mark the ancient coastline. Today, remains of houses, storage tanks, quarries, and roads (Fig. 8.11) are underwater, indicating sinking of the land. Harbors are sometimes found several kilometers

FIGURE 8.11 Photograph of partly submerged quarry and road surfaces along the Mediterranean coast. (From Fleming, 1969.)

inland, indicating uplift of the land. Utica (Fig. 8.12), which is located about 30 km northwest of Carthage, was a seaport about 2,000 years ago in what is now Tunisia. Because of the uplifting of land and siltation of the bay, the town is now about 10 km inland. These forces also caused the Medjerda River to be deflected about 20 km to the north.

ARTIFICIAL COASTAL STRUCTURES

Waves are sometimes capable of causing tremendous damage to coastal areas such as headlands which they can inundate in relatively short periods of time. Valuable property may be lost forever. Large storm waves and tsunamis, in addition to causing numerous deaths, are capable of running up the shore and damaging coastal installations. Furthermore, wave-produced longshore currents may cause deposition or removal of sediments along the coast, which for some areas may not be desirable. In order to prevent such changes, people often intervene—actually, interfere—along the coast by building various coastal structures, which often result in serious unwanted side effects. Generally, it is best to avoid unnecessary building on low-lying areas, cliffs, and narrow barrier islands, all of which may be subject to wave erosion.

Sea Walls

Sea walls are built to protect the coastal areas behind them from large waves as well as to maintain a permanent

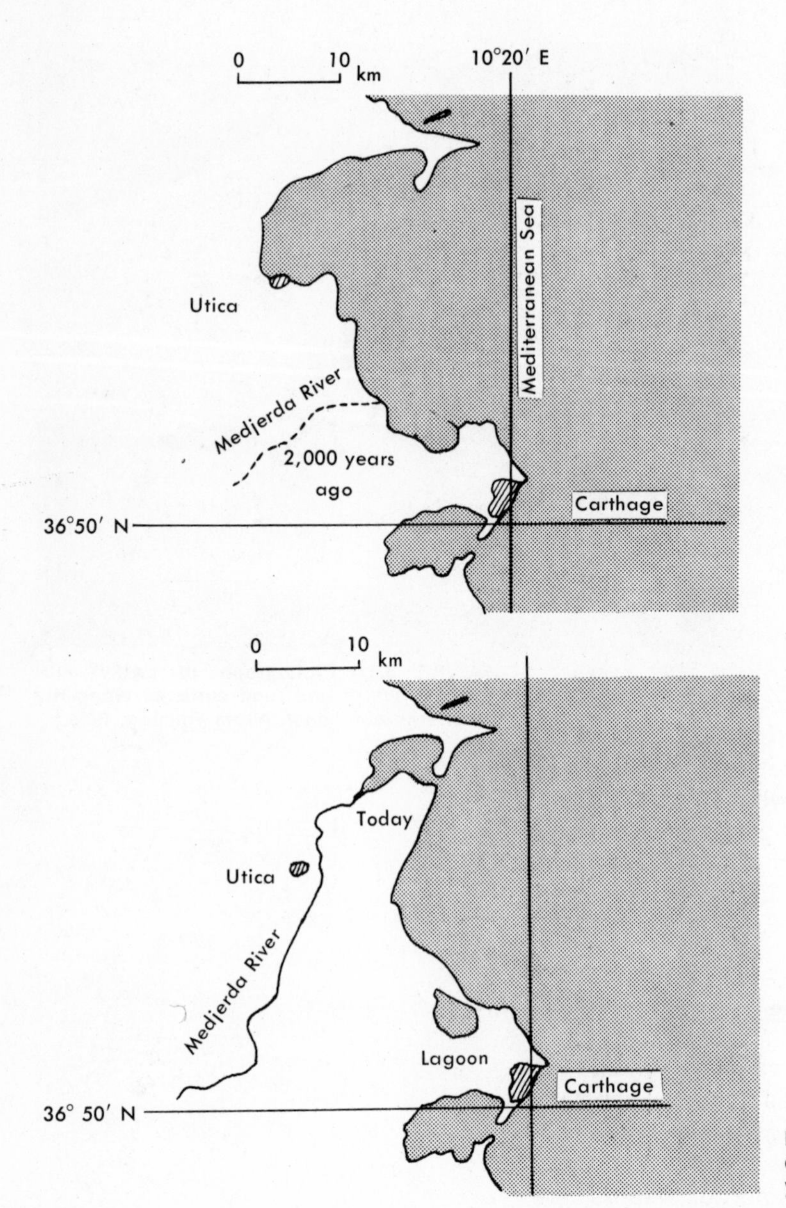

FIGURE 8.12 Changes in the coastline north of Carthage, Tunisia, during the past 2,000 years. (After Fleming, 1969.)

coastline. They are built along the coastline using concrete, boulders, or other materials. Large waves break on them or are reflected back to sea. They may be built in a variety of shapes (Fig. 8.13). The seaward sides of some sea walls have a parabolic design, which reflects waves more effectively than walls with sloping sides. Others may be built like steps, so that higher waves will break first, farther offshore. Some sea walls may have a drain at the top, a design which may someday be utilized to obtain power from large waves as well as from tidal fluctuations (Fig. 8.13E).

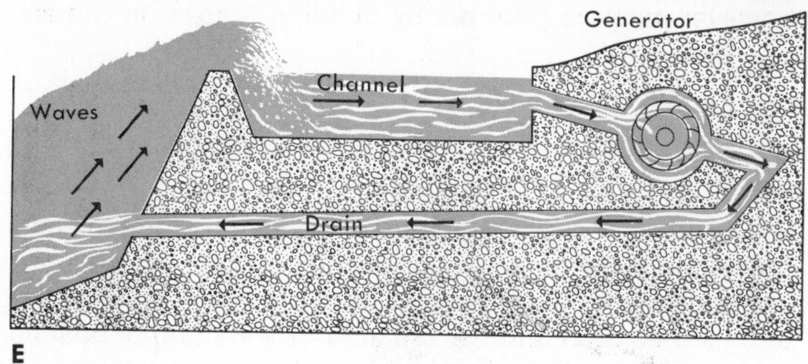

FIGURE 8.13 Different types of seawalls. *A*, Rocks, Fernandina, Florida; *B*, Parabolic, Galveston, Texas; *C*, Steps, Mississippi coast; *D*, Parabolic and steps, San Francisco, California; *E*, Channeled wall with possible application for power generation from waves and tides. (*A–D* from US Army Coastal Engineering Research Center, 1973, *E* after Smith, 1973.)

Jetties

Jetties are long structures constructed perpendicular to the coast and are usually built at the mouths of rivers and harbors. They may be single structures, or they may be in pairs, one on either side of the inlet. They are built to prevent sediment deposition in the inlet by longshore currents as well as to make stream and tidal flow favorable for navigation. In addition, jetties are capable of shielding the inlet from large waves arriving at an angle.

Although jetties often perform their functions well, they result in many unwanted side effects. Jetties some-

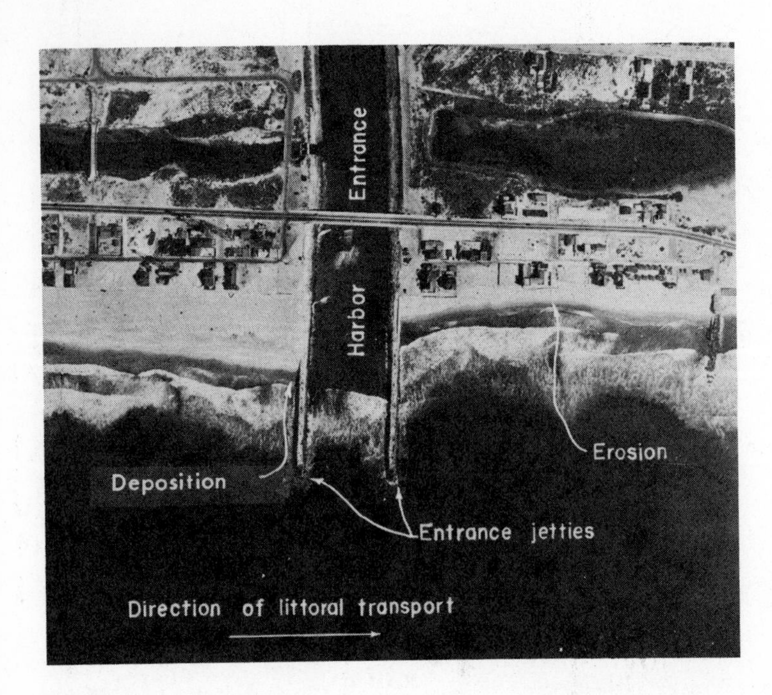

FIGURE 8.14 A pair of jetties protecting a harbor entrance at Ballona Creek, California. (From US Army Coastal Engineering Research Center, 1966.)

times cause deposition of sand on one side of the inlet and erosion on the other side (Fig. 8.14). In order to prevent erosion, sand must be continually supplied to this region—a costly and permanent operation. The erosion, however, may be retarded by building a groin in the area of erosion.

Groins

Groins are similar to jetties but are built along the open coast. They are usually shorter than jetties. The purposes of groins are many. They include the prevention or slowing of beach erosion and the creation or widening of beaches by trapping sediments from longshore currents. Like a jetty, a groin will cause erosion and deposition along the coast (Fig. 8.15). A second groin is often built to prevent the erosion caused by the first groin. The result is a chain reaction of groin construction along the coast. Many coasts, therefore, are littered with numerous groins. Groins certainly do their job of trapping sand but at the expense of the next beach.

Breakwaters

Breakwaters are offshore structures built parallel to the shore (Fig. 8.16), mainly to protect beaches and harbors from high waves. Waves either break on them or are

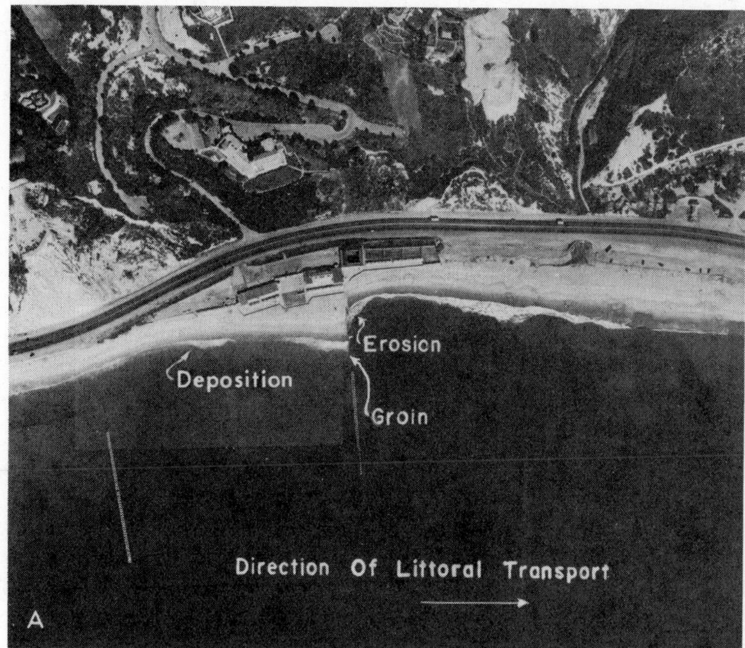

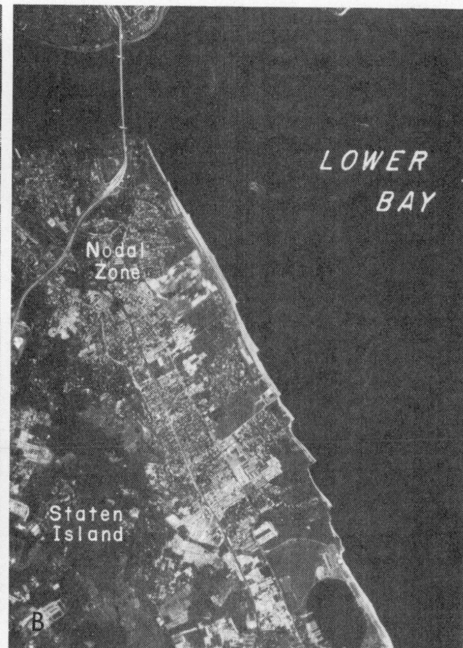

FIGURE 8.15 *A*, Groin at Santa Monica, California. *B*, Series of groins at Staten Island, New York. The "nodal zone" shown indicates that erosion and deposition are balanced here as a result of diverging longshore currents. (*A* from US Army Coastal Engineering Research Center, 1966; *B* from US Army Coastal Engineering Research Center, 1973.)

reflected back. Because of the reduction in wave energy behind the breakwater, the sediments suspended in the longshore currents are deposited here, and the result may be some erosion farther down the beach.

ESTUARIES

An ***estuary*** may be defined as the region where river water meets and mixes with seawater, resulting in inter-

FIGURE 8.16 A breakwater at Santa Monica, California. (From US Army Coastal Engineering Research Center, 1966.)

mediate salinities. The influence of the river on the sea may extend several kilometers into the sea and may be marked by a plume of muddy water extending out from the river mouth.

Historically, estuaries have been attractive as nuclei for the development of cities. Navigable water and proximity to the ocean make them ideal for seaport activities such as shipping. In addition, the natural load of land-derived nutrients carried by the rivers results in highly productive water, suitable for the rapid growth of many fish, crabs, and clams. Consequently, important fisheries have developed on and around estuaries.

As areas surrounding estuaries have become agriculturalized and industrialized, the water has become increasingly polluted and the natural estuarine resources threatened. In addition, oil tankers pose a constant threat of oil spills. However, many estuaries are too small for the new supertankers. In many parts of the world, offshore floating oil ports appear to be the only solution to this problem.

Types of Estuaries

The nature of an estuary is determined not only by how it is used but also by its dimensions, river flow, and tidal flow. When river flow is large, tidal flow is small, and the estuary is rather deep. The fresh water tends to flow over the seawater with little mixing (Fig. 8.17*a*). Thus, a ***stratified*** estuary is maintained. The deeper salty water forms a sort of "salt wedge" under the less saline, predominantly river water. In this case the salinity difference between surface and bottom water may exceed 20 ‰. The net flow of water is upstream near the bottom and seaward near the surface in a stratified estuary. The mouth of the Mississippi River is a good example of this type of estuary.

FIGURE 8.17 Types of estuaries, based on the distribution of salinity. Arrows indicate net flow of water.

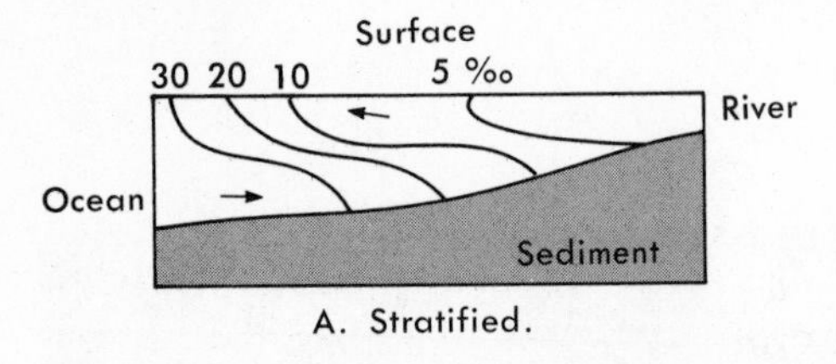

A. Stratified.

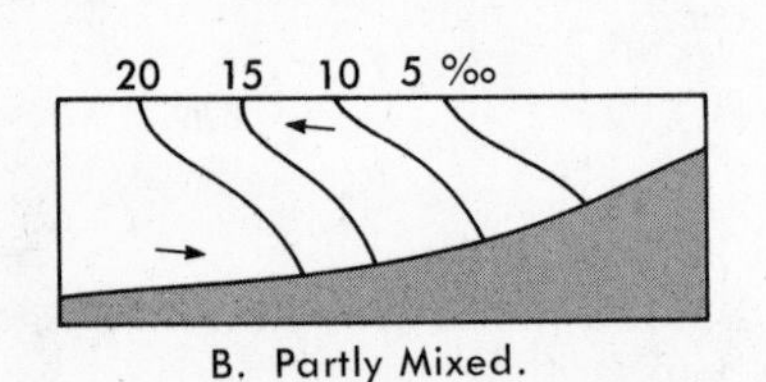

B. Partly Mixed.

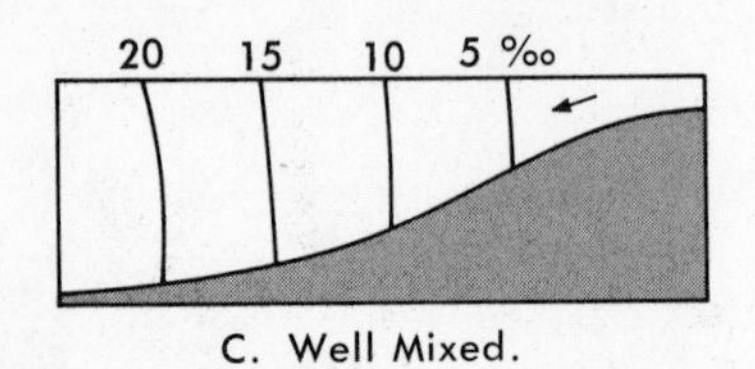

C. Well Mixed.

A ***partly mixed*** (or partly stratified) estuary occurs under conditions where depth and river flow are less and tidal flow is greater than in a stratified estuary. These factors all contribute to increased mixing of the fresh water with the salty water (Fig. 8.17*b*), and stratification is less. The salinity difference between surface and bottom may be from 4 to 20 ‰. Mixing causes some of the salty water to be brought into the surface water, where it flows back to the sea. Consequently, there may be a significant net flow upstream near the bottom as well as a seaward flow near the surface. The Columbia River estuary, near its mouth, is partly mixed when river flow is substantial.

If the river flow is small, the tidal flow is great, and the estuary is shallow and narrow, vertical mixing may be nearly complete, and the vertical difference in salinity generally will be less than 4 ‰. This is a ***well-mixed*** estuary (Fig. 8.17*c*), with a net outward flow of water at all depths. Many small estuaries—and some large estuaries during periods of low river flow—are of this type. For example, the Columbia River may be well mixed during the dry seasons.

Some very wide estuaries, such as the lower Chesapeake and Delaware estuaries, exhibit a horizontal (transverse) gradient of salinity owing to the effects of Coriolis force (see Appendix C for a discussion of the Coriolis force). In the Northern Hemisphere, this force causes salt water to be deflected to the right as it enters these estuaries (Fig. 8.18). The river flow is likewise deflected to

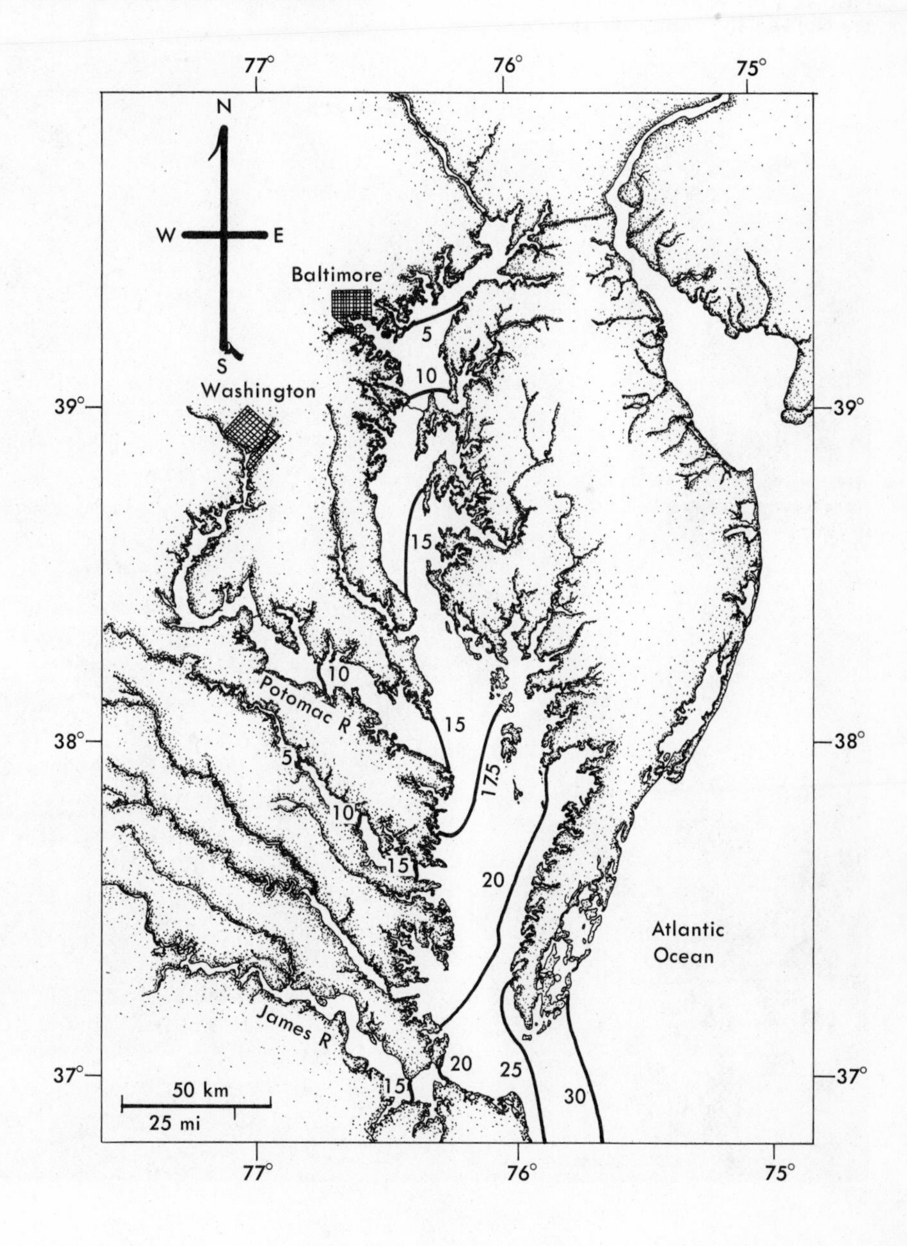

FIGURE 8.18 Distribution of salinity (in parts per thousand) in Chesapeake Bay. The shape of the surface salinity contours indicates the influence of Coriolis force deflecting the salty water to the right as it enters the estuary. It should be noted, however, that more fresh water enters the western side of the bay than the eastern side. (After Pritchard, 1952.)

the right. Consequently, one shore experiences higher salinities than the opposite shore.

The estuaries that we have discussed so far have been formed by the drowning of river mouths caused by rising sea level following the ice ages, or by the sinking of coastal areas. The lagoons or bays that are formed behind barrier islands are also considered to be estuaries, as they have many of the characteristics discussed above.

Another type of estuary is the fiord (Fig. 8.19). These have been formed by glacial erosion; they are quite deep (some are several hundred meters deep), and they have a shallow sill across the mouth. The sill marks the limit of seaward advance of the continental glacier. Circulation is

FIGURE 8.19 A Norwegian fiord. (Courtesy of the Norwegian Information Service.)

restricted in these estuaries, and thus they are quite distinct from the estuaries mentioned previously. They tend to be highly stratified at times, and circulation is restricted by the sill, which reduces the flow of seawater in and out of the fiord. Consequently there is a tendency for the deep water to become stagnant and anaerobic (deficient in oxygen). The anaerobic condition causes an increase in the concentration of hydrogen sulfide due to its production by certain anaerobic bacteria.

During a cold winter, freshwater run-off is reduced and surface temperatures are lowered. Consequently the stratification breaks down and the fiord becomes mixed. This causes a rather sudden drop in the concentration of oxygen and an increase in hydrogen sulfide at the surface. This combined effect may cause fish kills, and the hydrogen sulfide which is released into the atmosphere as a gas may drift to shore and cause houses to become blackened. (Recall that the presence of hydrogen sulfide in sediment is indicated by black color and a rotten-egg smell.)

Life in Estuaries

Since most estuaries are well nourished by nutrients that are carried by the rivers, they are frequently the sites of highly productive fisheries. Clams, oysters, shrimp, and many species of food fish thrive in estuaries. Many animals breed there, in waters rich in the food necessary for their developing young.

The distribution of estuarine organisms is highly dependent on the nature of the surrounding medium. In an estuary, attached organisms must tolerate changing salinities and temperatures as the tides move the salty water in and out. In addition, organisms in the intertidal (littoral) zone must tolerate periodic exposure to air. Barnacles and mussels can close their protective shells, trapping salty water within. Many attached seaweeds can tolerate periodic exposure and even drying of their surfaces. However, many bottom-dwelling worms and clams may burrow into the sediment in order to secure an environment that is very stable with regard to temperature and salinity. Plankton (drifting organisms) may be carried out to sea if they remain near the surface. Some plankton compensate for this danger by spending part of the time in deeper water, where the net water movement may be upstream. Estuarine fish can swim against currents as adults, and many have heavy eggs that sink to the bottom and thus avoid being washed out to sea.

The distribution of life in estuaries also depends on the distribution of salinity and various pollutants. Most truly estuarine organisms can tolerate wide ranges in salinity. Some, however, are restricted to the ocean end of the estuary. Often these are also found along the open coast. On the other hand, some organisms that are found in fresh water can also tolerate the low salinities found in the upper estuary. With increasing concentrations of pollutants in an estuary, species that are less tolerant to pollution are eliminated. Very rugged species, on the other hand, survive and often thrive.

Newark Bay is a well-mixed estuary in highly industrialized northern New Jersey. In the 1880s it was an important oyster-breeding ground, and the shad that were caught in the bay were highly regarded in the New York markets. Today, both fisheries are gone, having succumbed to the pressures of increased population and industrialization. However, some hardy organisms such as killifish and grass shrimp thrive.

Is it possible to clean up an estuary that has reached a highly polluted condition? Many estuaries are the sites of sewage and industrial waste disposal. If all pollution of water stopped today, it would still be many years before some of our estuaries returned to a condition suitable for fishing and recreation. Years of pollution have contaminated the sediments to the point that they would retain oil and grease for years. Heavy toxic metals such as mercury and lead may persist in the sediment for even longer periods, continuing to affect the water for many years.

Because many of the larger estuaries are important shipping ports, they are periodically dredged to maintain deep channels. This process effectively re-introduces pollutants from the sediment back into the water, either during the dredging operation itself or from the dredge spoils (the dredged-up sediment that is deposited outside the channel, frequently on shore). Of course, dredging may also remove contaminated sediments from the estuary, making it less polluted, but then the area in which the contaminated sediments are deposited becomes more polluted.

Human activities such as farming and construction that occur upstream may introduce sediments into rivers. These sediments are carried down to the estuaries, where they are deposited. When the shipping channels are dredged, the increased depth may affect the circulation of the estuary and may result in increased stratification of the water. In addition, construction of dams may reduce the flow of fresh water, increasing the mixing of the estuary. For example, a partly mixed estuary may become well mixed. Similarly, narrowing of estuaries by filling, or widening of estuaries may also affect their circulation. All of these changes may, in turn, have a significant effect on the distribution and survival of estuarine organisms.

Estuaries will continue to be under stress as long as urban centers and world trade are focused on them. However, the quality and utility of estuarine water may be improved considerably with care and patience.

BASINS

A ***basin*** is a semi-enclosed depression filled with seawater and usually having a narrow shallow opening to the ocean. The circulation in basins depends largely on the processes of fresh water input (precipitation and river run-off) and loss through evaporation.

If evaporation exceeds precipitation and river run-off, the basin becomes saltier than the ocean and a gradient of salinity occurs, with the less saline water located near the ocean. The Mediterranean Sea is an example of this type of basin. The sinking saline water causes the Mediterranean Sea to be well mixed rather than stratified. In this sea (Fig. 8.20), water flows in at the surface and saltier water flows out over the rim of the Gibraltar Sill and into the Atlantic Ocean. This salty warm Mediterranean water contributes to the deep circulation of the North and South Atlantic oceans.

If precipitation and river run-off exceed evaporation, a rather typical estuarine circulation occurs. In this case the excess fresher water tends to flow out of the basin near the surface, diluting the adjacent sea. Smaller amounts of

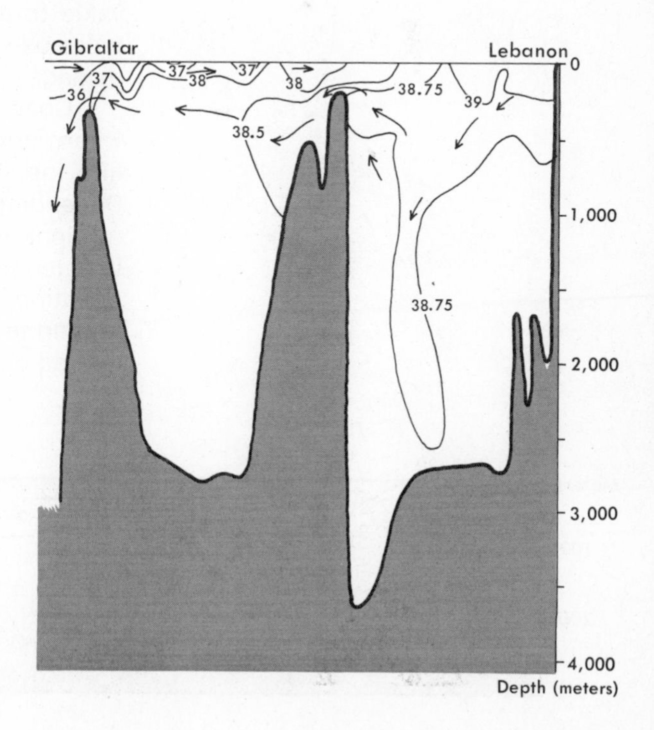

FIGURE 8.20 Circulation of water in the Mediterranean Sea. Salinity values are given in parts per thousand. (Modified after US Naval Oceanographic Office, H.O. 700, 1967.)

saltier water flow into the basin over the sill (Fig. 8.21). The Baltic Sea and the Black Sea are examples of this type of basin. The fiords, discussed earlier, may also be thought of as small basins of this nature. In silled basins of this type, there is a tendency toward isolation of stagnant salty water below the depth of the sill and, consequently, oxygen depletion and hydrogen sulfide build-up near the bottom of the basin. This effect is especially well illustrated in the Black Sea, where a sharp gradient of decreasing temperature coincides with the increase in salinity with depth. Here the plants and animals are restricted to the upper 200 meters or so. Below this depth only anaerobic microorganisms such as bacteria are found. The "rain" of dead organisms into the deep water provides energy for anaerobic bacteria, many of which release hydrogen sulfide into the water. The Baltic Sea is much shallower than the Black Sea, and consequently, anaerobic conditions exist only in the deepest areas, where stagnation occurs.

COASTAL POLLUTION

The coastal waters are most susceptible to pollution, because of the concentration of urban activity near the coasts and rivers which eventually enter the sea. Large cities and ports are located near the mouths of rivers. In fact, seven of the ten largest metropolitan areas of the world are situated near estuaries. Other coastal areas are popular as resorts. Coastal environments are of special value to us as a source of fuel and food. Because of intense use, our coastal oceans are in danger of misuse and damage. Their protection is the interest of all.

Coastal waters may become polluted by design, as when industrial or domestic wastes are dumped directly into the sea (Fig. 8.22); or by accident, as in an oil spill. Depending on the nature of the pollutant, whether toxic or nutrient, marine life may be either inhibited or stimulated. It is tempting to believe that "dilution is the solution to pollution." However, industrial wastes are frequently toxic to marine life near the source. Domestic sewage, whether treated or not, provides nutrients (such as phosphates) for

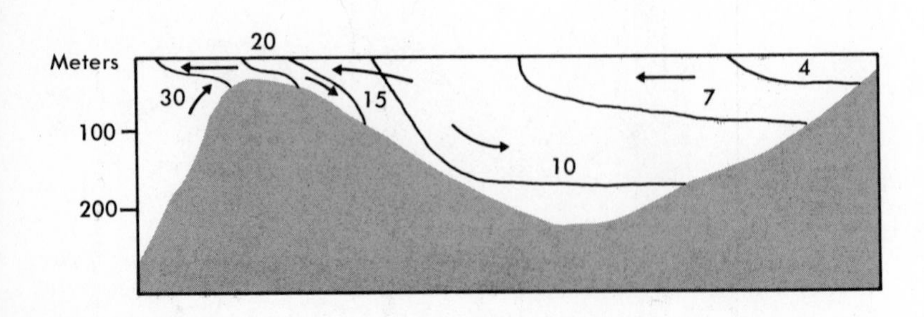

FIGURE 8.21 Circulation of water in the Baltic Sea. Salinity values are given in parts per thousand. (After Segerstråle, 1957.)

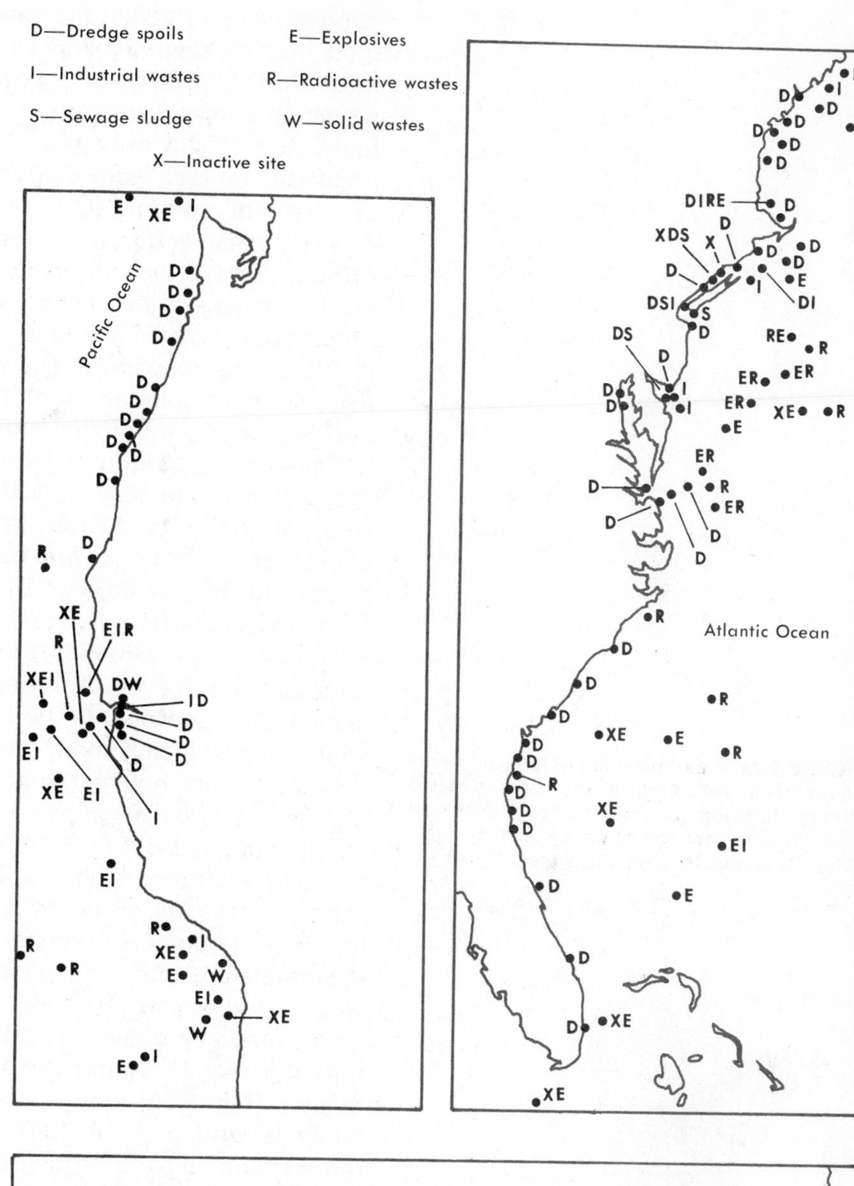

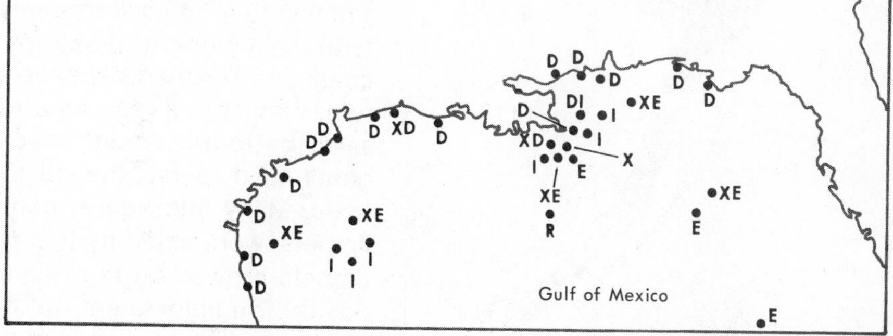

FIGURE 8.22 Known ocean dumping sites off US coasts. (Modified after Council on Environmental Quality, 1970.)

marine plants including the organisms that cause "red tides." (The role of red tides receives special consideration in Chapter 11.) In the case of oil on the sea, both effects may be seen. Oil kills some organisms, such as sea birds, but stimulates the growth of others, such as oil-digesting bacteria, which are slow acting and which certainly do not fully counteract the damaging effects of oil pollution. In fact, some of these bacteria secrete substances that are more toxic than the oil itself.

The ocean is frequently regarded as a handy dumping ground. Usually the municipalities and industries that dump wastes into the sea do so as close to shore as they are allowed by law (or sometimes even closer). The result is that certain areas, especially near metropolitan centers, become more polluted than others. In 1974, more than 70 per cent of the municipal sludge dumping in the United States and about 60 per cent of the industrial dumping took place in the New York Bight, in about 30 meters of water off the Hudson River. In this area, about five million cubic meters of sludge were being dumped about 25 kilometers from shore each year. Dredge spoils are dumped less than 15 kilometers from shore. In addition, industrial chemicals such as acids and other toxic materials are released into the sea. Heavier materials, such as sewage sludge, sink and pollute the bottom sediments (Fig. 8.23), killing or displacing many of the natural animal species. The wastes are not restricted to the area where they are dumped but are carried by currents. Lighter materials such as oil and plastics (Fig. 8.24) may even reach the shore.

FIGURE 8.23 **Lead pollution in the sediments of the New York Bight in the vicinity of the sludge dumping grounds. Concentrations of lead (in ppm) are indicated by curved lines. (After Carmody, Pearse, and Yasso, 1973.)**

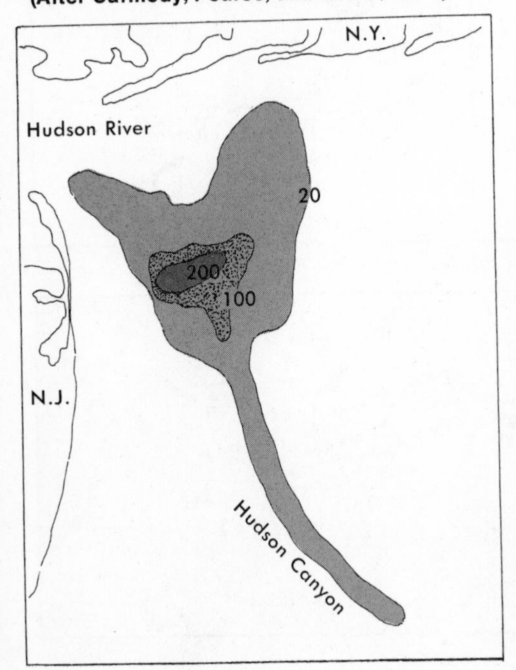

Oil pollution is a special problem in the coastal environment because oil floats on water and does not easily mix with it. Some oil seeps into the sea naturally from submarine oil fields. This oil is usually not a great problem since amounts are relatively small and some marine bacteria eventually digest the oil and render it harmless. On the other hand, drilling for oil or transportation of oil in tankers (Fig. 8.25) may occasionally result in large-scale oil spills. On March 18, 1967, the oil tanker *Torrey Canyon* ran aground off the British coast of Cornwall, releasing 118,000 tons of crude oil into the sea. Some of it was intentionally burned at sea, but most of it drifted along the coast and polluted beaches in Britain and France. In addition, many birds died as a result of the oil. In an effort to save the tourist trade, beaches were sprayed with detergents to disperse the oil. The detergent, however, was toxic. Many intertidal organisms, especially the snail-like limpets, were killed by the detergent. It is now known that limpets browse on the rocks, removing algae and oil deposits. The limpets eat the oil. It is unfortunate that many of these natural cleaners were killed in an attempt to

FIGURE 8.24 Plastics and oil pollution on a fine-mesh plankton net. (Courtesy of Harold Wes Pratt, National Marine Fisheries Service.)

disperse the oil. In fact it is probable that more marine life was killed by the detergents than by the oil itself. Ultimately, the disappearance of oil from the sea is mostly due to bacteria which decomposes the oil. Of course, this process takes years to accomplish the job.

More recently, in 1974, the tanker *Metula* (Fig. 8.26) ran aground in the Strait of Magellan, spilling 54,000 of

FIGURE 8.25 The supertanker *Atlantic Sun* with a capacity of about 250,000 tons. (Courtesy of the Sun Oil Co.)

FIGURE 8.26 The tanker *Metula,* aground in the Strait of Magellan. (Courtesy of the US Coast Guard.)

the 195,000 tons of oil that she carried. In 1975, the relatively small Liberian tanker, the *Spartan Lady,* broke in two in heavy seas off the New Jersey coast, spilling much of her more than 30,000 tons of fuel oil. Fortunately the oil drifted farther out to sea and no coastal damage occurred, at least none that was directly attributable to the *Spartan Lady.*

Although large, the *Torrey Canyon, Metula,* and *Spartan Lady* are not the largest tankers of today. The supertanker *Universe Ireland* carries almost three times the oil carried by the *Torrey Canyon.* In 1973, there were 26 supertankers, each with a capacity of 400,000 tons or more. What if some unforeseen disaster causes this tremendous cargo to be spilled onto the sea, particularly in cold, clean water such as in the Strait of Magellan or the Arctic?

On January 28, 1969, an oil well being drilled off Santa Barbara, California, blew out, sending more than three million gallons of oil (about 20,000 tons) into the sea. This is about as much oil as reached the Cornwall coast after the *Torrey Canyon* went aground. Again, beaches were coated with oil, birds died, and marine life suffered for a time. Eventually most of the oil was removed or was destroyed by bacteria.

Pollution of the coastal environment will continue. Someday it may not be necessary to dump sewage and industrial wastes into the water, as they may be recycled. Sewage waste may even provide nutrients to grow food for people, under controlled conditions (algae culture).

Oil will continue to be spilled accidentally, although one hopes that the chances of this disaster will be reduced. The value and scarcity of this fuel has resulted in special efforts to reduce the possibility of accidents and to facilitate the recovery of the valuable oil.

As our needs for power and development of coastal areas continue to grow, the coastal environment will be under increasing stress and perhaps in danger of even greater pollution despite stringent control measures. As the sea has a tremendous capacity to absorb heat, it is tempting to locate atomic power plants in or near the sea so that marine or brackish water can be used to cool the nuclear reactors. Of course, dangerous radioactive water pollution through accident or sabotage is a frightening possibility. A more likely source of pollution lies in the discharge of hot cooling water. The water will warm the local environment, causing some organisms to die and new communities of marine life to develop. These effects may be beneficial if they provide a greater diversity of life in the area, or they may be harmful if local fisheries are disrupted. The heat may also deflect marine fish from their normal patterns of migration to breeding grounds. Some fish may be attracted to the warm water around the power plant and fail to swim upstream to spawn. In addition, massive deaths of fish and other life may occur when the power plant is shut down and the water temperature drops suddenly.

SUGGESTED READINGS

Bascom, Willard. 1964. *Waves and Beaches.* Garden City, N.Y., Doubleday.

King, Cuchlaine A. M. 1959. *Beaches and Coasts.* London, Edward Arnold.

Lauff, George H. (ed.). 1967. *Estuaries.* Washington, D.C., American Association for the Advancement of Science (pub. 83).

Marx. Wesley. 1967. *The Frail Ocean.* N.Y., Ballantine.

U. S. Army, Coastal Engineering Research Center. 1966. *Shore Protection, Planning and Design,* 3rd ed. Washington, D.C., Government Printing Office.

______________. 1973. *Shore Protection Manual.* Washington, D.C., Government Printing Office.

A slug-like nudibranch crawling over encrusting sponges, which are also inhabited by sea anemones. (Photo by Harold Wes Pratt.)

CHAPTER 9

Marine Environments

Life in the oceans is unevenly distributed, both vertically and horizontally. Some areas are rich in life; others are nearly desert-like. Some of the important factors that limit an organism's distribution in the sea include its mobility and its relation to light, temperature, pressure, and nutrients (or food supply).

MOBILITY IN MARINE ENVIRONMENTS

Marine organisms may be classified with respect to their mobility in the sea. Thus we speak of ***plankton*** (drifters), ***nekton*** (swimmers), and ***benthic*** organisms (bottom dwellers).

Plankton (Fig. 9.1) drift at the mercy of currents in the sea. They drift with the water in which they live. They include plants (***phytoplankton***) and animals (***zooplankton***). Many of the zooplankton feed on phytoplankton. Some plankton actually float on the surface of the water, being held up by gas floats or by the surface tension membrane of the water. Most plankton are very small, less than 1 millimeter in length, but a few are quite large, such as jellyfish and *Sargassum* weed, the latter of which is actually more like its attached benthic relatives than most other phytoplankton.

Some organisms are planktonic during part of their life cycle and either nektonic or benthic at another stage. These plankton are called ***meroplankton***. For example, some jellyfish are planktonic during the sexually reproducing part of their life cycle and attached (sessile) during a stage that reproduces asexually by budding. Figure 9.1c illustrates the life cycle of a small jellyfish.

The larvae of most fish and invertebrates, including clams and crabs, are too small to swim against ocean currents and must be considered meroplankton. They mature into nektonic or benthic adults. In some cases the food of the larva differs considerably from the food of the adult.

The larvae of most fish that live near the surface are nearly transparent and are thus harder for predators to see. As the larvae mature, they develop the pigments characteristic of the adult and may also develop the vertical migration pattern characteristic of the adult of that species. It is interesting to note that the larvae of deep-

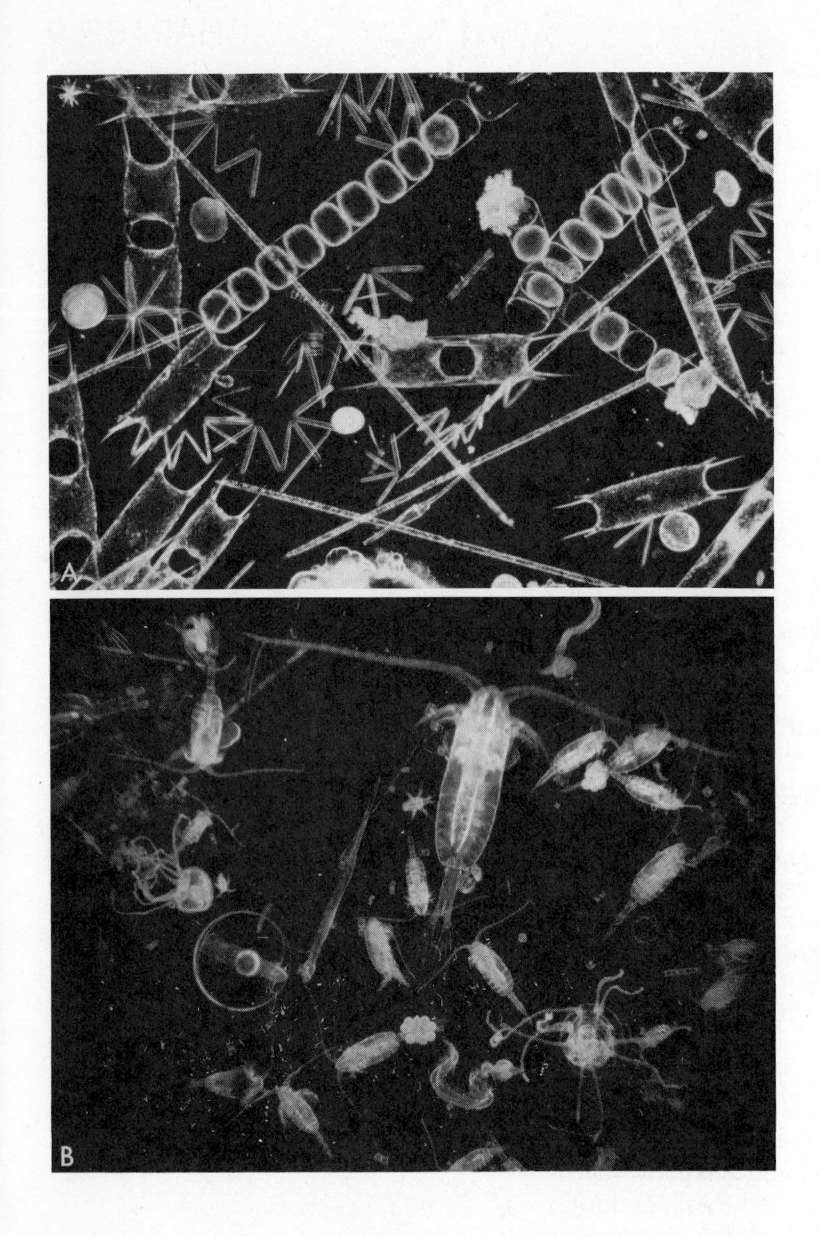

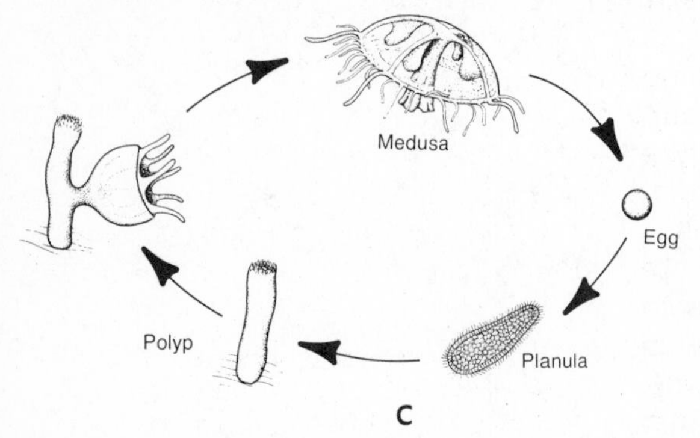

FIGURE 9.1 *A,* A variety of marine phytoplankton, mostly diatoms. 60 ×. *B,* Marine zooplankton. The largest animal is a copepod, as are the smaller similar ones. There are also two small jellyfish with tentacles, a straight arrow worm, two wormlike tunicates, and a spherical fish egg. 12×. *C,* The life cycle of a small jelly fish, indicating both planktonic (medusa) and benthic (polyp) stages. This type of life cycle is found in the fresh-water jellyfish, *Craspedacusta,* as well as in certain marine forms. (*A* and *B* photos © Douglas P. Wilson; *C* from Barnes, 1974.)

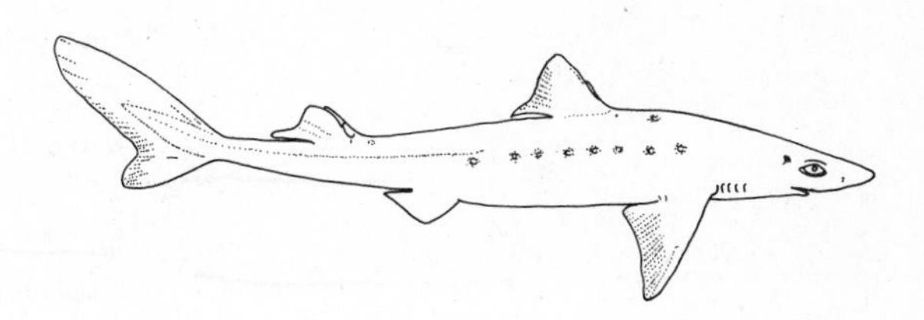

FIGURE 9.2 Sharks have upturned tails, which give them an uplift when they swim. They have no buoyant gas bladder as do most fish, so sharks tend to sink if they become motionless.

water fish, which live in the dark, usually have black pigmentation.

Animals that can swim rapidly enough to counteract ocean currents are called nekton. They include many fish, squid, porpoises, and whales. They may carry out extensive vertical migrations, as in the case of lantern fish, or much longer horizontal migrations as in the case of some eels and salmon. Nekton are adapted for swimming by the presence of fins, a generally flat tail, and some degree of streamlining.

Many fish have gas bladders that help maintain buoyancy, enabling the fish to remain above the bottom with little effort. Sharks, however, have no gas bladder and tend to sink when they are not swimming. In compensation for this tendency they have an upturned tail (Fig. 9.2). When the tail beats, it gives a downward and forward thrust to the rear and consequently an upward and forward thrust to the head, lifting the shark from the bottom. The shape of nektonic animals has much to do with their habitat and way of life (Fig. 9.3). Most fish that live near the bottom are slow moving and awkward looking but have great maneuverability at slow speeds. They usually feed on clams, crabs, and other sessile or slow-moving

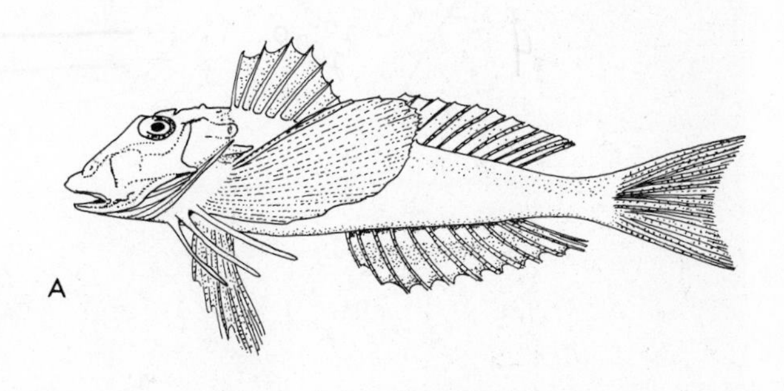

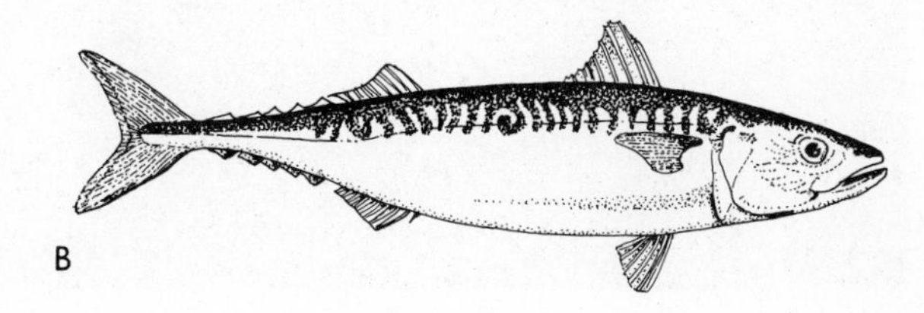

FIGURE 9.3 ***A,*** **A maneuverable but slow-moving fish, the common sea robin;** ***B,*** **A pelagic, rapid-swimming, streamlined fish, the mackerel.**

benthic species. Fish that live higher in the water tend to be much more streamlined and fast but are not very maneuverable.

Benthic animals may live on the bottom as ***epifauna,*** or within sediment as ***infauna.*** The epifauna may be attached, such as sponges and sea anemones, or they may move around, as do crabs and snails. The shallow-water epifauna are subjected to fluctuating temperatures and salinity. The infauna, such as various worms and some clams, are more protected from fluctuations in the environment. Even in shallow-water sediment, temperature and salinity vary less than in the water above it. There is little circulation of water in the sediment, so oxygen tends to be almost lacking. Many sediment-dwelling worms are adapted to this habitat, but the clams must receive oxygenated water from above, via an ***incurrent siphon***, a fleshy tube which also draws in small particles of food. Waste water is then pumped out through a separate ***excurrent siphon***.

FIGURE 9.4 The classification of marine environments.

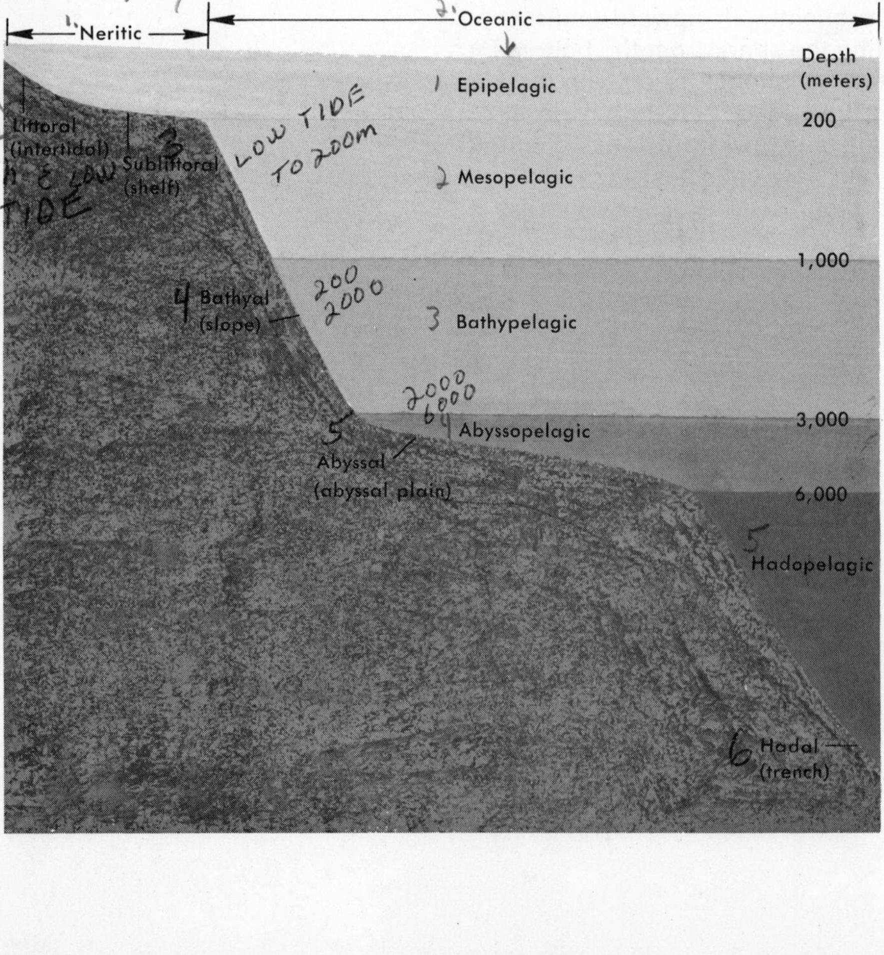

CLASSIFICATION OF MARINE ENVIRONMENTS

Pelagic Environments

The pelagic environment includes the entire ocean except the sea floor. The plants and animals that live in the open sea but which are not closely associated with the shore or sea floor are known as ***pelagic*** organisms. The pelagic environment is further subdivided into zones on the basis of water depth, light distribution, and temperature (See Figures 9.4 and 9.5).

Because light that is sufficiently intense for plant growth penetrates only a short distance into the sea (generally less than 200 meters in the clearest open ocean water and only a few meters in cloudier coastal water), plants—which require light for photosynthesis—are found only in the upper layer of the sea. Essentially all of the conversion of light energy into the stored energy of organic matter occurs in the well-lighted water, the ***euphotic zone***. The production of plant life in this zone provides food for various plant-eating animals (herbivores). Some of this stored organic matter may eventually settle to the sea floor, become buried, and be converted into oil and gas. This process is especially prevalent over the continental shelf, where production tends to be greater than in areas farther from shore. The deeper, dark water, the ***aphotic zone***, is cold; the only light found there is produced by some animals. Between the euphotic and aphotic it is convenient to include a ***disphotic zone***, where light exists but is not intense enough for effective production of plants. This intermediate zone may extend as deep as 1,000 meters or so in the clearest ocean water. Figure 9.4 illustrates the classification of marine environments.

Neritic Pelagic

The ***neritic*** pelagic region comprises all the water that lies over the continental shelf, generally extending downward to a depth of about 200 meters. This area includes the most productive parts of the oceans, because it is close to rich supplies of nutrients from land as well as from upwelling of deeper nutrient-rich water. It is continuous with the surface water farther from shore, and the processes that apply here apply also in the water beyond the shelf. Note that the depth of light penetration (Fig. 9.5) is less in the neritic water than in the open ocean, farther from shore. This effect is primarily due to the greater amounts of suspended sediment and marine life in the coastal water.

Oceanic Pelagic

The ***oceanic*** pelagic region includes all the water beyond the continental shelf, from the surface down to

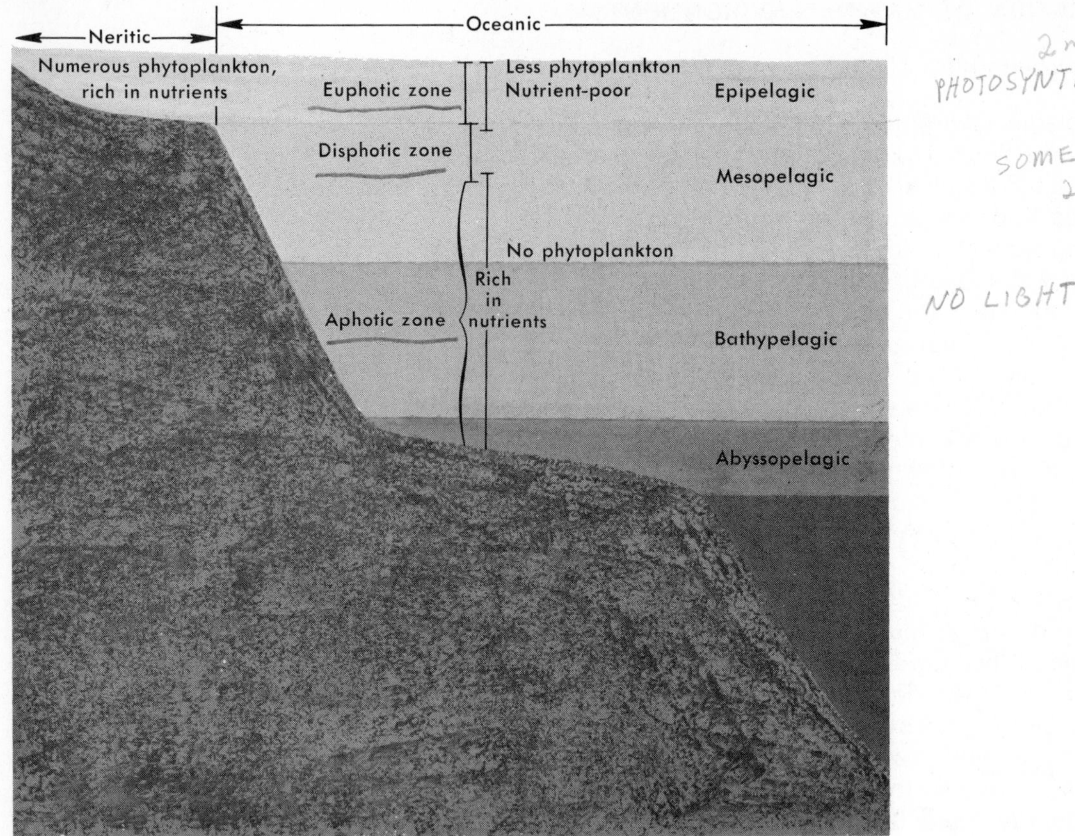

FIGURE 9.5 Distribution of phytoplankton, light, and nutrients in neritic and oceanic water. Penetration of light is much less in the turbid neritic water than in the clearer oceanic water.

the greatest depths found in the trenches. These waters are generally much poorer in nutrients than the neritic water; hence the production of life is much less.

Pressure in the sea increases at a rate of about one atmosphere for each ten meters of depth. Thus at a depth of 2,000 meters, the pressure is about 200 times that at the surface. In spite of the tremendous pressures in the deep sea, animals exist at all depths. Apparently the distribution of marine life is less affected by pressure than by other factors such as nutrient availability, food supply, light, and temperature.

Oceanic plant life is restricted to the euphotic upper layer of the sea, the ***epipelagic*** zone. The animals which eat the plants, however, are not so restricted and may take part in daily vertical migrations upward as the sun sets and down again in the morning.

The migrant animals may be eaten by carnivores at any of the depths at which they are found. This may be near the surface, or at depths of 400, 500, or even 700 meters or more in clear water. Organic energy-rich matter in living organisms is thus carried down into the ***mesopelagic zone,*** which extends from about 200 to 1,000 meters and is sometimes called the "twilight zone" (disphotic) because of the reduced light in this zone.

Another way that energy may be transferred to deeper zones is by the sinking of ***detritus*** (dead organic matter). This may be in the form of dead plants or animals or fecal material. Some detritus even reaches the bottom in the trenches, below 6,000 meters, and may be the main source of food for deep sea organisms.

The mesopelagic and the deeper ***bathypelagic zone*** (1,000 to 3,000 meters) are isolated from the surface waters and the benthic environments except where they touch the continental slope. Thus, the organisms in this zone depend on migration of other animals and on descending detritus for food.

The animals of the ***abyssopelagic*** (3,000 to 6,000 meters) and ***hadopelagic*** (deeper than 6,000 meters) zones are closely associated with the animals of the deep-sea floor. These deep-sea pelagic animals, along with the ever-present detritus, provide food for the deep-sea benthic animals. The most striking features of these zones are the monotonously constant cold temperature (below 4°C) and darkness.

Benthic Environments

The benthic environment is the sea floor, and the plants and animals that live there are called benthic organisms. Benthic organisms require a proper substrate such as rock, sand, wood, or any other material upon or within which they may survive. Great variety in the nature of substrate materials results in a great variety of niches. A ***niche*** consists of all the requirements (physical and biological) of an organism. Most clams, for example, require a sandy or muddy bottom. They also require a water current from which they can filter phytoplankton food. Different species tolerate different temperature ranges, salinity ranges, oxygen concentrations, and so forth. No two species may have exactly the same environmental requirements, and thus no two species occupy the same niche. If two species with similar environmental needs compete for the same place on the substrate, only the one best suited to the total environment will survive. A rocky coast generally has a greater number of niches than a sandy coast or the deep ocean floor. Thus the diversity of benthic species tends to be greater on the rocky coast than in the more uniform environments.

Littoral Zone

The part of the sea most readily seen by people is the intertidal or ***littoral*** zone. Organisms at the edge of the sea are subject to periodic wetting by waves and tides. The pattern of waves and tides varies from place to place, depending on the shape of the ocean basin, weather, and other factors discussed in the chapters on waves (Chap. 5) and on tides (Chap. 6). Most coasts have semi-daily tides, with two high tides and two low tides each day. Some areas, for example along the Gulf of Mexico, experience daily tides with one high and one low tide per day. Some examples of tidal patterns are shown in Figure 9.6.

Casual observation of a piling or rock jetty shows a vertical distribution of attached organisms, in which discrete zones or stripes of color and pattern are seen (Fig. 9.7). The organisms found above the high-tide levels are only occasionally covered by the splashing of waves (***splash zone***) and are adapted to greater exposure and less water coverage. Those found between the high and low tides (***littoral***) are immersed in water at least once a day, but they must also be able to survive exposure to warm and cold air temperatures. They may accomplish this by closing their shells, as do barnacles and mussels, or they may live near the substrate, covered by a layer of seaweed. Organisms living at levels below low tide (***sublittoral***) are always covered.

There are three general ways in which organisms adapt to life in the intertidal zone: (1) Some organisms can adapt to alternate periods of submersion and exposure to air. (2) Others migrate with the tides, so that their

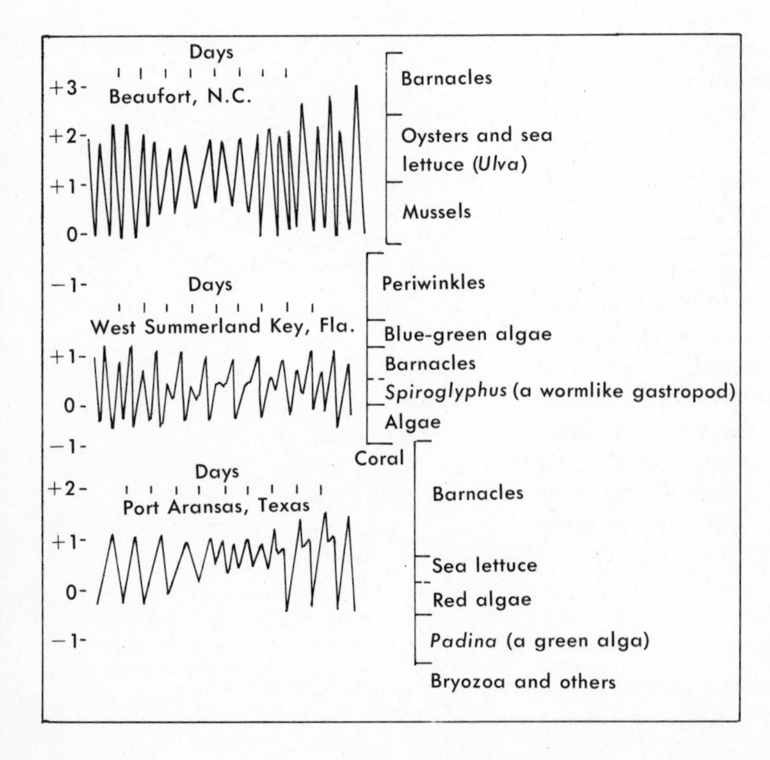

FIGURE 9.6 Tidal patterns in relation to the distribution of attached organisms. The scales shown on the left are in feet (3.28 feet = 1 meter). (After Hedgpeth, 1953.)

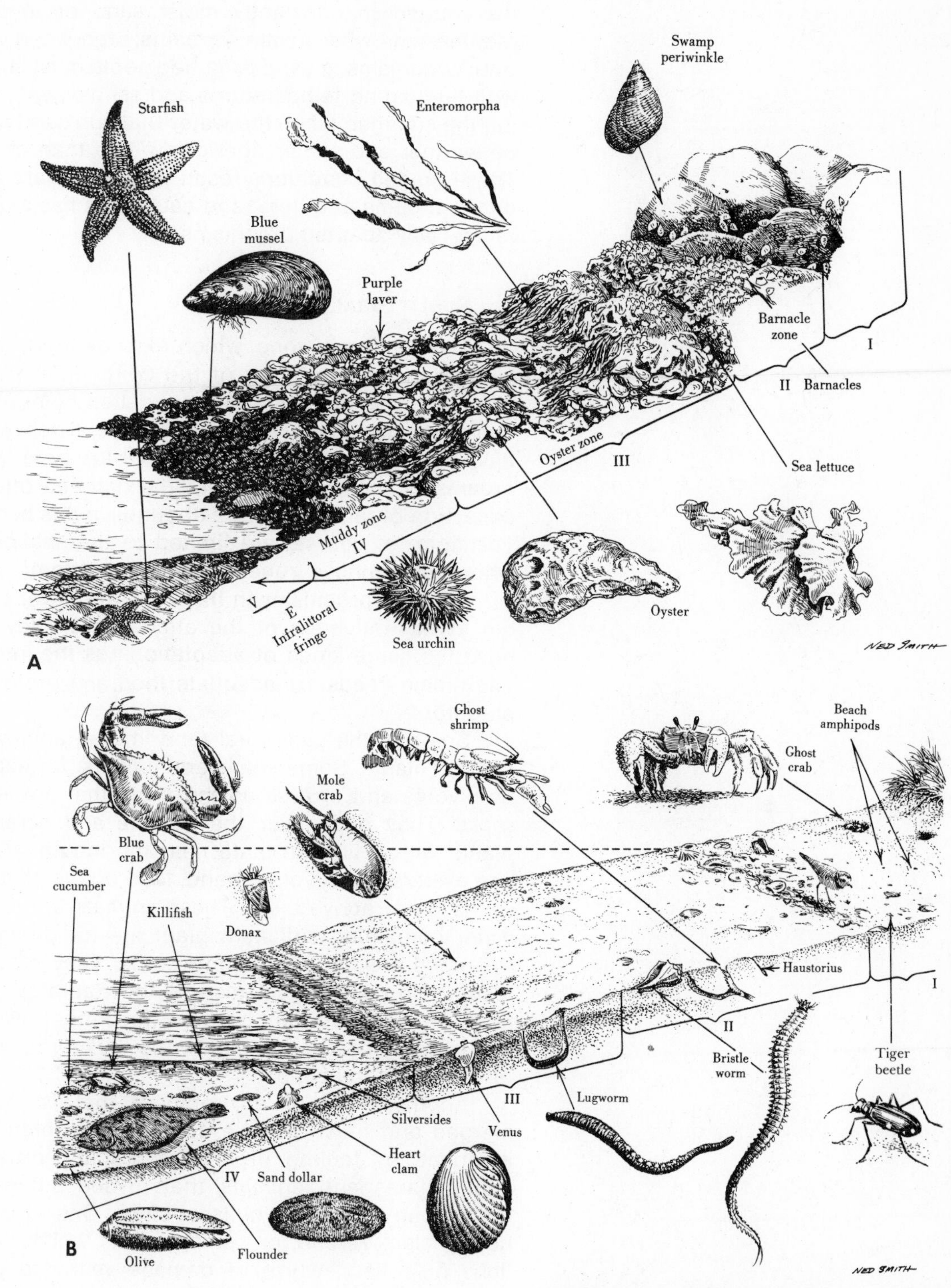

FIGURE 9.7 *A,* Intertidal zonation on a rocky surface at low tide. *B,* Sandy beach zonation. Dashed line indicates high tide. (From Smith, R. L.: Ecology and Field Biology. New York, Harper and Row, Publishers, 1966.)

environments remain fairly stable. (3) Still others live either buried in constantly moist sand or in tide pools (depressions where water remains even when the tide is out). Organisms in sand or in tide pools must still contend with fluctuating temperatures and salinity, especially during the summer, when the water that surrounds them may be warmer and saltier during low tide than at high tide. These graded conditions result in zones characterized by different fauna and flora, and each zone has a community of specially adapted organisms.

SUBLITTORAL ZONE

The sublittoral zone, which extends from the low tide levels down to the edge of the continental shelf (about 200 meters) is generally well nourished by nutrient-laden rivers and upwelling of deeper nutrient-rich water from below, as well as by the rain of detritus from the pelagic organisms above. Consequently, a rich and often diverse collection of fauna and flora abounds in this benthic zone. The flora, of course, are limited to the shallower areas, where the euphotic zone reaches the bottom.

Because organisms in this zone are never exposed to the drying influence of the atmosphere, they need not have the same kinds of adaptations as the intertidal life. Their main needs are adequate food and protection from predators.

Some of the sublittoral (and littoral) animals feed on benthic plants. Some snails and limpets, for example, are herbivores and browse on the algae that are attached to rocks. They glide over the surface and scrape off the plants with a file-like plate near the mouth. Many clams and oysters, on the other hand, feed on the phytoplankton in the water above them. These animals are restricted to water containing sufficient plant life for growth. Not so restricted are the scavenging crabs. They crawl around, feeding on dead organisms and invertebrates on the sea floor. They in turn may be food for bottom-dwelling fish, including some sharks. Figure 9.8 illustrates some views of the sublittoral zone.

Another kind of environment that is always submerged and must also be considered sublittoral is that inhabited by fouling organisms. Fouling organisms include plants and animals that invade pilings, floating docks, buoys, and the hulls of ships. One of these is the boring clam *Teredo*, or "ship worm," which causes millions of dollars' worth of damage yearly to docks and other wooden structures in the sea. Also included are certain types of barnacles, which feed on plankton but may attach to the hulls of ships and slow them down owing to the increased frictional drag of the ship against the water.

FIGURE 9.8 *A,* Sublittoral environment dominated by sea stars, brittle stars, and sea lilies, at 92 meters on the continental shelf off Antarctica's Palmer Peninsula. *B,* An Australian coral reef. *C,* Sublittoral environment on the New England shelf, dominated by sea urchins. Note also the sea star and hermit crab. (*A* courtesy of Smithsonian Institution, Oceanographic Sorting Center; *B* courtesy of Australian Tourist Commission and Qantas Airlines; *C* courtesy of Bruce Reynolds, National Marine Water Quality Laboratory, Narragansett, Rhode Island.)

Antifouling paint, however, can be used to prevent the young larval forms from attaching and becoming established.

DEEPER BENTHIC ZONES

The deeper benthic zones comprise the entire ocean floor deeper than the continental shelves, and they lack sufficient sunlight for plant growth. The amount of organic matter is much reduced in these zones as compared with that in shallower environments. The principal sources of food for the animals here are sinking detritus from above and migrating pelagic animals.

The ***bathyal zone*** is the benthic environment of the continental slope, about 200 to 2,000 meters in depth. Temperatures in this zone are usually less than 10° C and

FIGURE 9.9 Sea cucumbers and sea lilies at about 600 meters on the continental slope off Antarctica. (Courtesy of Smithsonian Institution, Oceanographic Sorting Center.)

are nearly unchanging, resulting in a very monotonous environment. One view of the bathyal environment is shown in Figure 9.9. A wide variety of animal life exists in this zone, including many species of sponges, soft corals, sea lilies, sea stars, sea cucumbers, crabs, shrimp, and fish.

The ***abyssal zone***, which comprises the abyssal plains and hills, represents more than 80 per cent of the sea floor. Less food is transported onto this zone than onto the shallower benthic environments; hence the abundance of animal life is greatly reduced. There are also fewer species here as compared to shallower zones. Most of the abyssal animals (Fig. 9.10) are either mud-eaters, such as sea cucumbers and brittle stars, or predators, such as sea stars and certain grotesque fish with large mouths. The distribution of life in the abyssal zone is not uniform. It may be abundant in one area and very sparse in others, depending largely on the productivity in the epipelagic zone above. In addition to darkness, this zone is characterized by temperatures of less than 4°C.

The ***hadal zone*** consists of long, narrow oceanic trenches located close to many island chains or continents. With temperatures generally ranging from 1.2 to 3.6°C and pressures in excess of 600 atmospheres (depths greater than 6,000 meters), this area rates as the most extreme of ocean habitats. Larger predators such as fish, sea stars, and crabs are generally absent in the deep trenches, apparently because of the low concentration of food organisms. The concentration of life may be more than 1,000 times as great on the shelf as it is in the hadal benthic environment. The dominant species tend to be slow-moving mud-eaters such as sea cucumbers and other animals that are able to make the most of the sparse

FIGURE 9.10 Life in the abyss. *A,* Sea cucumber (right) and trail of feces left by a worm (left) at about 4,000 meters in the southern Indian Ocean. *B,* Sea urchin and brittle star at about 4,000 meters between South America and Antarctica. (Photos courtesy of Smithsonian Institution, Oceanographic Sorting Center.)

food available. These include sea anemones and strange worms with no mouth or stomach (the pogonophorans). These worms can absorb dissolved food through their skin.

SAMPLING THE MARINE LIFE

Sampling of ocean life may be simple, as in scraping organisms from a measured area of a rock or piling; or it

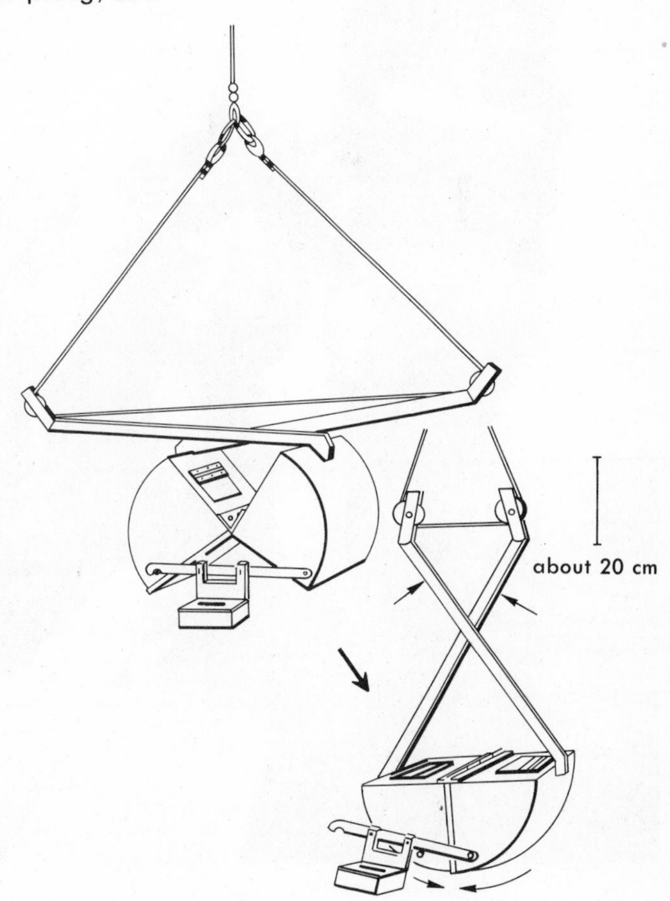

FIGURE 9.11 A van Veen grab, open on the way down (above), and closed on the way up, with a sample of the mud (below). (After Hedgpeth, 1957.)

may be very difficult and time consuming, as in sampling the deep-sea benthos. In a day's time, many samples may be collected in the littoral and sublittoral zones, but it may take months to collect an equivalent number of samples from the abyss. The equipment used in deep water is frequently very heavy and must be lowered slowly. For example, if a deep-sea grab-sampler (Fig. 9.11) is lowered to 4,000 meters at 50 meters per minute and retrieved at 30 meters per minute, it will take about four hours to get a single bottom sample. This figure would be increased to about 10 to 12 hours if a dredge (Fig. 9.12) is used to sample the epifauna or infauna because much more line must be let out (at an angle), for the sampler may be dragged nearly horizontally. Use of a plankton net (Fig. 9.13) or a mid-water trawl (Fig. 9.14) to sample the deep pelagic zones involves similar complications.

FIGURE 9.12 A large dredge, used to collect large samples of ocean sediment and infauna. (Courtesy of Woods Hole Oceanographic Institution.)

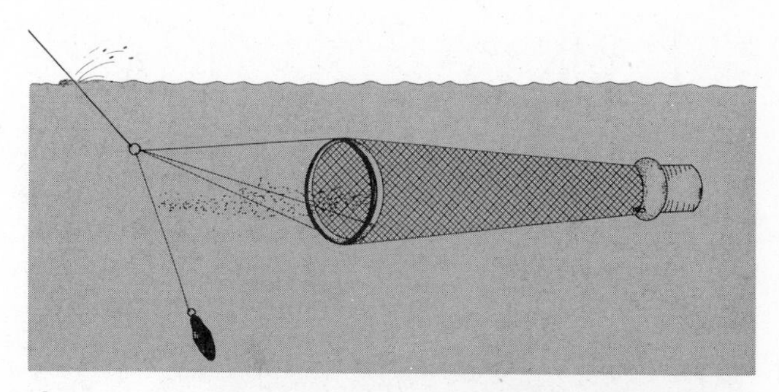

FIGURE 9–13 *A,* **Plankton net in tow.** *B,* **"Bongo" nets, which can be used to sample plankton in deep water. (Photo by David Stern, Oregon State University.)**

Thus, the time and expense of obtaining ecological data increase dramatically as one works in deeper and deeper water. Neritic waters have been studied most extensively because of this factor and also because of the presence of economically important food resources in this zone.

An economical way around the need for specialized expeditions to sample the epipelagic plankton was devised by Sir Alister Hardy in 1925. His ***continuous plankton recorder*** (Fig. 9.15) may be towed behind a commercial vessel traveling across the ocean. The plankton are caught, preserved, and continuously wound between two gauze bands and stored until they are "unwound" and identified. This method provides continuous sampling of

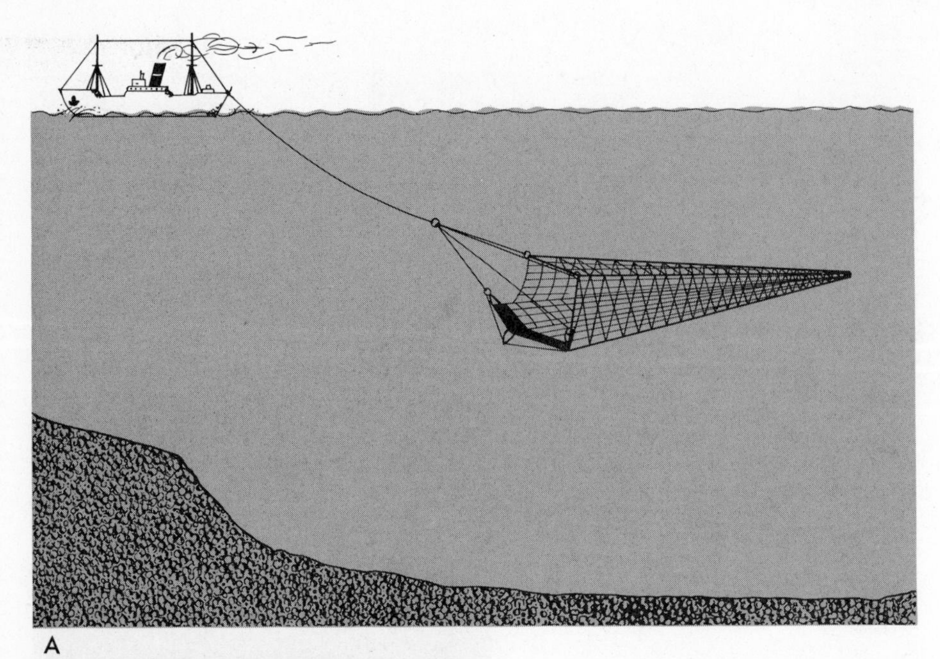

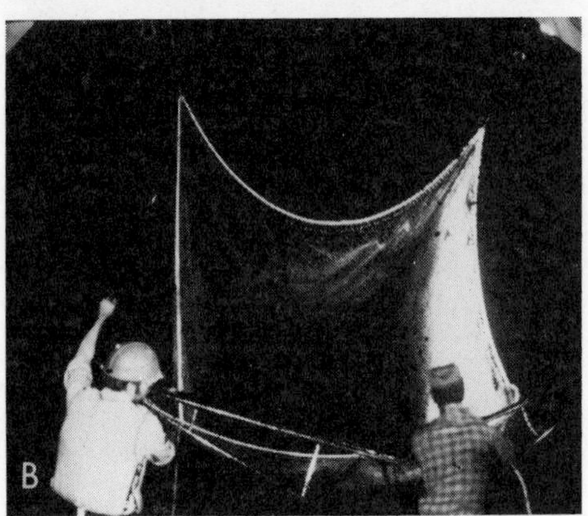

FIGURE 9.14 The midwater trawl (A) being towed; (B) being hauled in. (Photo by David Stein, Oregon State University.)

the plankton between two points. Of course, many of the organisms, especially the zooplankton, become crushed, making identification more difficult. Consequently, experts in identifying crushed plankton have evolved out of these studies.

BIOLUMINESCENCE

Although sunlight is the most obvious source of light in the sea, it is not the only one. At night and in disphotic and aphotic regions, the phenomenon known as bioluminescence is also present. This other light source can have a profound influence on many aspects of life from vertical migration to reproductive behavior to predation.

Actually, even bioluminescence depends indirectly on the sun, because sunlight is the source of energy for the

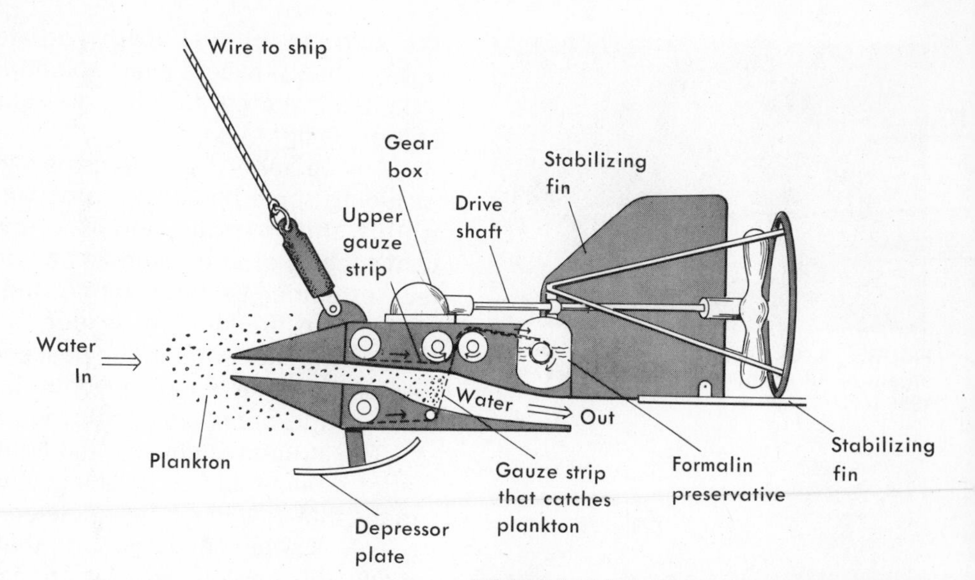

FIGURE 9.15 The Hardy continuous plankton recorder, which collects the plankton on a gauze band that is continuously wound into a tank of preservative. (After Hardy, 1956.)

plant life which make organic matter out of inorganic nutrients. Bioluminescent animals may eat plants or other animals that had eaten plants and thereby attain energy to live. Some of this energy is released in certain organisms as a cool blue-white light. Little heat is given off. Bioluminescence also provides light in surface waters at night, when it frequently lights up the surf or the wake of a ship.

Pliny and other ancients had observed luminescence in some of the larger jellyfish. But the glow sometimes seen in the surf and in the wakes of ships was still a mystery in the seventeenth century, when it was thought by some that the sea absorbed the sun's energy by day and gave it back at night. The British chemist Robert Boyle, in the seventeenth century, thought that it was caused by friction of waves against the air or the side of a ship. This light was observed only at night and frequently on the tips of waves or in the wake of a ship or along its hull. In the eighteenth century, it was discovered that the glowing of waves and surf is often due to a tiny one-celled dinoflagellate, *Noctiluca*, which means "night light." Although an individual is less than one millimeter in diameter, the presence of great numbers of these organisms causes the water to glow.

The reaction involved in bioluminescence is oxidation of a substance referred to as ***luciferin*** in the presence of an enzyme called ***luciferase***. The exact composition of the raw materials and the nature of the reactions involved vary with the kind of organism. A typical reaction might be something like this:

$$2LH_2 + O_2 \text{ (in the presence of luciferase)} \rightarrow L + H_2O + \text{light}$$

where L = oxidized luciferin. This reaction is much like

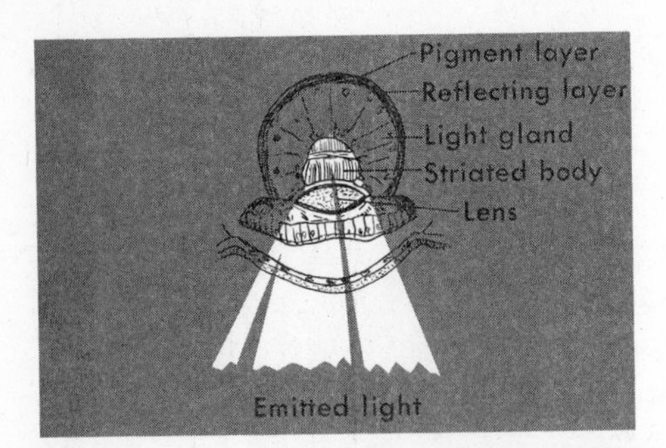

FIGURE 9.16 Photophore (light-producing organ) of an euphausid shrimp. (After Hardy, 1956.)

the burning of fuel (carbohydrate + $O_2 \rightarrow CO_2 + H_2O$ + light + heat), except that bioluminescence releases very little heat, whereas burning results in a great deal of heat as well as light.

A wide variety of organisms exhibit bioluminescence, including some bacteria, protozoans, jellyfish, polychaete worms, shrimp, squid, and fish. However, few bioluminescent species exist on land as compared to the sea. Only a few terrestrial bacteria, fungi, and insects (such as the well-known firefly) light up during the night. The nature and distribution of bioluminescence in the sea are so diverse that one wonders about its significance. Of what value is it to organisms that produce it?

The structures causing the light range from the single cell of a bacterium or dinoflagellate to the complex light organs that very much resemble an eye found in some fish. In fact, it was once thought that these complex light organs, or ***photophores***, such as those found on the shrimplike euphausid (krill), some squid, and lantern fish, actually represented eyes that were adapted to giving off light rather than receiving it (Fig. 9.16). However, the nerve pattern and development of eyes differ greatly from photophores, so this theory has been abandoned.

An interesting relationship exists between the lantern fish, *Photoblepharon,* and the luminescent bacteria that it maintains in pouches on its cheeks. The fish provides food and lodging for the bacteria. The bacteria produce light for the fish. The light may help the fish by illuminating the nearby surroundings, by attracting a mate (Fig. 9.17), or by attracting small fish which may be eaten by *Photoblepharon* (Fig. 9.18). It might even attract a larger fish which might eat *Photoblepharon* (Fig. 9.19).

Some angler fish (Fig. 9.20) have a similar relationship with luminescent bacteria which they carry in a light-organ (photophore) on the end of a spine extending over the mouth. The photophore resembles a large, luminescent copepod, a favorite crustacean food for many small

FIGURE 9.17 Photophores of *Photoblepharon* may aid in mating.

FIGURE 9.18 *Photoblepharon's* light may also attract prey.

ocean fishes. Small fish may be lured by the light to within grabbing distance for the angler fish.

Some animals produce colors other than blue-white by covering the light source with a colored filter. For example, the deep-sea squid *Thaumatolampas diadema* (Fig. 9.21) produces lights of red, white, and blue (leading some to suggest that it should be our "National Squid"). Structurally, the light organs closely resemble those of some fish and euphausids. This similarity is an example of convergent evolution as these three groups of animals are not otherwise closely related. Most of the lights of *Thaumatolampas*, euphausids, and lantern fish are on the undersides. What might the function of the light be? (Note that red light penetrates only a short distance in water.)

MARINE MIGRATIONS

Many marine animals move at one time or another from one environment to another, quite often either to breed or to feed. These migrations may seem mysterious at first because they can involve considerable distances and great navigational precision. In fact, humans would need quite sophisticated navigational instruments to duplicate some of these remarkable journeys. Marine animals, unlike land animals, can migrate either horizontally or vertically, and we will discuss each type of migration in turn.

FIGURE 9.19 The light produced may attract predators.

Horizontal Migration

Horizontal migrations cover distances of many miles. The direction is often well defined and accurately aimed at a target destination. Birds seasonally migrate thousands of miles north and south, without the use of compass, clock, sextant, or map. They may rely on landmarks part

FIGURE 9.20 Some angler fish have luminescent lures that resemble copepods, which are food for fish.

of the time, but often they must migrate over unfamiliar territory, or even over clouds or the sea. A bird may be able to use wave patterns on the sea as an indicator of constant direction for a short time, perhaps long enough to notice with its keen eyesight the direction and rate of the sun's movement and its height above the horizon. With this information it might be able to estimate its latitude at a particular time of year during the migration. If the bird has a strong "biological clock" which can tell it when noon should be, back home, it can note the time that the sun reaches its greatest height above the horizon on a particular day and by the time difference, it can determine the longitude. If the bird drifted to the east from its starting point, the sun would reach its peak (noon) earlier, according to "home time." Remarkably enough, it appears that some birds actually have sufficiently powerful eye-sight, biological clocks, and instincts to perform such complicated tasks.

FIGURE 9.21 *Thaumatolampas diadema,* the amazing squid with red, white, and blue lights. (After Hardy, 1956.)

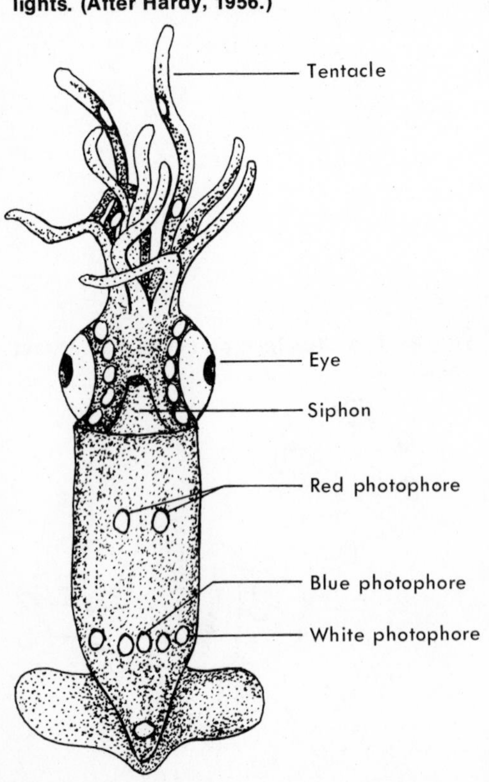

Some migrations of fish may be related to chemicals in the water. The chemicals washed from the land by a particular stream may be characteristic and may "label" that stream as unique. Anadromous fish, which return from seawater to fresh water in order to spawn, can probably recognize their home stream (the place of their birth) by a chemical sense similar to that of smell. In the case of anadromous salmon, migration involves journeys into fresh water and up rivers to specific breeding grounds (Fig. 9.22). Most of the feeding, however, takes place in the sea, perhaps thousands of miles from the home stream. This migration is made especially difficult because of the physiological adaptations required for the organism to be able to tolerate contact with different waters, having a salinity difference of about 35‰.

Chemical pollution may someday pose a problem to organisms that rely on their sense of "smell" for migrational orientation. Thermal pollution might misguide some fish into a detour on their journey to breed, causing a

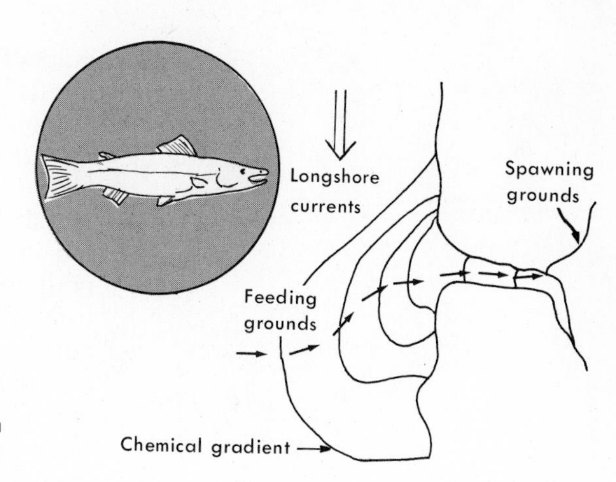

FIGURE 9.22 Migration pattern of an anadromous salmon in relation to a chemical gradient emanating from a stream.

decrease in the production of a species. Offshore power plants may heat water locally, causing not only the formation of a new community around the plant but also the deflection of some marine migrants from their course.

Perhaps some migration patterns, even though they are no longer really needed by a species, are imprinted from past times when they were necessary. For example, the American and European eels migrate from the east coast of North America and from Europe, respectively, and meet in the Sargasso Sea (in the West Central Atlantic), an area of weak currents and low productivity, to breed separately (Fig. 9.23). The larvae return to the freshwater home of their parents without ever having made the trip before. What mechanism could these eels, especially the larvae, use to navigate such distances? Is it the earth's

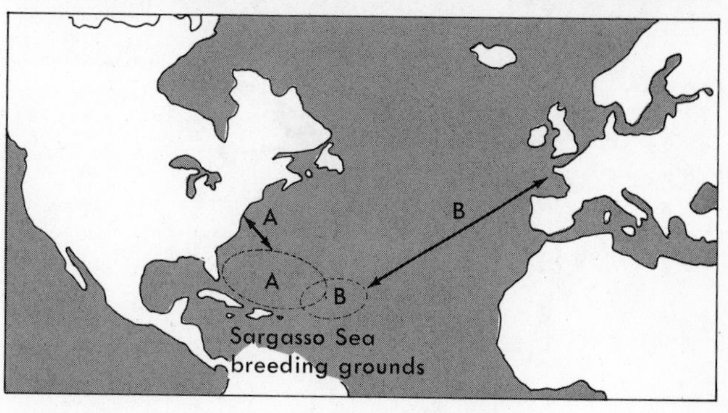

FIGURE 9.23 Migration routes and breeding grounds of the American eel (*A*) and the European eel (*B*).

magnetic field, sun arcs, chemicals, or some other agent? Whatever the factors, they are definitely part of the hereditary make-up of the eels.

One theory postulates that when the continents were joined together, the Sargasso Sea was a fresh-water lake and that the two types of eels were a single species which fed in tributaries that emptied into the lake. When the continents of Europe and North America drifted apart, the tributaries were separated, and the Sargasso Sea became part of the Atlantic Ocean. Through evolutionary adaptation, the eels became two species, reproductively isolated and adapted to life in both fresh and salt water. The larvae retained an instinct to migrate into the tributaries of Europe and North America, respectively, but the adults still migrate back to the Sargasso Sea to breed. The process of continental drift is very slow, requiring millions of years, which is plenty of time for a gradual adaptation, by many steps, to the new migratory pattern. The migration itself takes about three years, during which the larvae mature. The eels remain in the fresh water for 7 to 15 years. The reproductive organs do not mature until the return trip, when the adults are 10 or more years old.

Some sea turtles, seals, and whales also migrate great distances. The navigational skills of these animals are no

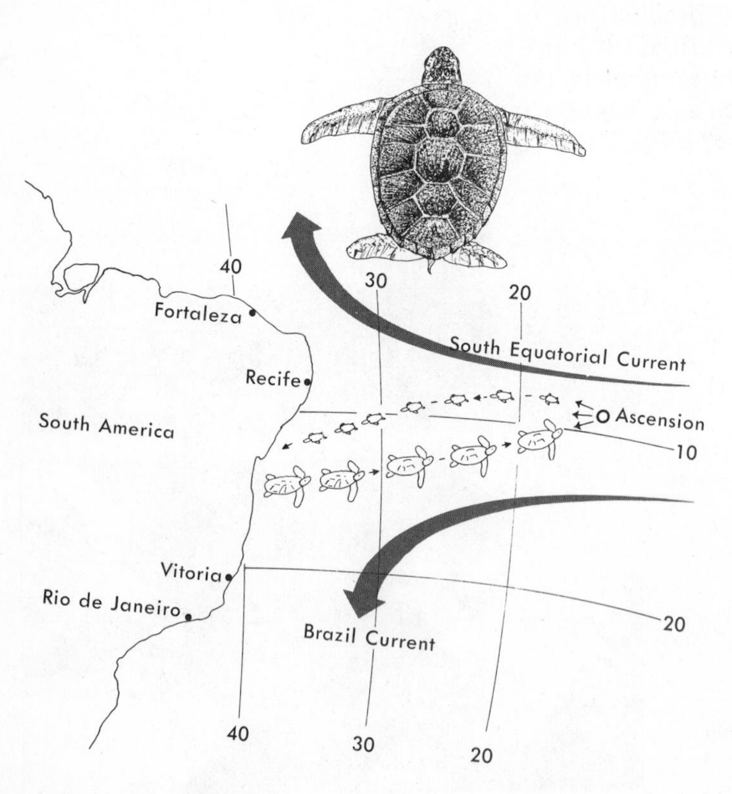

FIGURE 9.24 Migration route of the green turtle in the South Atlantic. (After Carr, 1965.)

less amazing than those of birds and fish. Some green turtles (*Chelonia mydas*) are known to migrate about 2,000 kilometers from their feeding grounds along the coast of Brazil to their breeding grounds on Ascension Island in the South Atlantic (Fig. 9.24). This journey is all the more amazing because Ascension Island is only about eight kilometers wide and the adult turtles must swim against the South Equatorial Current. The young turtles, however, are aided by the current during their westward migration back to Brazil.

The baby turtles head for the sea as soon as they hatch. Somehow they know where the water is, even when sight of it is obscured by barriers such as sand dunes. Once they find the sea they swim away from land until they are picked up by the South Equatorial Current and carried to Brazil, where they feed and mature. As adults, they migrate back to Ascension Island every two to three years to breed. Their primary navigational aid appears to be the sun. The height of the sun at noon, the arc made by the sun as the earth turns, plus a seasonal sense cued by temperature changes, could provide sufficient information to determine latitude. Although such capabilities have not been proven for the green turtle, if the turtle can find the latitude of Ascension Island (about 8° S), it can migrate due east, using the arc of the sun as an east-west guide. Once the turtle is within range of a chemical gradient emanating from Ascension, it can swim toward the island, using a sense of smell or taste until the island can be seen. Mating in coastal waters, visual and chemical orientation guide the female turtle to a specific beach that is suitable for egg laying.

FIGURE 9.25 *A,* **Vertical distribution of the copepod *Calanus* at selected times during a 24-hour period. The width of the patterns shown indicates abundance at a particular depth. (After Russell, 1927.)** *B,* **Continuous representation of (*A*), where the darkest slading indicates the greatest abundance of *Calanus*.**

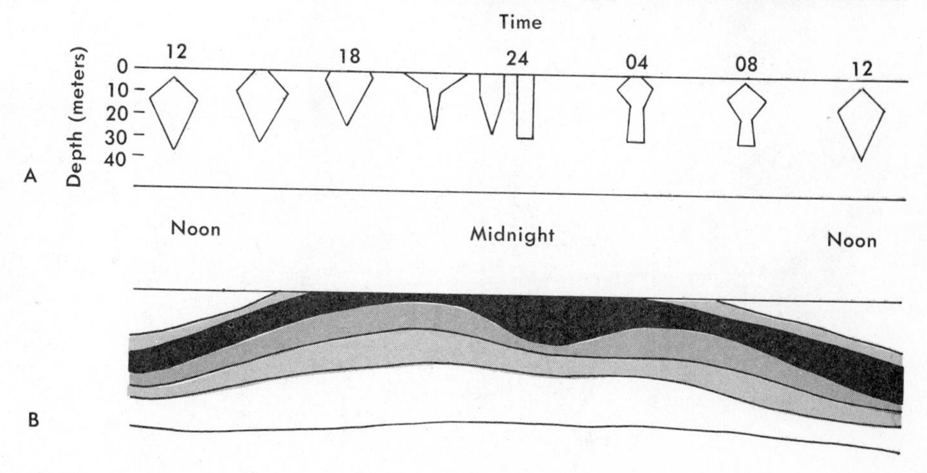

Vertical Migration and the Deep Scattering Layer

Almost as soon as biologists began systematic sampling of plankton, in the nineteenth century, it was observed that more zooplankton were collected at the surface during the night than during the day. It was suggested by some that the plankton avoided the nets during the day because they could see the nets (this is frequently a problem when sampling for larger animals such as fish). It was subsequently found, however, that deep samples contained more animals during the day than during the night (Fig. 9.25), suggesting that the zooplankton were migrating downward at dawn and upward at dusk. The vertical migration of animals has now been proven experimentally.

In the 1930s Sir Alister Hardy and his co-workers invented a "plankton wheel" (Fig. 9.26), which they used to measure the speed of migration of various small plankton animals in response to various light conditions. Most of the zooplankton studied actively swam toward a dim light and away from a bright light. In other words, they tended

FIGURE 9.26 The Hardy plankton wheel in use. (After Hardy, 1956.)

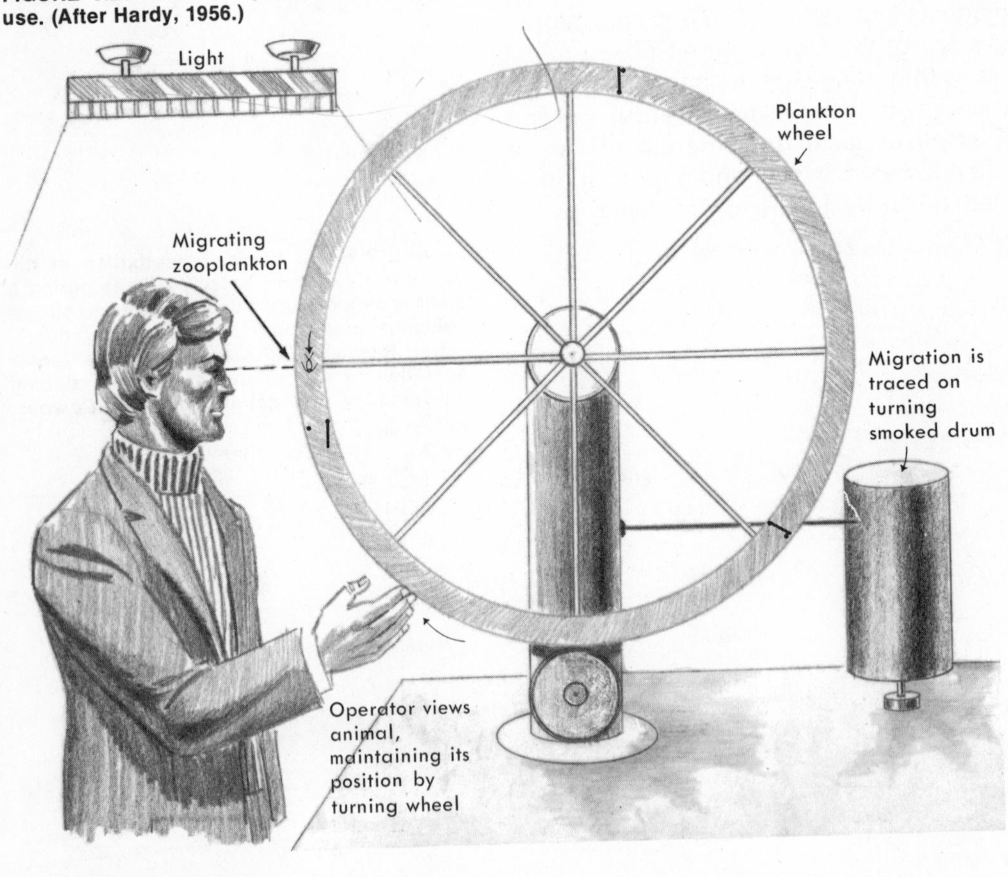

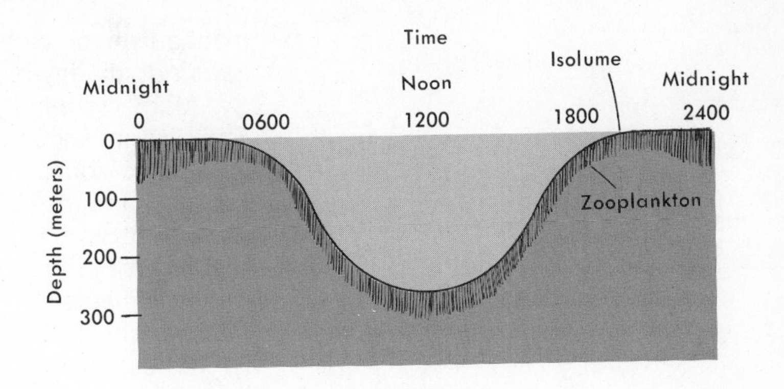

FIGURE 9.27 Path of vertical migration in relation to light intensity.

to migrate toward an area of optimum light (a movable "twilight zone"). Each species was found to have its own light optimum and consequently its own path of migration. They discovered that copepods could migrate as fast as 30 meters per hour (about 0.8 cm/sec). At that rate, a copepod could migrate 180 meters in six hours. The shrimplike euphausid, which is much larger than a copepod, can migrate at a speed of 135 meters per hour, and the polychaete worm *Tomopteris* swam in the wheel at a rate of more than 200 meters per hour. These are all upward speeds; downward rates were frequently even faster.

Some crustaceans may migrate 800 meters or more during their trek. Others may travel just a few meters. The range of migration generally depends on the turbidity of the water, the brightness of light, and the sensitivity of the animal to light, or its own light optimum. Even within the same area, one species may descend to depths several times greater than another since they follow different light intensities (Fig. 9.27).

During maximum darkness, some migrants stay near the surface, others tend to become scattered by random swimming, and some even sink to slightly lower depths during what Hardy refers to as "midnight sinking." As dawn approaches and a light optimum is re-established, the animals re-aggregate in the "twilight" of the sea. Then they follow this light into the depths. This is probably a

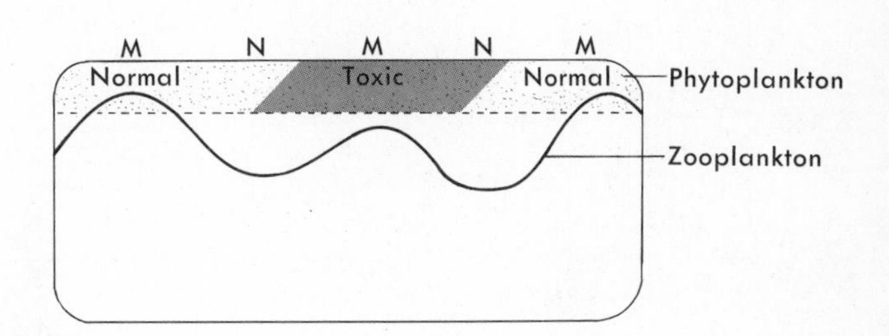

FIGURE 9.28 The effect of a toxic phytoplankton bloom on vertical migration. Note that the zooplankton do not rise to the surface layer when it is occupied by poisonous phytoplankton. (M = midnight; N = noon.)

mechanism of vertical migration for many, though perhaps not all, day-night migrants.

Many zooplankton require phytoplankton as food. The phytoplankton are restricted to the upper waters, the epipelagic zone. Therefore the zooplankton that migrate between the epipelagic and mesopelagic zones are in water that is rich in food during the night and in water that is poor in food during the day. They could not survive if they remained in the deeper water; they must come up to the surface water to feed.

It has been observed that some zooplankton will not migrate into extremely dense phytoplankton blooms but will engage in a shorter migration route until the bloom has diminished or drifted away (Fig. 9.28). The phytoplankton secrete chemicals that may be toxic or perhaps merely repulsive to the zooplankton. In support of this hypothesis, it has been shown that "old" phytoplankton raised in the laboratory can cause the water-flea *Daphnia* to die, apparently because of a toxic effect of the plants. Vertical migrations can be advantageous in that different water is sampled by the zooplankton on succeeding days if the currents are different at the surface as compared to deeper water. It is also possible that some phytoplankton secrete more toxins during the day, when they are photosynthesizing, than during the night.

FIGURE 9.29 Multiple Deep Scattering Layers, descending at dawn (0700 hours) and rising at dusk (2000 hours) in the open sea west of Africa. The lower trace indicates the ocean bottom. (From Lowrie and Escowitz, 1969.)

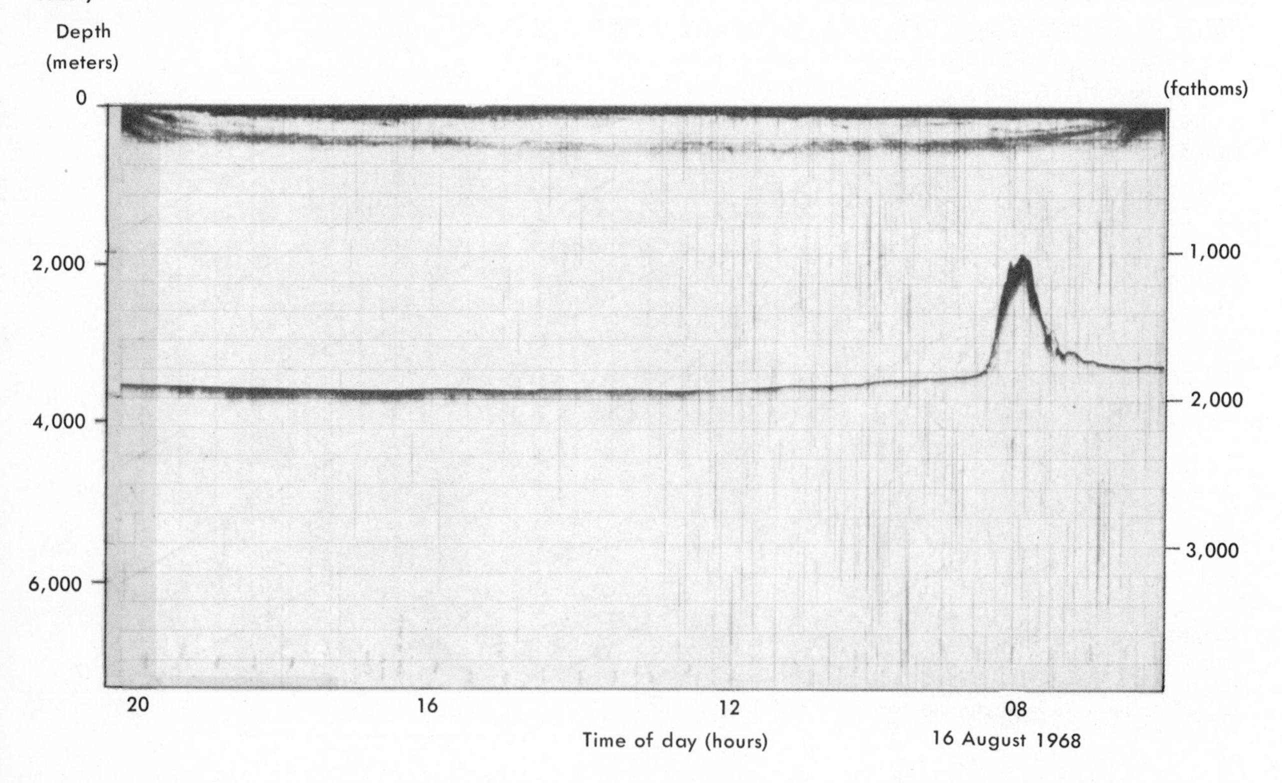

It has also been suggested that by migrating to deeper darker water during daylight hours, the zooplankton avoid being eaten by predators that could see them if they were in bright light. Interestingly enough, however, some zooplankton are followed by their predators, usually small fish, during their migrations. These fish may be following a light optimum or the prey itself.

Layers of migrating fish are among those animals that may be responsible for "false bottoms" recorded on echo sounders. When echo sounding replaced sounding lines as a means of determining water depth, mariners were puzzled by the appearance of reflections from something in the water above the bottom. Occasionally these echoes were thought to be the bottom itself and were reported as isolated reefs. Later, when continuous echograms were made over long distances and over many hours, the "false bottom" or ***deep scattering layer*** (D.S.L.) was observed to rise in the evening and descend during the morning. The D.S.L. could be differentiated from the true bottom as a soft trace rather than the hard trace produced by the sea floor.

Now it is known that the D.S.L.'s are actually layers of vertically migrating animals that swim toward the surface as night approaches and descend in the morning. In fact, several D.S.L.'s may be seen, one over the other (Fig. 9.29). Many marine animals, including copepods, shrimp, krill (shrimplike food of whales), and lantern fish engage in vertical migrations. Copepods are probably too small to produce distinct echoes, but the larger shrimp, krill, and fish probably cause D.S.L. traces or echoes. Different species may migrate to different depths during the day, producing multiple D.S.L. traces.

Fish with gas bladders (swim bladders) are presumably the most effective sound reflectors owing to the difference in density between the gas and the seawater. Fish that are mesopelagic and have swim bladders, such as

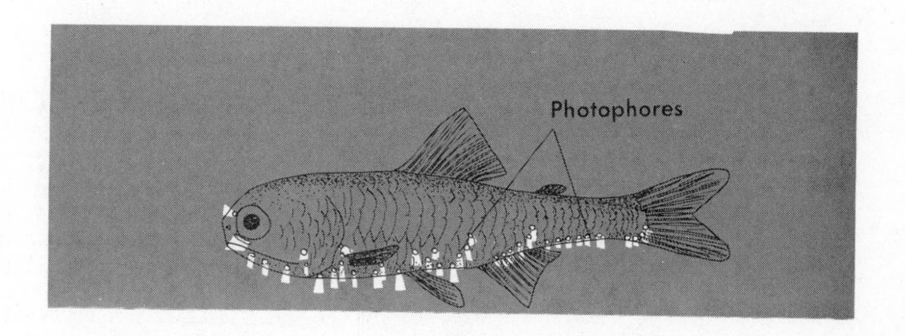

FIGURE 9.30 A lantern fish (*Diaphus theta*) countershaded (camouflaged) in the "twilight zone" during its vertical migration. White areas indicate bioluminescence. The fish is less readily seen from above or below than from the side.

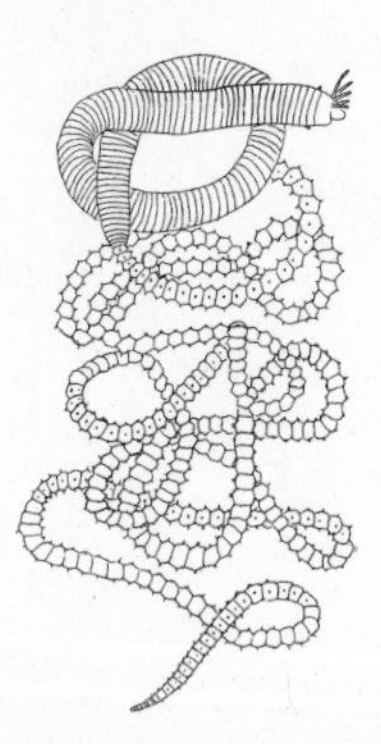

FIGURE 9.31 The palolo worm (*Eunice viridis*), whose reproductive tail sections (lower part) migrate vertically in response to the light of the moon. (From Barnes, 1974.)

lantern fish, probably are among the most important animals in causing D.S.L. traces.

Lantern fish carry their own light source as they trek to the epipelagic regions. They cover a migratory pattern similar to that of the copepods and other small crustaceans that they eat. At first it might be thought that the lights would attract predators that eat the lantern fish, but on a closer examination of the fish, it is noticed that most of the lights are directed downward (Fig. 9.30). Larger fish from above, looking down would see nothing, since the dark upper surface blends in with the dark, deeper water. The same predator viewing the lantern fish from below would probably still see nothing. Perhaps the predator's eyesight is not too sharp, and the glow produced by the lantern fish tends to blend in with the "twilight" glow from above. The light, therefore, provides the lantern fish with a type of camouflage.

Some vertical migrations are dependent not on the sun but on the moon. An example of this type is the migration of the Palolo worm of the South Pacific (Fig. 9.31). It lives most of its life among rocks in clear tropical water, emerging at night to feed on zooplankton. During the eighth to tenth day after the full moon in October or November, the Palolo worms, male and female, lose their reproductive hind ends, which contain sperm and eggs. These tail sections, which can swim on their own, even without a head, migrate to the surface of the water, where they burst, allowing the sperm to fertilize the eggs. These develop into new Palolo worms. The reproductive stages of the worm are regarded as a delicacy by the natives of Samoa and other islands of the area. Once a year they have quite a feast of Palolo worm tails, which they catch in nets at night. The most remarkable aspect of this process is the timing of the migration. Why is it 8 to 10 days after the full moon in October or November every year? The most likely stimulus is the lunar light itself. A plausible suggestion is that the worms, which are exposed to moonlight at night, mature to a certain point, then stop until the bright light of the full moon triggers the final stage of maturation, which lasts for 8 to 10 days and culminates in the sexual migration.

SUGGESTED READINGS

Carr, Archie. 1965 (May). The navigation of the Green Turtle. *Scientific American.*
Dietz, Robert S. 1962 (August). Deep scattering layers. *Scientific American.*
Ekman, Sven. 1953. *Zoogeography of the Sea.* London, Sidgewick and Jackson.
Limbaugh, Conrad. 1961 (August). Cleaning symbiosis. *Scientific American.*
Hardy, Alister C. 1956. *The Open Sea, the World of Plankton.* Boston, Houghton Mifflin.
——————. 1959. *The Open Sea, Fish and Fisheries.* Boston, Houghton Mifflin.

Harvey, E. N. 1952. *Bioluminescence.* N.Y., Academic Press.
Hedgpeth, Joel W. (ed.). 1957. *Treatise on Marine Ecology and Paleoecology, I: Ecology.* N.Y. Geological Society of America (memoir 67).
Southward, A.J. 1965. *Life on the Seashore.* Cambridge, Mass., Harvard University Press.
Tait, R. V., and R. S. De Santo. 1972. *Elements of Marine Ecology.* N.Y., Springer-Verlag.
Yonge, C. M. 1949. *The Seashore.* London, Collins.

A stalked sea pen at 5,066 meters on the abyssal plain west of Africa. (Photo courtesy of Naval Oceanographic Office.)

CHAPTER 10

Life in the Sea

The oceans contain a bewildering assortment of living things—bewildering because of the sheer variety and also because we simply do not see most marine creatures in our everyday lives. How can we study these organisms? One way to tackle the problem is to look at the roles played by various organisms in marine ecosystems.

Perhaps the most convenient approach is to classify organisms according to the way in which they acquire or manufacture food. Thus, they may be ***producers, consumers***, or ***decomposers***. Plants produce organic matter using the sun's energy and inorganic nutrients, whereas animals consume organic matter, deriving their energy from the organisms that they eat. Bacteria and fungi function as a part of the decomposition process, whereby dead organic matter (detritus) is broken down into inorganic nutrients that may be used by plants in producing new organic matter.

The role or function of an organism in the sea (its niche) depends on its relationship to the physical environment and to other organisms through the ***food web*** (feeding relationships, discussed in the next chapter) and ***symbiotic*** relationships (close interactions among two or more species).

Symbiotic relationships may take several forms. If one species benefits and the other is harmed, the relationship is called ***parasitism,*** as in the case of a roundworm living in the intestine of a fish. In a ***mutualism,*** both species benefit. An example of this is the relationship of certain single-celled algae to some sea anemones and corals. The algae live within the tissues of the host animal and supply it with organic substances that stimulate its growth. The anemone, in turn, gives the algae a home and inorganic nutrients. In ***commensalisms,*** one species benefits and the other is neither harmed nor helped. For example, some marine worms form a burrow which is shared by other species such as shrimp or crabs. The worm is not harmed, but the others have a free home and may even share the worm's food.

All of the interactions among organisms constitute a closely interrelated community of producers, consumers, and decomposers, the composition of which is largely determined by the physical environment.

THE ELECTRON MICROSCOPE

Several of the photographs in this chapter were produced with the aid of electron microscopes. Most of these organisms are less than a millimeter in diameter, and some, such as bacteria, may be less than 0.001 mm (1 μ) in diameter. Most conventional light microscopes effectively magnify only as much as 1000 times (1000×). Greater magnification is needed to show the internal structure of simple organisms such as blue-green algae and the surface features of the minute plankton. Electron microscopes have the advantage of using a beam of electrons, instead of light, to magnify objects as much as tens of thousands of times. The transmission electron microscope passes the electrons through a thin piece or slice of the object and focuses on a fluorescent screen or a photographic plate. This is an excellent technique for showing internal structure. The scanning electron microscope scans a beam of electrons over the surface of the object and causes other electrons to be thrown off the surface and produce an image of the organism on a television screen. Then a photograph is taken of the television image. This is ideal for showing great surface detail of small opaque organisms, such as the plankton, and the skeletons of ooze-forming organisms, such as diatoms, foraminifera, and radiolaria.

In this chapter, we will discuss some of the more familiar or important kinds of marine organisms. Familiar species are not necessarily the most important to the ecosystem, because they may have low productivity or they may not be available to marine food webs. The tall plume grass *Phragmites,* a common plant of disturbed salt marshes, is of questionable value. Some researchers regard it as worthless except as a source of detritus; others point out its value as a nesting ground for birds. On the other hand, certain species may be unknown to most people, but they are of tremendous importance to life in the sea. For example, the bristlemouth *(Cyclothone),* a small oceanic fish, is one of the most common marine fish, but few people have seen one. Some organisms, such as bacteria and diatoms, are minute and hence rarely noticed except when present in numbers great enough to discolor the water. Bacteria are essential to the recycling of nutrients used by plants. Diatoms, which are tiny single-celled plants, are important producers of organic matter in the sea.

PRODUCERS

Plants are frequently called the primary producers, since they incorporate light energy with carbon dioxide, water, and inorganic nutrients to make organic matter, which directly or indirectly becomes the food for the rest of the organisms. This is true in the oceans as well as in fresh water and on land. In littoral environments, simple marine plants (seaweeds or algae) often flourish on rocks, pilings, and other solid objects. In the open waters of the euphotic zone, many kinds of drifting plants (phytoplankton), mostly microscopic in size, fill the role of primary producer. In addition, flowering plants such as salt-tolerant grasses are found growing in shallow bays, estuaries, and salt marshes. They also provide food and shelter for a wide variety of animal life. The major groups of plants are categorized into ***divisions*** (equivalent to the term ***phyla,*** as applied to animals) based on their structure, mode of reproduction, and the nature of their pigments. The marine plant divisions may be roughly classified into the seaweed divisions, the phytoplankton divisions, and the flowering plants.

Seaweed

Seaweeds are marine plants familiar to most who visit a rocky coast. These are generally simple attached plants. The colors of the seaweeds are quite varied, suggesting terms such as "brown algae" (Division Phaeophyta), "red algae" (Division Rhodophyta), and so on. All seaweeds contain the green pigment, chlorophyll, required in photosynthesis, although other pigments may be present in large enough amounts to mask the green chlorophyll and give the seaweed a brownish, reddish, or blue-green color. Although they are usually attached to solid objects and to the shallower parts of the sea floor, some such as the brown alga *Sargassum* drift on the surface of the open sea and hence may be regarded as phytoplankton. The Sargasso Sea is a region of the central Atlantic Ocean famed for the large masses of *Sargassum* that drift there.

BLUE-GREEN ALGAE (DIVISION CYANOPHYTA)

Blue-green algae are commonly seen as very dark encrusting masses of tiny filaments in the splash zone of rocks or wood pilings. They may appear black to the naked eye, but closer examination of a thinned-out sample under an ordinary light microscope reveals a blue-

FIGURE 10.1 *A,* **Structure of** ***Anabaena,*** **a typical blue-green alga.** *B,* **Electron micrograph of ultrathin longitudinal section of** ***Anabaena.*** **(*B* courtesy of G. B. Chapman, Georgetown University.)**

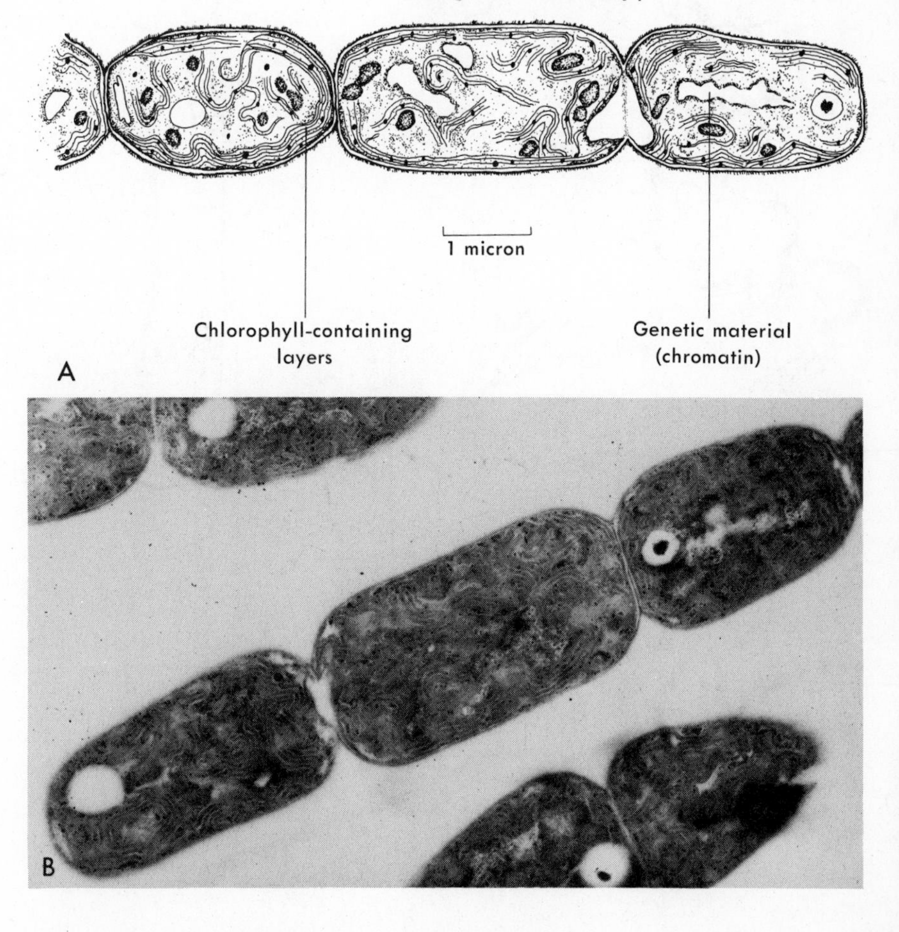

green color (in most species). No internal structure is seen, but often a protective gelatinous sheath surrounding the chains of cells can be observed. In order to see the internal structure of the cells (Fig. 10.1), it is necessary to use a powerful electron microscope. Because of their simplicity, they are thought to be similar to the first plants to inhabit the earth.

Green Algae (Division Chlorophyta)

The members of this group (Fig. 10.2) that are perhaps the most obvious appear as green sheets (sea lettuce, *Ulva*) or tubes (sea intestines or *Enteromorpha intestinales*) or clumps of thin filaments (for example, *Cladophora*). *Ulva* may grow in sheets up to a meter or more in length. It is sometimes dried and may be used in seaweed soups. It may also be an important food for intertidal animals such as amphipods and sea urchins.

FIGURE 10.2 *A,* **The green alga,** ***Ulva*** **or sea lettuce.** *B,* ***Enteromorpha,*** **a tubular shaped green alga.**

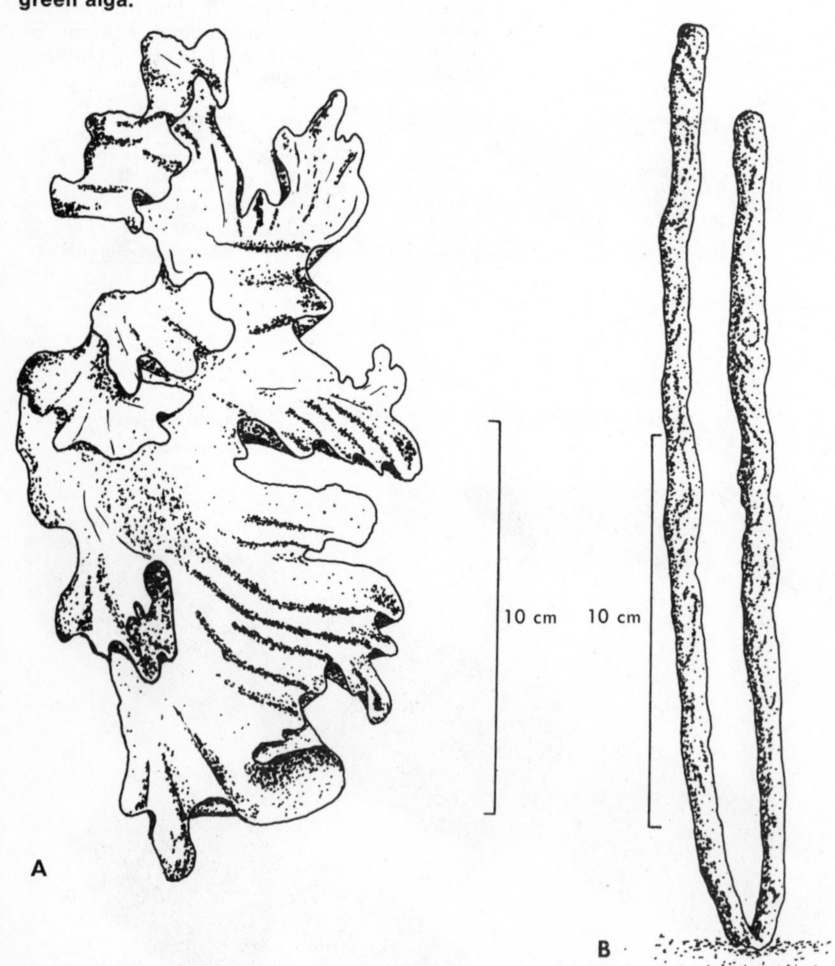

Brown Algae (Division Phaeophyta)

"Brown algae" (Fig. 10.3) are really more brownish-green than brown. The largest of the algae belong to this Division, including the kelp *Macrocystis* (up to 60 meters in length), which is important as a source of "algin," used as a smoothing agent in commercial ice cream. Bladder wrack *(Fucus)* is common on rocky coasts and in harbors, where it grows in great masses on pilings. Some species of *Sargassum* are the dominant algae floating in the Sargasso Sea.

Large brown algae such as those mentioned above have gas-filled floats built into their blades, keeping the plants well exposed to light near the water's surface. Thus they become the dominant seaweeds in water too deep for other algae.

Brown algae are important as a source of food for certain snails living on their blades. Also, a wide variety of other algae and animals live among the kelp or are attached to them. The food and protection offered by kelp attracts a tight little community of great diversity.

Red Algae (Division Rhodophyta)

The red algae, which are generally quite sensitive to exposure, are usually found only at lower levels on intertidal rocks or in tide pools. They get their characteristic color from the red pigment phycoerythrin, which masks

FIGURE 10.3 *A,* **The giant kelp,** ***Macrocystis; B, Sargassum; C,*** **The common rock-weed,** ***Fucus.*** **(*B* and *C* are drawn roughly natural size.)**

A

B

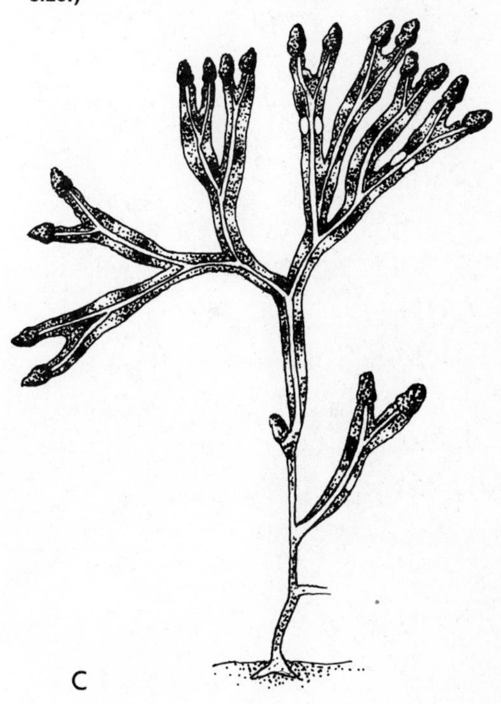

the green chlorophyll. These algae vary from delicate lacy forms to flat sheets as in *Porphyra*. Some species that are common in tide pools and on coral reefs have a tough calcareous covering.

Two species of red algae are of special economic importance. *Gelidium* is the source of agar, a non-nutritious medium used in culturing bacteria. *Porphyra* is cultured in the Orient as a source of food. It resembles the sea lettuce *Ulva* and is also used in "seaweed soups." Red algae and other seaweeds frequently provide a home for small marine animals (Fig. 10.4).

Phytoplankton

Even though the most easily observed marine plants are the seaweeds mentioned above, the most abundant with regard both to number and biomass are the phytoplankton, minute in size (usually less than 0.1 mm), widely distributed, and tremendous in numbers. The phytoplankton discussed in the following sections belong to two divisions, the Chrysophyta (diatoms, silicoflagellates, and coccolithophores) and the Pyrrophyta (dinoflagellates).

CHRYSOPHYTA

Diatoms. The most prominent marine plants are generally the diatoms (Fig. 10.5), sometimes called the

FIGURE 10.4 The red alga *Ceramium*, harboring two unusual, attached jellyfish. The animal on the left is about two centimeters in diameter. (Photo © Douglas P. Wilson.)

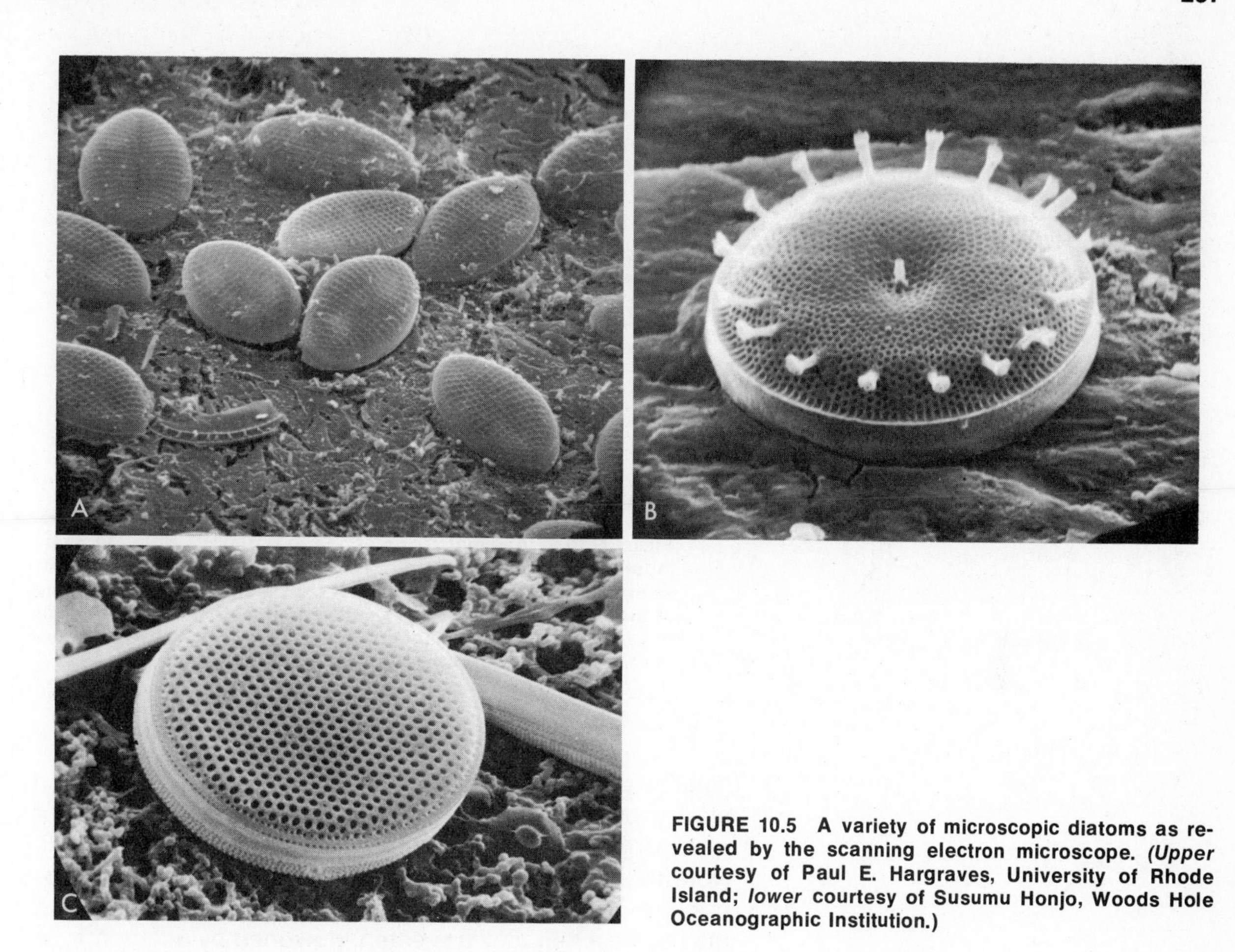

FIGURE 10.5 A variety of microscopic diatoms as revealed by the scanning electron microscope. *(Upper* courtesy of Paul E. Hargraves, University of Rhode Island; *lower* courtesy of Susumu Honjo, Woods Hole Oceanographic Institution.)

"grass of the sea" because of their importance as a food for some small shrimplike herbivores (copepods), which are called the "cattle of the sea." (Perhaps fish should be called the "wolves of the sea" since many feed on herbivores.)

Diatoms are brownish in color and are surrounded by a glassy "shell" made of silicon dioxide. The shell remains after death and contributes to the composition of the sediment. If diatoms are the predominant constituent, the sediment is called ***diatomaceous ooze***. These deposits may become 1,000 meters thick and may contain more than five million shells per cubic millimeter.

The "shell" of a diatom is built much like two Petri dishes (Fig. 10.6). One half is small enough to fit inside the larger half. The walls are porous, allowing dissolved gases and nutrients to pass through. When the cell divides asexually (the usual means of reproduction), each original cell becomes the outer half of the new shell. A new smaller half-shell is secreted. The average size of the cells that are produced decreases until they can get no

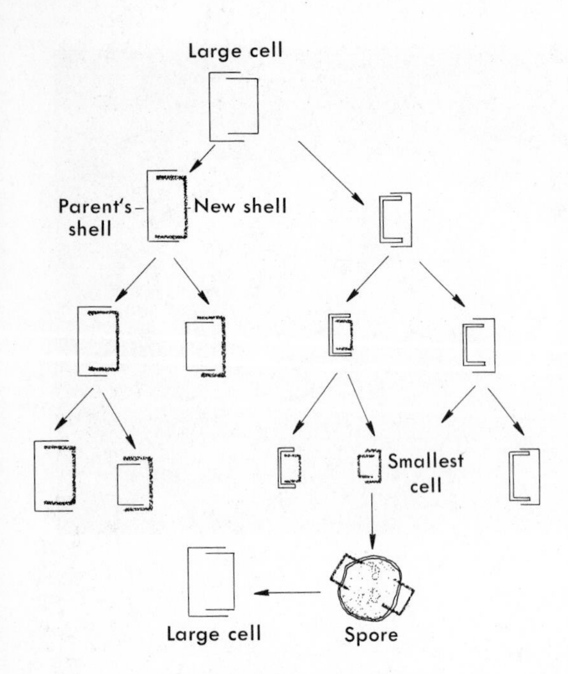

FIGURE 10.6 Asexual reproduction in a diatom. Note the overlapping half of the shell and decreasing size of successive "offspring" owing to the production of new shells inside the margins of the parents' shells. When a cell is produced that is too small to divide, it forms a spore that grows and produces a new large cell.

smaller yet still live. These small cells form a "spore" and start growing, shedding the old shell. Then they form a new, larger shell and continue as a renewed "large" cell. Note that there is no natural death in this type of reproduction. These organisms are, in a sense, immortal. Death may come when they are eaten, poisoned by pollutants, or sink below the photic zone. This feature is characteristic of most phytoplankton that reproduce principally by asexual means.

Long ornamented spines help to retard the sinking of some diatoms, especially in warm water, which is less viscous than cold water. Some diatoms attach to other organisms. Others form a slimy coating on rocks and wood pilings.

FIGURE 10.7 The silicoflagellate *Distephanus.* (After Hardy, 1956.)

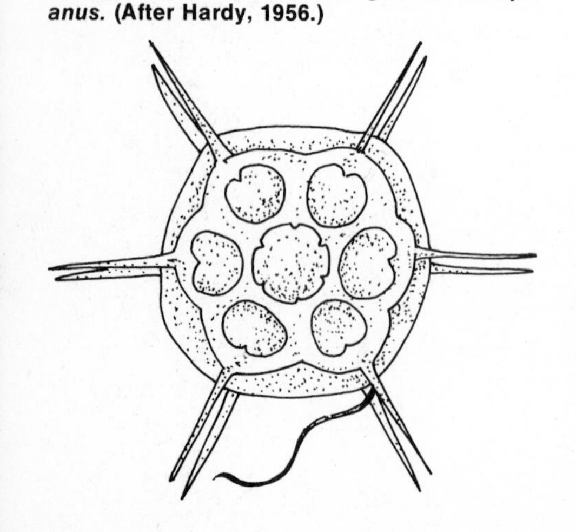

When nutrients become sparse or conditions become unfavorable for the support of great diatom populations, other phytoplankters become dominant. These are frequently small cells with whiplike flagellae that beat and move them through the water. Three major groups of flagellates are discussed below.

Silicoflagellates. These organisms are widely distributed as a minor constituent of plankton and may be quite abundant locally (Fig. 10.7). They contribute their glassy SiO_2 skeletons to the sediment when they die. In the North Sea, these chrysophytes may be the dominant primary producers among the plankton.

Coccolithophores. These tiny round flagellates are usually less than 10 microns (0.01 mm) in diameter with minute, buttonlike plates made of calcium carbonate (Fig.

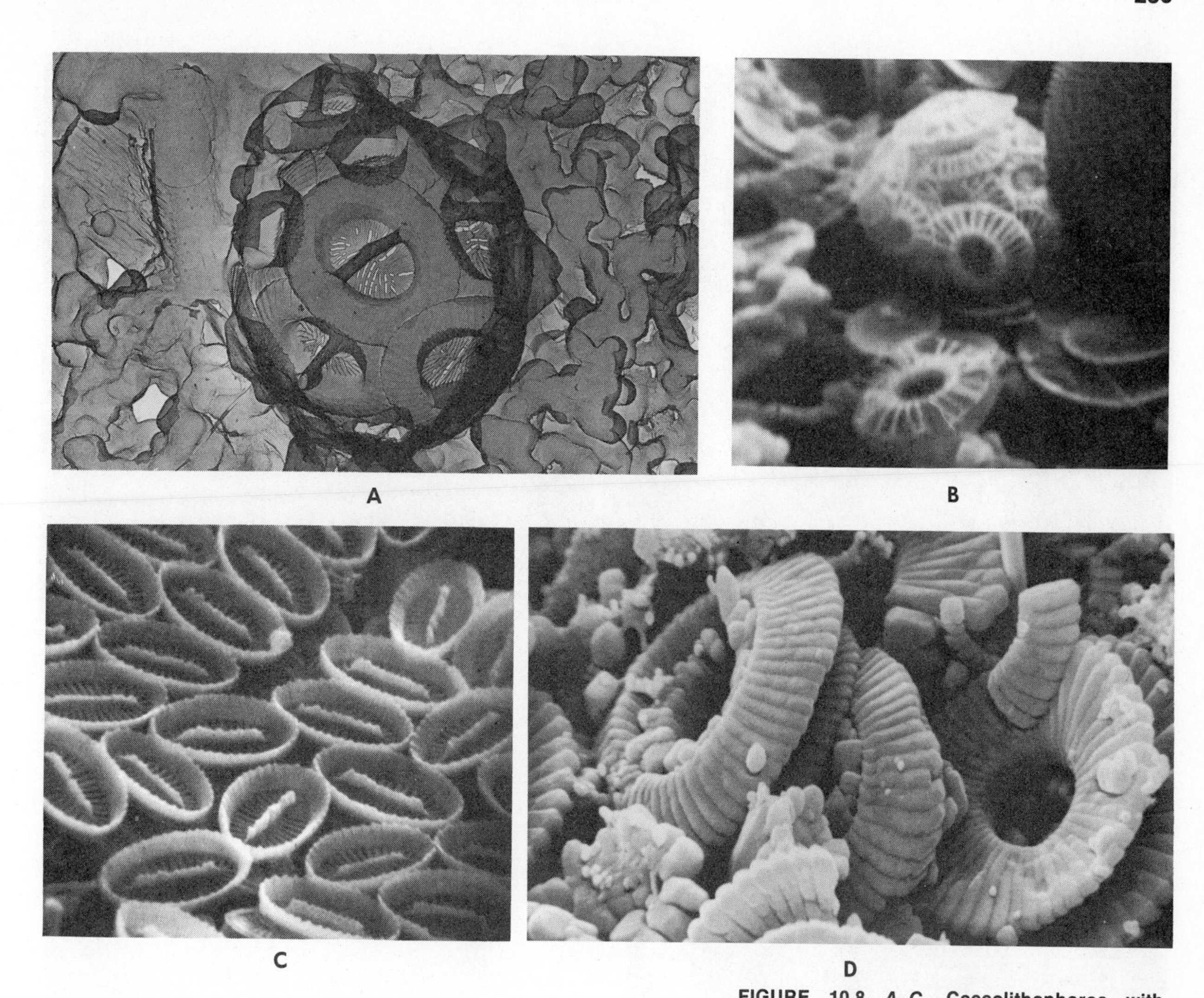

FIGURE 10.8 *A–C*, Coccolithophores with many calcareous plates, or coccoliths, sometimes referred to as "collar buttons." *D*, Collar buttons of *Coccolithus* sp. (*A*, transmission electron micrograph; *B* and *C*, scanning electron micrographs, courtesy of Susumo Honjo, Woods Hole Oceanographic Institution. *D* courtesy of Sea Library/SIO.)

10.8). These "buttons" may be abundant in the sediments of warm shallow water (calcium carbonate tends to dissolve in cold or deep water), reflecting their sometimes great abundance in the water above.

PYRROPHYTA

Dinoflagellates. Among the most abundant of the phytoplankton are the dinoflagellates (Fig. 10.9). They may be naked or armored with heavy, often ornamented cellulose plates. They move about by means of a pair of flagellae. They are more abundant in warm water but may grow into blooms in temperate water when conditions are just right and may cause "red tides," some of which are poisonous. Red tides are discussed in detail in Chapter 11.

Some dinoflagellates are bioluminescent (see Chap. 9), producing their own cool blue-white light. This light is frequently observed illuminating the surf or the wake of

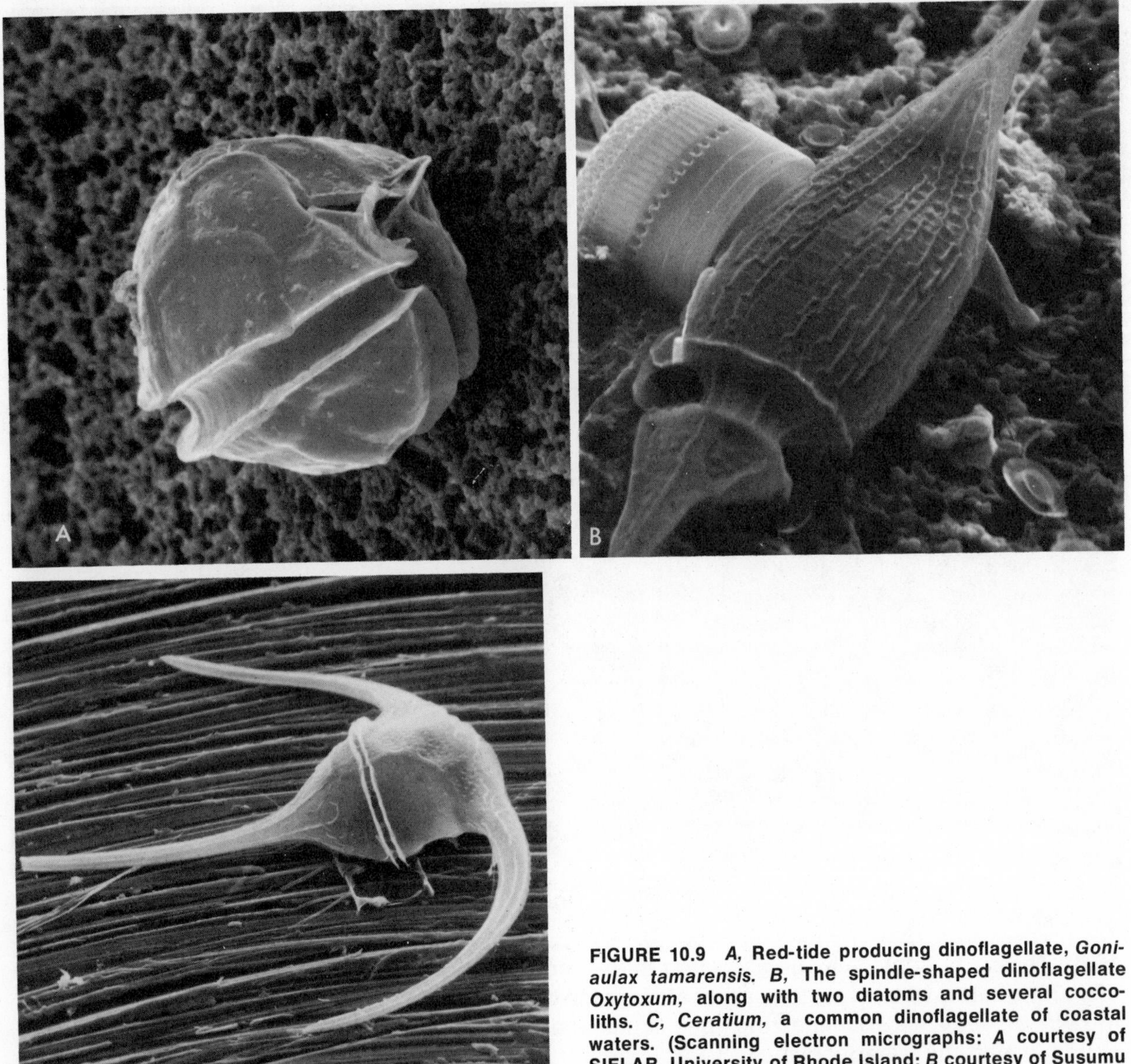

FIGURE 10.9 *A*, Red-tide producing dinoflagellate, *Goniaulax tamarensis*. *B*, The spindle-shaped dinoflagellate *Oxytoxum*, along with two diatoms and several coccoliths. *C*, *Ceratium*, a common dinoflagellate of coastal waters. (Scanning electron micrographs: *A* courtesy of SIELAB, University of Rhode Island; *B* courtesy of Susumu Honjo, Woods Hole Oceanographic Institution; *C* courtesy of Paul E. Hargraves, University of Rhode Island.)

ships on moonless nights. Their division name, Pyrrophyta ("fire plants"), is especially appropriate for these light-giving phytoplankton.

Flowering Plants (Division Anthophyta)

Flowering plants are absent in the open ocean, as they require a substrate for attachment and light for photosynthesis. However, some flowering plants may be found in the marine environment. Eelgrass is frequently seen in rather protected shallow coves with sandy or rocky bottoms. Salt marshes are usually covered with grasses that are partly submerged by the high tides. These plants contribute much detritus to the food webs of shallow estuarine bays.

CONSUMERS

Animals are consumers; that is, they require a source of organic food that they themselves cannot make. This is in contrast with the plants, which can utilize the sun's energy and inorganic nutrients for growth. Animals that feed on plants are termed herbivores (***secondary producers***). Those that eat other animals are termed carnivores.

For classification purposes, the animals are divided into ***phyla*** (similar to the divisions of plants) and the phyla into ***classes***, based mainly on structure and reproduction and in some cases on the method of feeding. Consumers vary considerably in their mode of feeding. Some of the animals in the sea are ***filter-feeders***: they strain the food (phytoplankton, zooplankton, and detritus) out of the water with featherlike appendages, gills, or tentacles around or in the mouth. These animals may be tiny, such as the copepods, or huge, as are the baleen whales. The ***scavengers***, including many crabs, feed on the dead remains of other organisms. ***Deposit-feeders*** eat the sediment and digest the organic food, living and dead, contained in the bottom materials. Many of the so-called worms survive in this manner. ***Browsers***, such as many snails, cover the surface of rocks and pilings, scraping off and eating the algae that grow thereon. ***Predators*** consume other living animals. They may actively pursue their prey, as many fish do, or they may engulf prey that come their way by chance. For example, jelly fish paralyze the unwary organism with stinging cells located on the tentacles and then eat it.

The Simple Life

These organisms, including protozoans, sponges, and the so-called jelly fish, are among the simplest animals. They have no actual brain and no blood. Therefore their behavior is relatively simple, and feeding is semiautomatic, responding to simple stimuli such as food, touch, and light.

SINGLE-CELLED ANIMALS (PHYLUM PROTOZOA)

These little one-celled animals (Fig. 10.10) are abundant in the sea, feeding on detritus, bacteria, and proba-

FIGURE 10.10 Single-celled marine animals *A*, Stalked, ciliated protozoa around the openings of "moss animals" (bryozoa) 120×. *B*, Planktonic spiny foraminiferan. *C* and *D*, Foraminifera with calcareous skeleta. *E* and *F*, Radiolaria with siliceous (opal) skeleta. (*A* courtesy of J. McN. Sieburth, from Sea Microbes. Williams and Wilkins Co., Baltimore, Md., in preparation for 1976 publication. P. W. Johnson and J. McN. Sieburth, scanning electron micrograph; *B*, scanning electron micrograph courtesy of Allan W. H. Bé, Lamont-Dougherty Geological Observatory of Columbia University; *C–D*, scanning electron micrographs courtesy of James Kennett, University of Rhode Island; *E–F*, scanning electron micrographs courtesy of ETEC, Corp.)

bly even dissolved organic matter. Ciliates are fairly common among grains of sand and among detritus debris and masses of hydroids and other benthic organisms. They move by waving tiny whiplike threads called cilia on the surface of the cell. Radiolarians and foraminiferans are amoebalike protozoans that are common in the plankton. Foraminiferans are also found in the sediment. Radiolarians have SiO_2 skeletons that resemble cages, and foraminiferans have porous $CaCO_3$ shells. Both of these kinds of skeletons or shells contribute to the formation of the sediment in the areas where they abound. (See also Chapter 2.)

SPONGES (PHYLUM PORIFERA)

Sponges are familiar to us in the form of "bath sponges," although natural sponges have been replaced almost totally by synthetic types. We use only the flexible but tough spongin skeleton as a bath sponge. The living parts are removed by drying and washing. The cells, called amebocytes and collar cells, are similar to amoebas and flagellates. The collar cells have whiplike flagellae (similar to cilia but much longer than them) that beat, driving water through canals in the sponge. As the water flows through, food is filtered and transferred to the amebocytes, which engulf and digest it.

Sponges are quite variable in shape and size, ranging from thin encrusting forms, which may be yellow, orange, red, or bluish-green, to large grayish masses one meter in diameter. Some may be quite prickly owing to tiny, sharp, calcareous or glassy spicules, which, along with spongin, make up the structural support for sponges. In the glass sponges, the spicules form an elaborate network (Fig. 10.11).

JELLYFISH AND THEIR KIN (PHYLUM COELENTERATA)

Coelenterates are a diverse group of simple animals with radial symmetry, tentacles, and soft, often jellylike bodies. They are carnivorous, stinging their prey with

FIGURE 10.11 The skeleton of the glass sponge *Euplectella*, about one half natural size. (Courtesy of the American Museum of Natural History.)

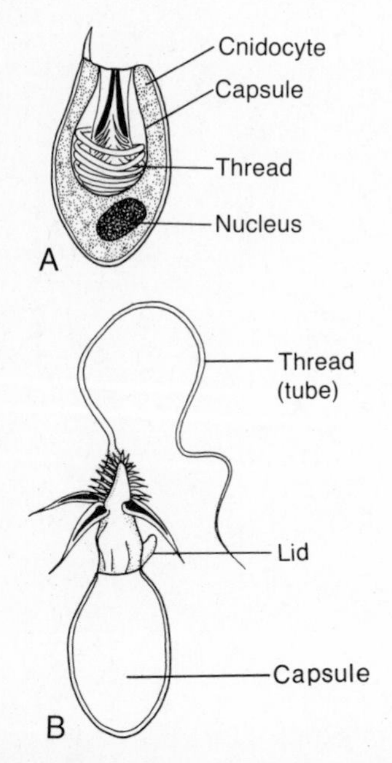

FIGURE 10.12 A Coelenterate nematocyst, its "poison dart" undischarged (*A*), and discharged (*B*). The nematocytes are microscopic and are most abundant on the tentacles and around the mouth of the coelenterate. (After Hyman, 1940.)

poison darts called nematocysts (Fig. 10.12) and then manipulating the food into their blind digestive tract with their flexible tentacles. They have two basic forms, medusae (jellyfish) and polyps (hydroids, sea anemones, and corals), as shown in Figure 10.13. A polyp is like an upside-down medusa or vice versa. Many small jellyfish have a sessile polyp stage. They have an alternation of asexual (polyp) and sexual (medusa) generations. In other words, the jellyfish that we see drifting in the water is the sexual generation. On the other hand, sea anemones and corals have only the polyp stage, which may reproduce asexually or sexually.

Coelenterates are found in all oceans and at all depths. They feed on small crustaceans, fish, or any other animals that they can ingest. Because they feed on the same food as many fish, they may be serious competitors and may deplete the supply of the fish's food. Some fish eat coral polyps, and a few arthropods such as "sea spiders" eat hydroid polyps. Otherwise they are not eaten by many animals.

Some drifting or floating jellyfish and Portuguese men-of-war produce such a strong sting that they may cause painful welts on human swimmers. The cubomedusa jellyfish, found off the coast of Australia, may even cause death.

Corals are close relatives of the sea anemones. They differ, however, in that they form a protective calcareous exoskeleton around their polyps. Many corals grow in tremendous colonies that form reefs (Fig. 10.14). Others may grow as isolated polyps.

Reef-building coral generally requires warm (greater than 18°C) clear water such as is found in the upper 20 meters of the tropics. These coral polyps contain symbiotic algae, which apparently supply dissolved organic matter that stimulates the growth of the coral. The algae benefit in having a home and nutrients excreted by the coral. The algae require light, and therefore so do the corals.

The living reef-building coral is found only at the outer edge of the mass of coral rock; the interior is made up of dead coral skeletons. On or near the surface of dead coral rock grow other algae, some filamentous, some fleshy, and others with calcareous "skeletons" of their own. This concentration of epibenthic life results in a characteristic environment dominated by the reef life but including a wide variety of fish that are adapted to feeding on the reef plants and animals.

Comb Jellies (Phylum Ctenophora)

Comb jellies are frequently seen as small globs of jelly, a few centimeters across, washed ashore by the surf. They are small, round or oval jellyfish-like animals with

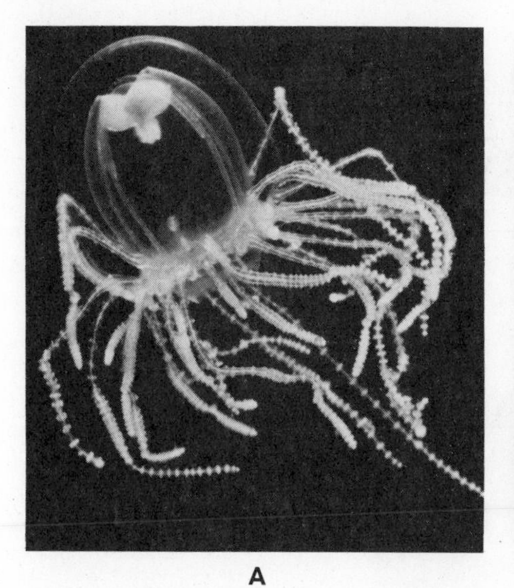

A

B

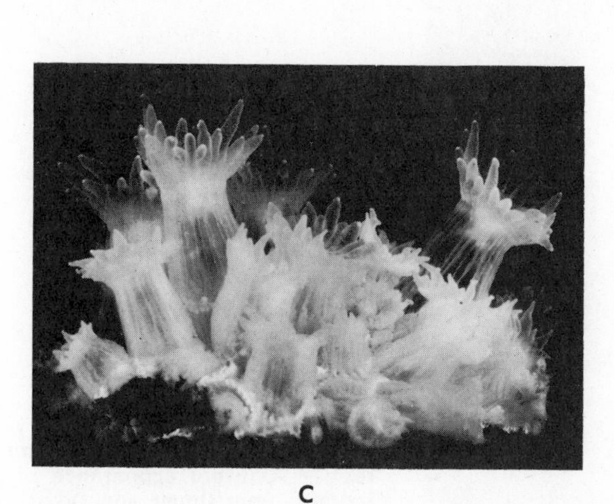

C

D

FIGURE 10.13 A variety of coelenterates. *A,* Hydromedusa, *Gonionemus. B,* A common sea anemone, *Metridium senile. C,* Polyps of the star coral. *D,* The Portuguese man-of-war, *Physalia,* eating a fish. (*A* photo © Douglas P. Wilson; *B* photo by Harold Wes Pratt; *C* courtesy of American Museum of Natual History; *D* courtesy of New York Zoological Society.)

eight vertical rows of ciliated comb-plates, their locomotive organs (Fig. 10.15).

Comb jellies are similar to medusae, ecologically. They are voracious carnivores, eaten by few other animals. They differ from the true jellyfish in that they tend to be roundish rather than bell-shaped. They have no sessile polyp stage and no stinging cells but only sticky-cells with

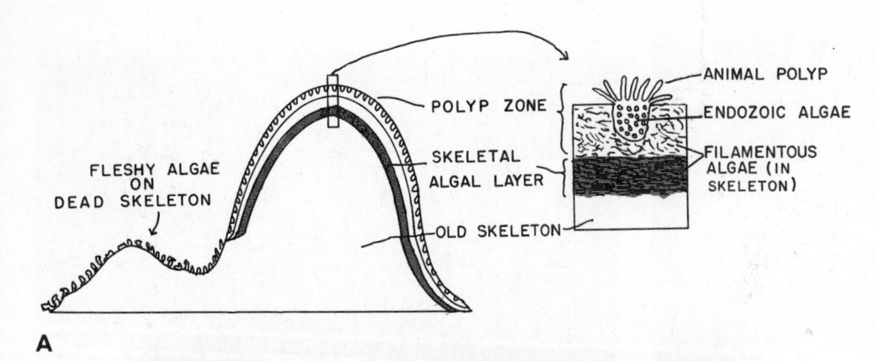

FIGURE 10.14 Coral reef. *A*, Structure of reef; *B*, Outward appearance of reef. (*A* redrawn from Odum and Odum, 1955; *B* courtesy of Amikam Shoob, Tel Aviv, University of Israel.)

which they ensnare their prey. Their distribution is very patchy, and hence they appear to be absent one day and abundant the next. When they are abundant, they may foul fishing nets and decimate the food of fish.

Worms

What we commonly call "worms" really comprise a number of phyla that share similar shape and habitats.

FIGURE 10.15 A comb jelly, *Pleurobrachia.* This one is also known as the sea-gooseberry. (Photo by Harold Wes Pratt.)

Most are long and slender and live in the mud or sand. Some live in tubes that they construct out of sand or other materials. Several of these phyla include worms that are parasitic or have entered into commensal relationships with other animals. Their shape is well adapted for crawling in and out of small spaces. A few, including some flat worms, segmented worms, and arrow worms, have abandoned their benthic ways and have assumed a pelagic swimming life.

FIGURE 10.16 The flat worm *Bdelloura,* a commensal on the gills of the horseshoe crab *Limulus.* (Courtesy of Ward's Natural Science Establishment.)

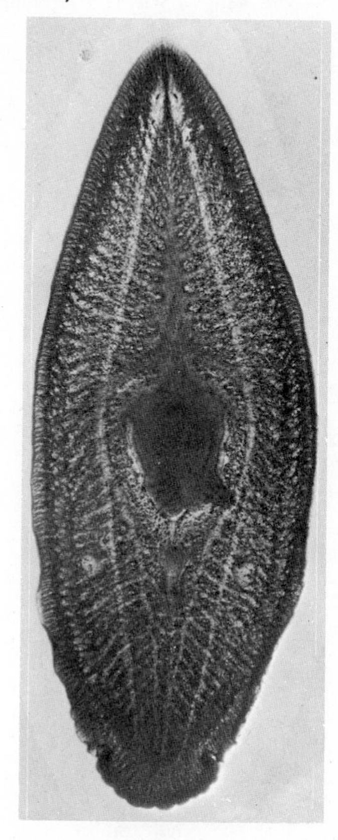

Flat Worms (Phylum Platyhelminthes)

Flat worms (Fig. 10.16) are fairly common on or under rocks, in shallow water, and on the sediment. They also crawl around benthic growths of hydroids and other small attached organisms such as moss animals and tunicates. Flat worms are mostly carnivorous and may eat dead fish, clams, tunicates, and other soft-bodied sessile animals. One genus, *Bdelloura* lives on the gills of the horseshoe crab as a commensal. Most flat worms are small, but they may reach five centimeters in length and some may swim through the water by undulating their leaf-shaped bodies. Although many flat worms live in fresh water, more are marine. Most of the marine species, however, live as parasites in fishes and other marine animals.

Round Worms (Phylum Nemathelminthes)

These are small, round, unsegmented worms with ends that are generally pointed. They survive well under conditions that would be lethal to most animals. Round worms are important in the decomposition of detritus and are common in sediments that are rich in organic matter but which contain little oxygen. Many species of round worms are parasites of other organisms.

Segmented Worms (Phylum Annelida)

Most of the segmented worms (Fig. 10.17) in the sea fall into the Class Polychaeta, having many spines that emerge from lateral projections on each segment. They may be predators, filter-feeders, or deposit-feeders. The predaceous worms have strong jaws on the head. Filter-feeders may use nets composed of a ring of featherlike tentacles to capture their prey. Hence they have popular names such as "plume worms, "fan worms," or "feather duster worms." The plume worms live in mucus-lined burrows or in tubes that the worm makes out of calcium carbonate, sand, or debris cemented by a mucus that it secretes. Some tube worms form sandy tubes that accumulate in cemented masses and form distinctive reefs which provide homes for many other organisms.

Deposit-feeders (Fig. 10.18) consume detritus that may be found on or in the sediment. They may use tentacles to pick up the food or may eat the sand or mud and then digest the food that it contains. Some, such as the lug worm *Arenicola,* leave coils of sandy feces at the rear end of the burrow. It pumps water in at the tail end of the burrow and forces it through the sand at the head end. Food is filtered out by the sand, which is then eaten by the worm. Then the worm backs up to the burrow opening to defecate.

Some polychaetes share their burrows in a commensal relationship with other organisms. One such relationship exists among the tube-dwelling, filter-feeding polychaete *Chaetopterus* and small crabs that share the worm's burrow and food. The worm apparently is neither benefited nor harmed by the crab. The movements of the worm (Fig. 10.19), however, bring in plenty of food for both the worm and the crab.

Arrow Worms (Phylum Chaetognatha)

Arrow worms are among the most voracious predators for their size (about 1 to 2 cm). They may compete with small fish for their copepod food or may even eat fish larvae that are as long as the arrow worms themselves (Fig. 10.20). They can dart forward and grasp their prey with strong hooks attached to their head. They then manipulate the prey into the stomach, sometimes bending a fish larva in half. They are common in the epipelagic zone of most ocean water.

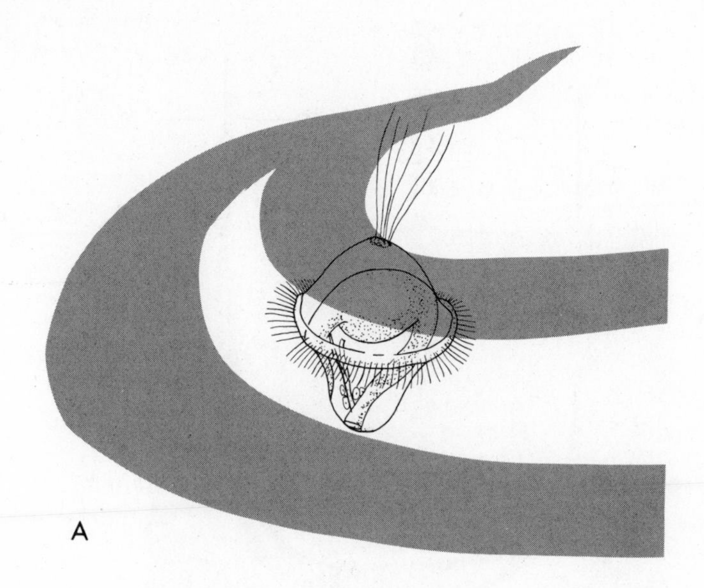

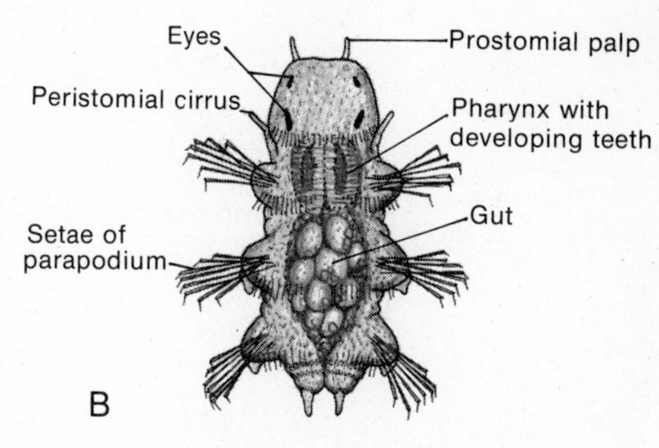

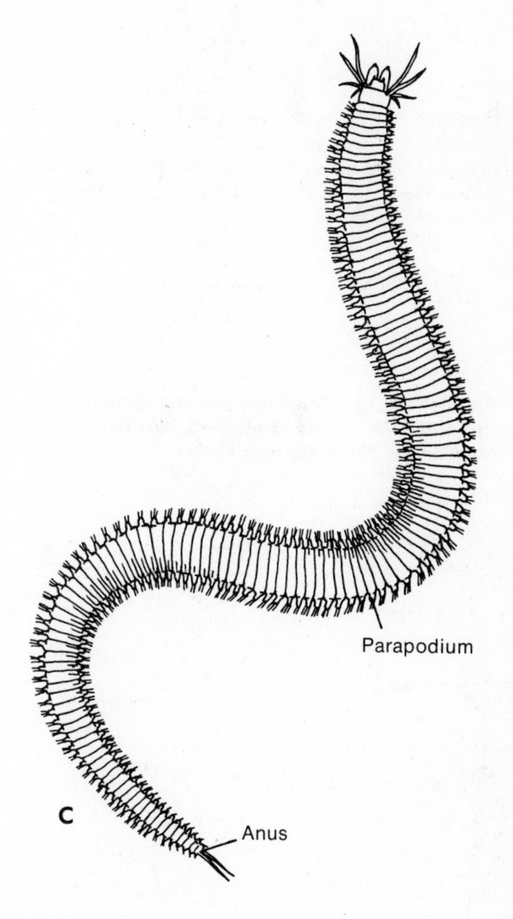

FIGURE 10.17 Larvae and adult of a common predaceous polychaete worm, *Nereis,* the "clamworm." *A,* The trochophore larva develops into *B,* A nektochaete larva, which matures into *C,* An adult polychaete. *D,* A closer view of the head. *E,* Same, with jaws extended. (From Villee et al., 1973.)

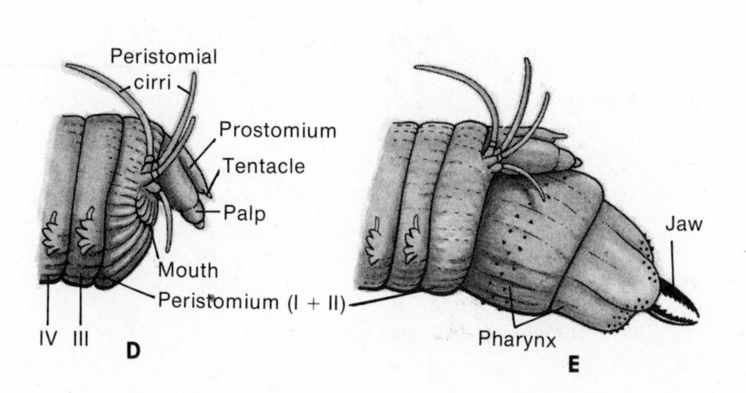

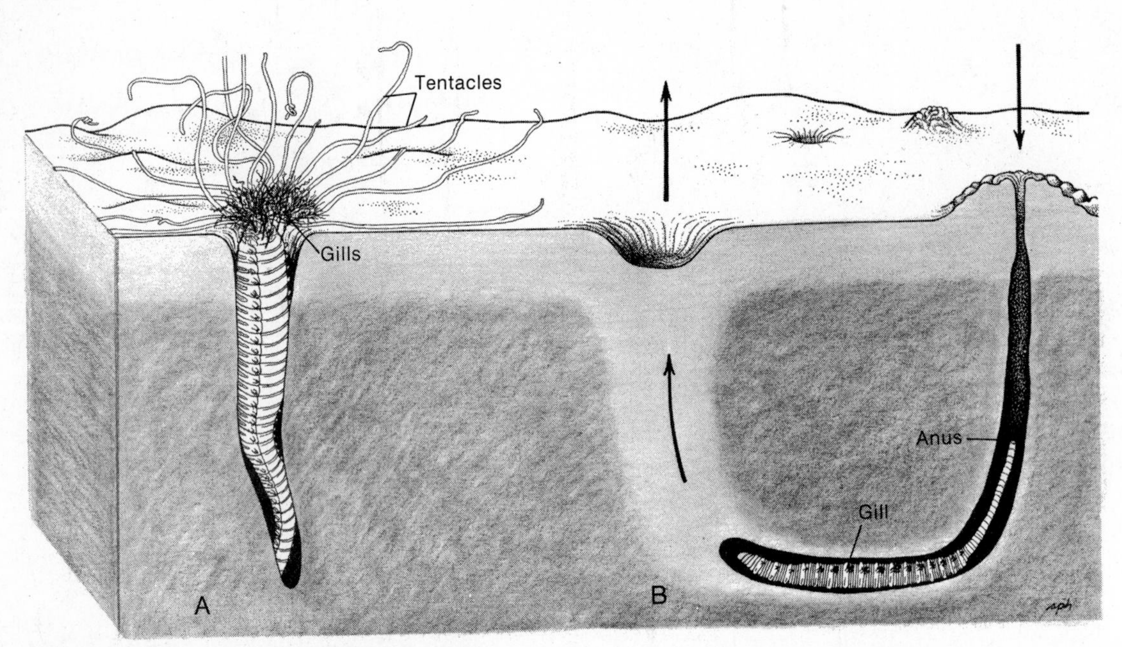

FIGURE 10.18 Two deposit-feeding polychaetes. *A, Amphitrite; B,* The lugworm, *Arenicola.* Arrows indicate the direction of water flow. (*A* modified from Wells; *B* after Villee et al., 1973.)

FIGURE 10.19 The filter-feeding polychaete *Chaetopterus* in its U-shaped burrow. (After MacGinitie. From Barnes, 1974.)

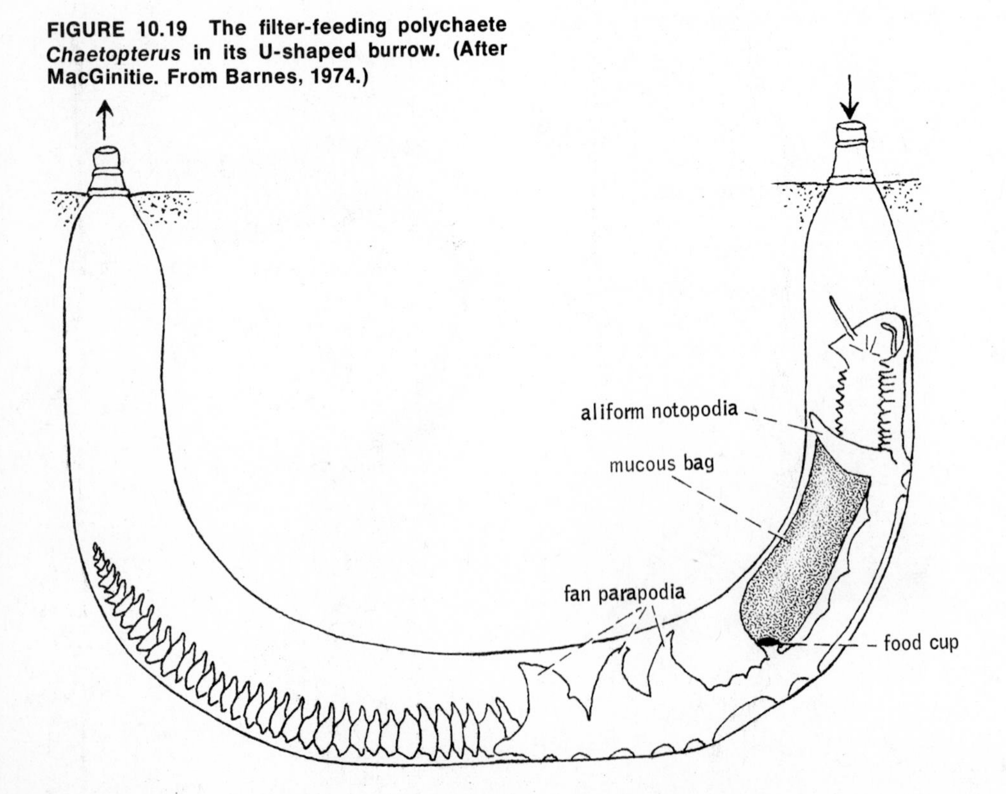

Sausage Worms (Phylum Echiuroidea)

Most of these saclike worms are found in shallow water, but some may inhabit depths greater than 8,000 meters. The majority feed on detritus trapped by a mucus secretion. They have no eyes or sense of hearing. Even though they live at a wide range of depths, including the deep aphotic zones, they have sexual reproduction. Some species have evolved an interesting way of getting the male and female together (Fig. 10.21). If a larva settles in an area of the bottom that has a moderate to low carbon dioxide (CO_2) concentration, it develops into a female. If it settles in an area of high CO_2 concentration, it develops into a male. An actively respiring female would have a "cloud" of CO_2 produced around it. This causes a settling larva to develop into a male. The male grows and becomes parasitic on the female sausage worm, ensuring a high probability of reproductive success.

Peanut Worms (Phylum Sipunculida)

Peanut worms are fairly common, burrowing in mud and sand or under rocks, from the intertidal zone to depths greater than 4,500 meters. Most feed on detritus trapped in mucus on their tentacles, but a few species are predators on other worms. Most peanut worms are less than 30 cm in length (Fig. 10.22).

Acorn Worms (Phylum Hemichordata)

Acorn worms are mostly wormlike animals that burrow in mud and sand or hide under rocks. They are thought by some to be related to the chordate phylum, which includes fish, because of the presence of gill slits. They are usually between 10 and 50 cm in length, with a stubby proboscis followed by an overlapping collar, giving the forward end the appearance of an acorn (Fig. 10.23). The mouth is located between the proboscis and the collar. Plankton, detritus, and sediment are trapped by mucus on the proboscis and carried to the mouth by the beating of cilia.

Larger Invertebrates

Phylum Mollusca

One of the characteristics shared by this phylum is a soft body frequently covered by a protective shell of $CaCO_3$. A tissue called the ***mantle*** enshrouds the internal organs of the animal and produces any shell it may have. Most mollusks have a muscular "foot" that provides them

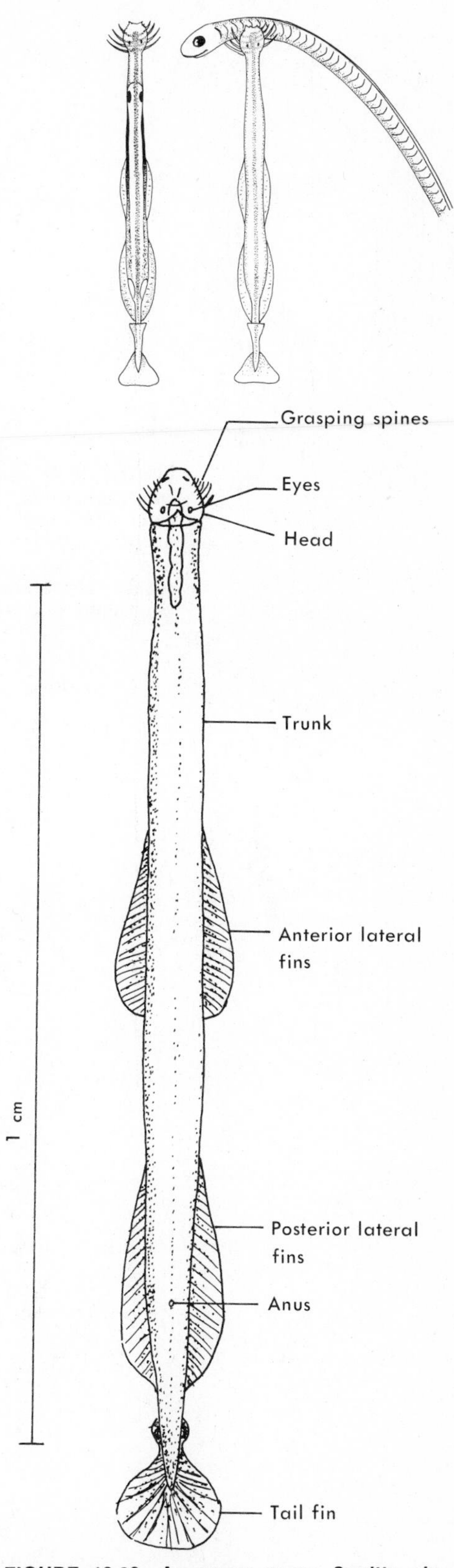

FIGURE 10.20 **An arrow worm, *Sagitta elegans*. (After Hardy, 1956.)**

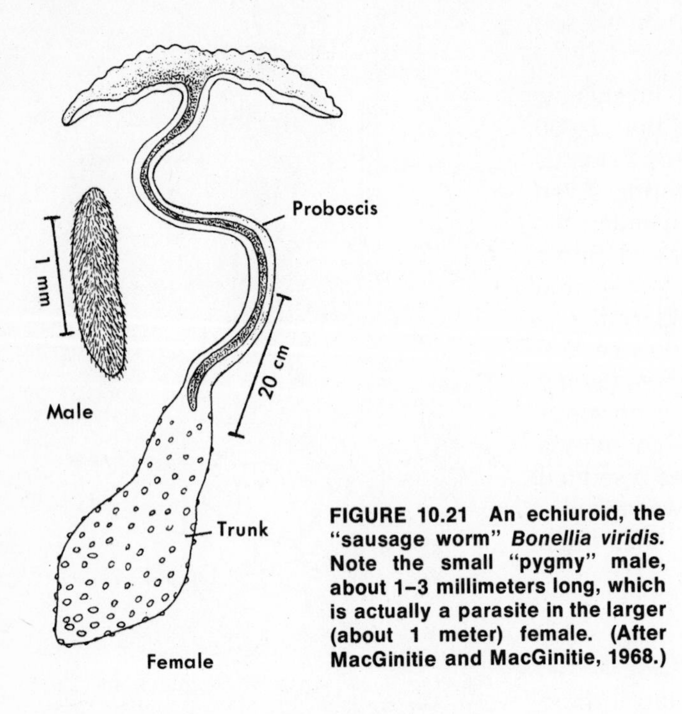

FIGURE 10.21 An echiuroid, the "sausage worm" *Bonellia viridis.* Note the small "pygmy" male, about 1–3 millimeters long, which is actually a parasite in the larger (about 1 meter) female. (After MacGinitie and MacGinitie, 1968.)

with locomotion. Most are adapted to life in the benthos, where they burrow into sediment, crawl over the bottom, or are attached to rocks and other substrata (Fig. 10.24). However, in the squid and octopi, the foot has been modified into arms and a jet propulsion siphon, which enables them to swim through the water.

In Table 10.1 are presented characteristics of the most common classes of mollusks.

Snails, limpets, and chitons are generally browsers, crawling over the substrate, scraping food into their mouths with a file-like radula (Fig. 10.25). Some snails, such as the oyster-drill, use their radula to bore through the shells of oysters and clams and then devour their flesh. Others, such as the mud snail *(Nassarius obsoletus),* are scavengers and deposit-feeders. They thrive in great numbers on shallow mud flats.

Most clams are benthic filter-feeders, drawing in water through an inhalant siphon and expelling it through an exhalant siphon (Fig. 10.24). As the water passes over the gills, food particles such as phytoplankton are filtered out. Most of the economically important clams and oysters live on sandy or muddy bottoms and are thus easy to collect in large numbers. Many others, such as mussels, live on rocky substrates. Yet others have evolved the ability to burrow into rock or wood, thus achieving a great deal of protection.

The wood-boring shipworm (Fig. 10.26) has lost its ability to filter-feed and now eats the sawdust that it produces as it bores into wood. Shipworms may become abundant in untreated wood, causing millions of dollars' worth of damage.

In the cephalopods, which include the octopus, squid, cuttlefish, and nautilus, the foot has evolved into tentacles or arms with suction cups (Fig. 10.27). Of these, only the nautilus has an external shell. In the cuttlefish, the shell has been reduced to a calcareous internal "cuttle bone." Cuttle bone is often hung in the cages of pet birds for the birds to peck and sharpen their beaks. The squid has a thin flexible "pen" and the octopus has lost all traces of a shell. The cephalopods are thought to be the most advanced of the mollusks. They have well-developed brains and eyes and are generally predaceous. Squid and even the clumsy-looking octopus (Fig. 10.28) may become quite active in pursuit of their prey.

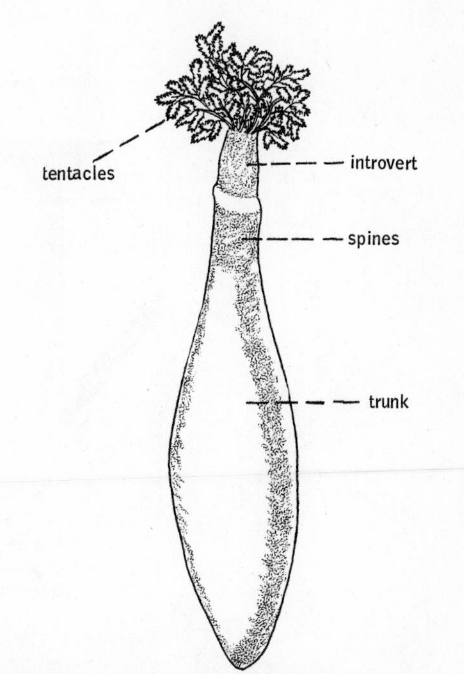

FIGURE 10.22 A typical peanut worm, *Dendrostomum*, about 12 centimeters long. (After Fisher, from Tétry, 1959.)

PHYLUM ARTHROPODA

The arthropod phylum occupies one of the crucial links in the marine environments. The greatest diversity is found in the Class Crustacea, which includes the copepods, shrimp, lobsters, crabs, and sand fleas. They inhabit all regions of the seas and live as herbivores, carnivores, scavengers, and parasites. Along with the mollusks and fish, they are of great importance to us as a source of food.

Arthropods are characterized by jointed appendages and a hard protective exoskeleton covering their bodies. The life cycle of many arthropods involves several distinct stages (Fig. 10.29). When crabs molt, or shed their old skeleton, they lose their protection and are especially vulnerable and become ready prey for other animals. Many compensate in part for this vulnerability by living in the sand or among rocks.

Copepods, the "cattle of the sea" (Fig. 10.30*a* and *b*), are mostly less than 2 millimeters in length, but some are as long as a large grain of rice, about 5 millimeters. In spite of their small size, they are among the most important herbivores in the sea, providing a key link between the phytoplankton which they consume and the fish that eat them.

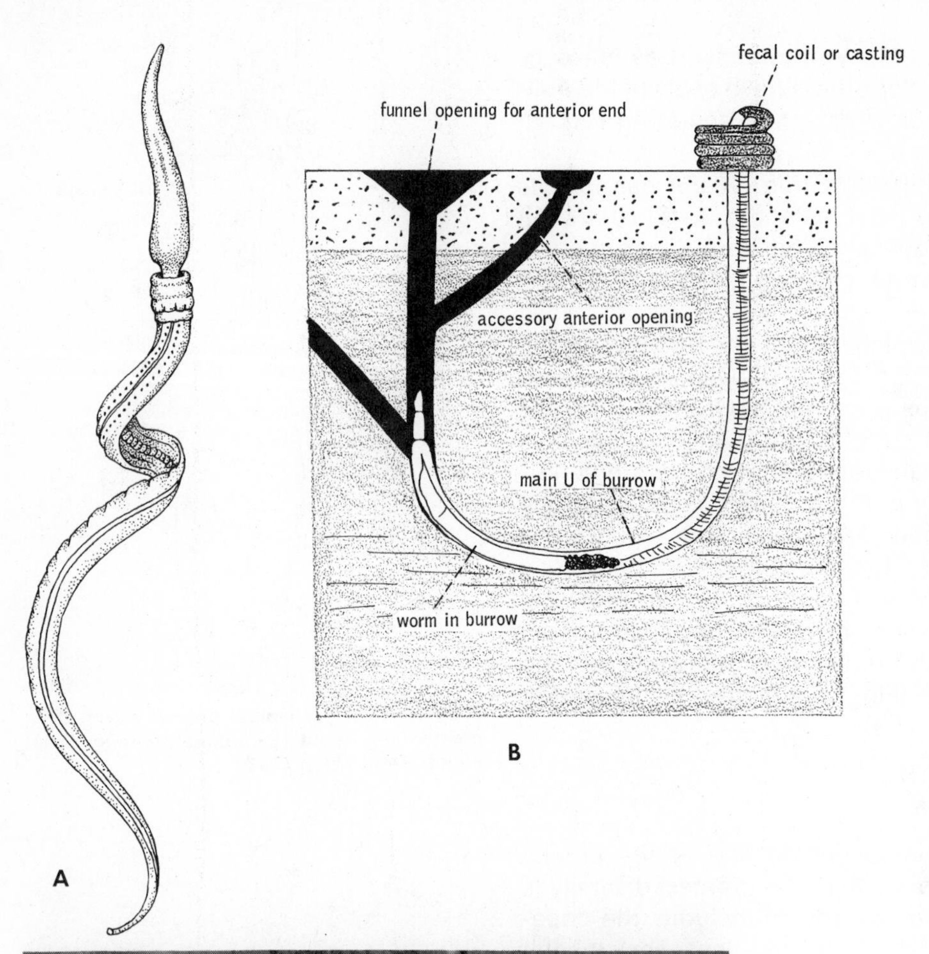

FIGURE 10.23 *A,* A shallow-water acorn worm, *Saccoglossus;* B, the burrow system of an acorn worm; *C,* An abyssal acorn worm with its trail of feces. (*B* after Stiasny from Hyman, 1959; *C* courtesy of Lamont-Doherty Geological observatory of Columbia University.)

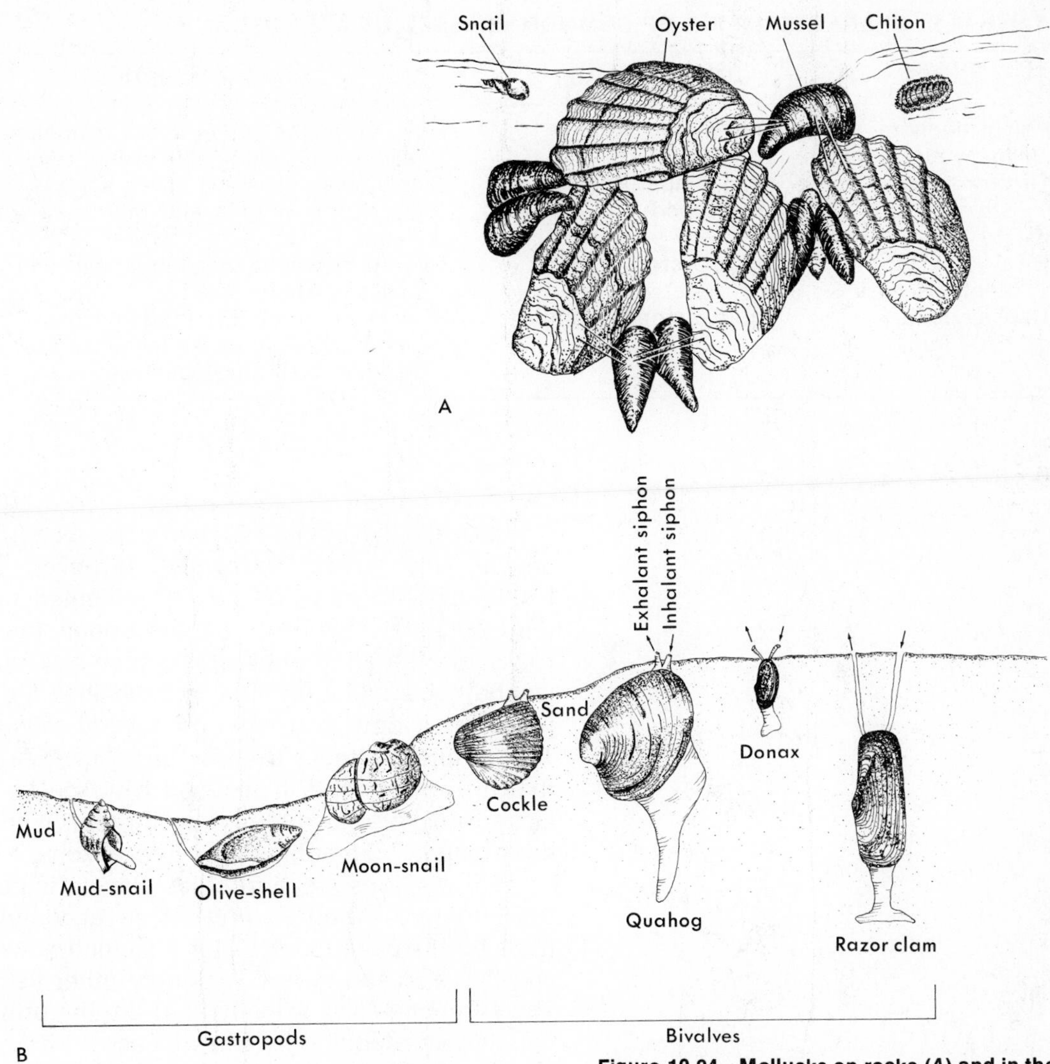

Figure 10.24 Mollusks on rocks (*A*) and in the sea-floor sediments (*B*). (*A* modified after Villee et al., 1973; *B* after Pearse, Humm and Wharton, 1942.)

Copepods swim through the water with jerky movements caused by their thoracic appendages (those on the midsection, directed downward in Fig. 10.30*b*). When they stop swimming, the antennae project outward and away from the body, thus retarding sinkage. Feeding is usually accomplished by filtering the water with the smaller hairy appendages around the mouth. These legs set up water currents that draw plankton, including diatoms, dinoflagellates, and the tiny larvae of other animals, toward the mouth. The food is caught in the hairs of some of the legs and transferred by others into the mouth. Although most copepods are herbivores, some may be carnivores when phytoplankton is scarce. Other copepods that live in the less productive oceanic water or in the deep sea rely on zooplankton and detritus as a source of food.

Table 10.1 CHARACTERISTICS OF COMMON CLASSES OF MOLLUSCA

CLASS	COMMON NAME	CHARACTERISTICS
Polyplacophora "many plates"	chiton	Row of plates along back, sometimes covered by mantle; gills along side of foot
Gastropoda "stomach foot"	snail, limpet, sea slug, pteropod	Single shell, usually a coiled spire (snail), or uncoiled (as in the limpets and pteropods) or missing (sea slugs); digestive tract twisted in most types
Bivalvia "double shell"	clam, oyster	Two shells, joined by a hinge; gills and foot enclosed by shells when closed
Cephalopoda "head foot"	squid, octopus, nautilus	Shell coiled (nautilus), internal (squid), or missing (octopus); eight or ten tentacles with suction cups; generally well-developed eyes

As with the copepods, barnacles begin their pelagic life as tiny larvae. Barnacles, however, are destined to mature into attached adults cemented to a substrate (Fig. 10.30*c*). This may be a wooden piling, intertidal rocks, or even another organism such as a whale. The typical barnacle has a hard calcareous shell around it with a hinged trap door consisting of several calcareous plates. The barnacle is really a rather ordinary crustacean except that it is positioned in its shell lying on its back, and it kicks food into its mouth with its feet. It functions as a filter-feeder, commonly in the littoral zone.

Shrimp, lobsters, and crabs play an important role as scavengers, helping to prevent an accumulation of detritus on the ocean floor. They also feed on worms, clams, and fish and are, in turn, eaten by other fish and people. The shrimplike krill (Fig. 10.31*a*) are the principal food of the plankton-feeding baleen whales.

The isopods and amphipods (Fig. 10.31 *b* and *c*) are common small crustaceans, usually less than 2 centimeters in length. They occupy similar niches in the benthic environments, especially in littoral communities. They differ in that the isopods have legs that are similar in size and shape (like the terrestrial pill bug), whereas the amphipods have two kinds of legs, dissimilar in size and shape. Some of the amphipod legs are aimed forward and some are aimed backward, enabling amphipods such as sand fleas to hop about on the beach. The food of many isopods and amphipods consists of algae and soft-bodied invertebrates such as hydroid polyps, among which they crawl. Some isopods, such as the wood-boring gribble (*Limnoria*), may cause damage to marine pilings. This is a role similar to that of the molluscan shipworm.

Sea spiders (pycnogonids) are not true spiders but resemble them in a superficial way, having long legs that enable them to crawl over and prey upon colonies of hydroids and bryozoans that abound on littoral rocks (Fig.

THE PROLIFIC COPEPOD

Some widely distributed species of copepods may produce one to five generations per year, making them very prolific herbivores. In fact, a male and female of the copepod *Tisbe furcata* could produce four generations in 100 days, resulting in a total of more than one billion progeny *if* all of the offspring survived. However, most of them die or are eaten by carnivores such as small fish and jellyfish. Thus copepods such as *Tisbe* are an important link between the phytoplankton and the pelagic carnivores.

Copepods may be very abundant in highly productive surface waters. There may be more than 25,000 of these little animals per cubic meter of neritic ocean water. They provide a ready source of food for the small carnivores such as anchovies and herring.

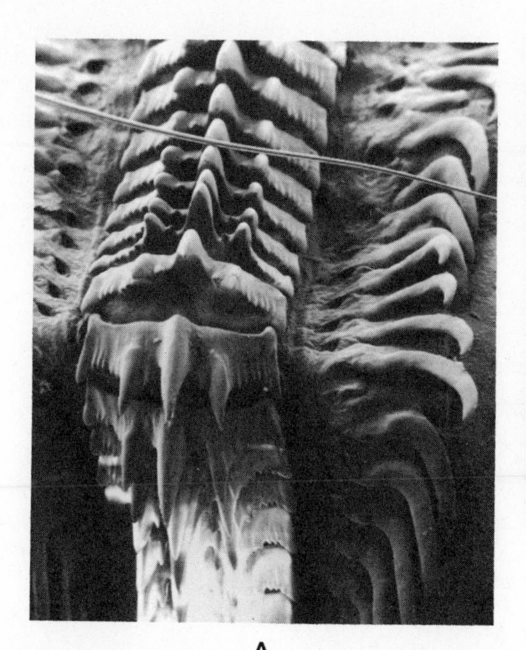

A

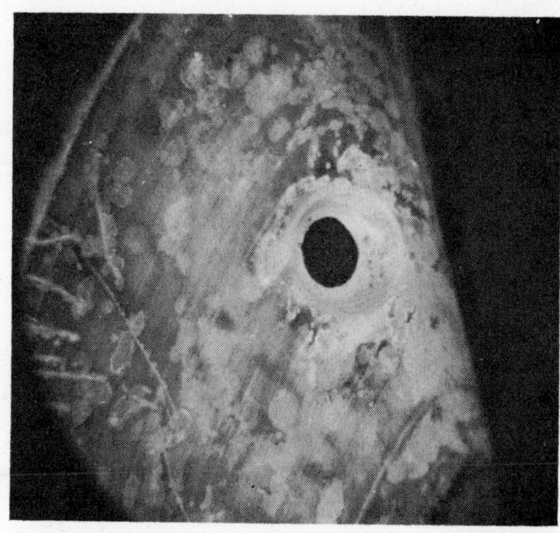

B

FIGURE 10.25 *A,* The file-like radula of a boring snail. *B,* A hole in a bivalve shell, drilled by a boring snail. (*A* scanning electron micrograph from Carriker, 1969; *B* photo by Betty M. Barnes.)

10.31*d*). The littoral species are usually less than a centimeter in leg span. Those found in the abyssal zone, however, may have a leg span of over 60 centimeters (two feet).

The common Atlantic Coast horseshoe crab *(Limulus)* is not a true crab (Fig. 10.31*e*). It is probably more closely related to spiders, because of similarities in their feeding appendages. It is another scavenger in the littoral zone, also feeding on worms and small clams. Horseshoe crabs bear a striking resemblance to the early arthropods such as the extinct trilobite, suggesting that they have evolved little over a period of millions of years. Because they are not economically important and have few natural enemies, they may survive for millions of years more if their home is not destroyed by human activities.

PHYLUM ECHINODERMATA

Members of this phylum are generally found in shallow water, but there are echinoderms at all depths, including the trenches. They usually have spiny skins, radial symmetry, and five sets of short arms. However, the body shape and location of arms vary considerably from one class to another (see Table 10.2 and Fig. 10.32).

Sea stars are especially familiar to visitors of intertidal rocky areas. They are basically carnivorous. Most feed on living or dead animals such as crabs, clams, snails, or dead fish. Some can even remove hermit crabs from their

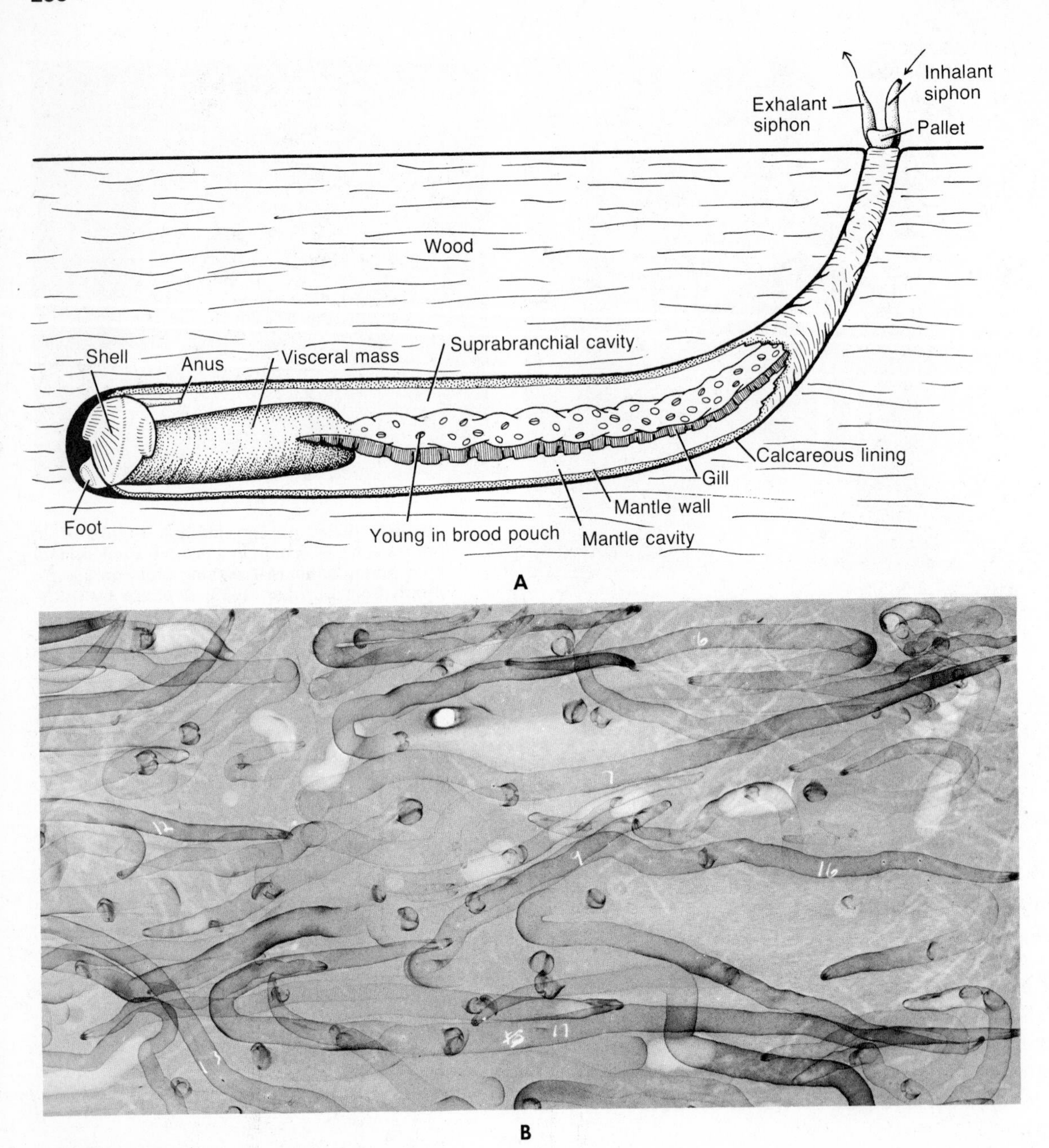

FIGURE 10.26 The work of a wood-boring shipworm. *A*, **Detail of the bivalve mollusk** ***Teredo. B,*** **X-ray of the shipworms in wood. (*A* from Villee, et al, 1973; *B* photo by C. E. Lane.**

shells and eat them. Many, however, feed on organic matter in the sediment or even on suspended detritus and plankton. They are found over a wide range of habitats, especially along rocky coasts, where they may feed on clams and mussels.

Sea urchins and sand dollars have no arms as sea stars do, but their calcareous shell is covered by many

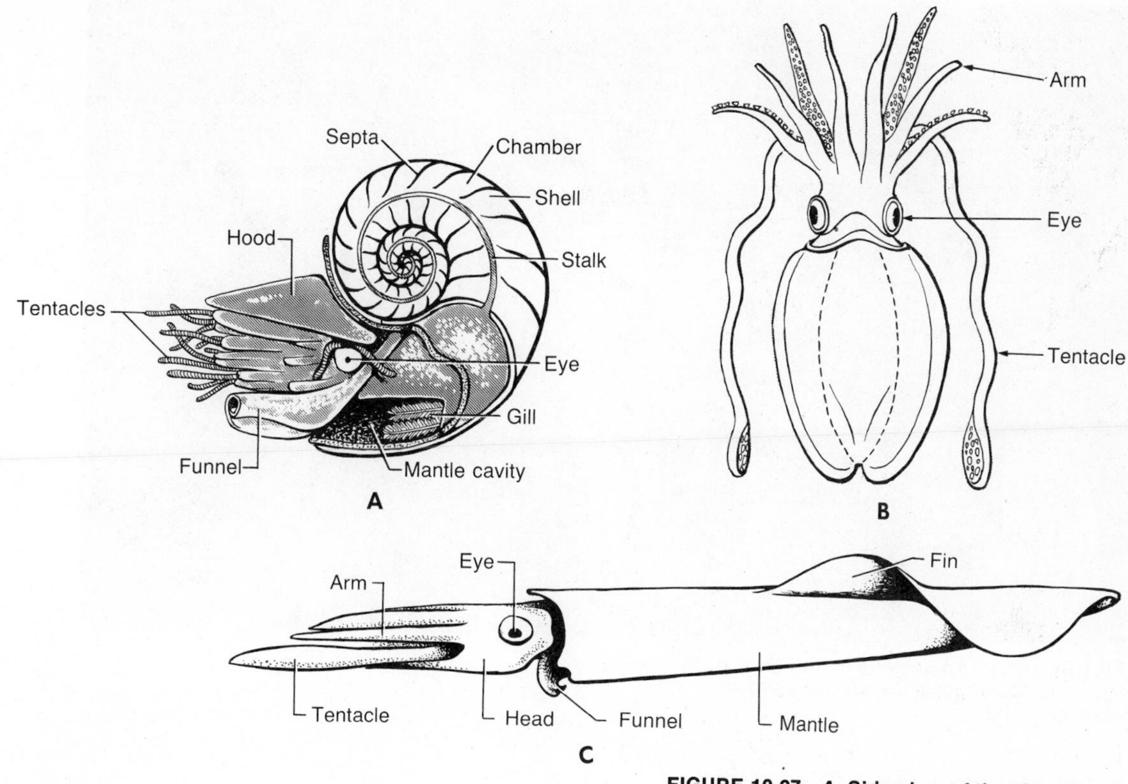

FIGURE 10.27 ***A,*** **Side view of the chambered nautilus, showing internal anatomy.** ***B,*** **Top (dorsal) view of a cuttlefish.** ***C,*** **Swimming squid. (From Villee et al., 1973.)**

spines, which are used in locomotion over the sea floor. Sea urchins in shallow water usually feed on dead or living algae and animals living on rocky substrates. Sand dollars and deep-sea urchins usually eat detritus from the sea floor.

Brittle stars have long, slender arms used in locomotion and feeding, and small mouths surrounded by five jaws, used in chewing. They usually feed on small benthic animals and detritus, although some with branched arms can capture relatively large and active crustaceans. A few even capture suspended organic matter by waving their arms about and then eat the food that adheres. Some of the largest forms are found in the deep sea, where they browse on the detritus on the sediment surface.

Sea cucumbers may be found at any depth, from the littoral zone to the deep sea. They have been found on the floor of the Peru-Chile Trench at over 6,000 meters. At depths of more than 8,000 meters in the Kurile Trench, sea cucumbers may constitute over 80 per cent of the fauna by weight. Sea cucumbers may be scavengers, mud-eaters, filter-feeders, or browsers. Many browsers of the deep sea crawl over the sediment and pick up detritus

FIGURE 10.28 An octopus in pursuit of a crab (By Fritz Goro, courtesy of *Life* Magazine, © 1955 Time, Inc.)

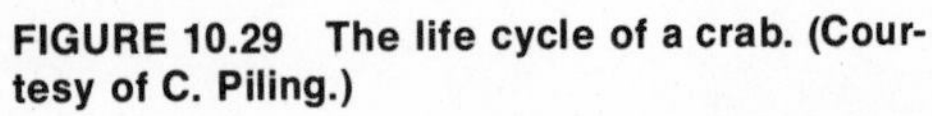

FIGURE 10.29 The life cycle of a crab. (Courtesy of C. Piling.)

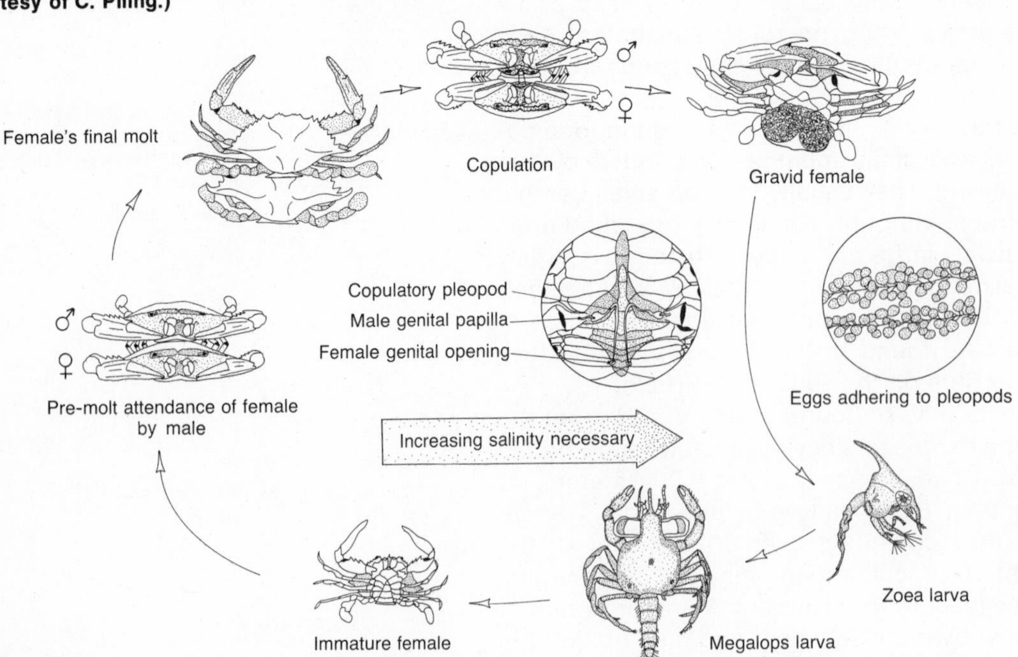

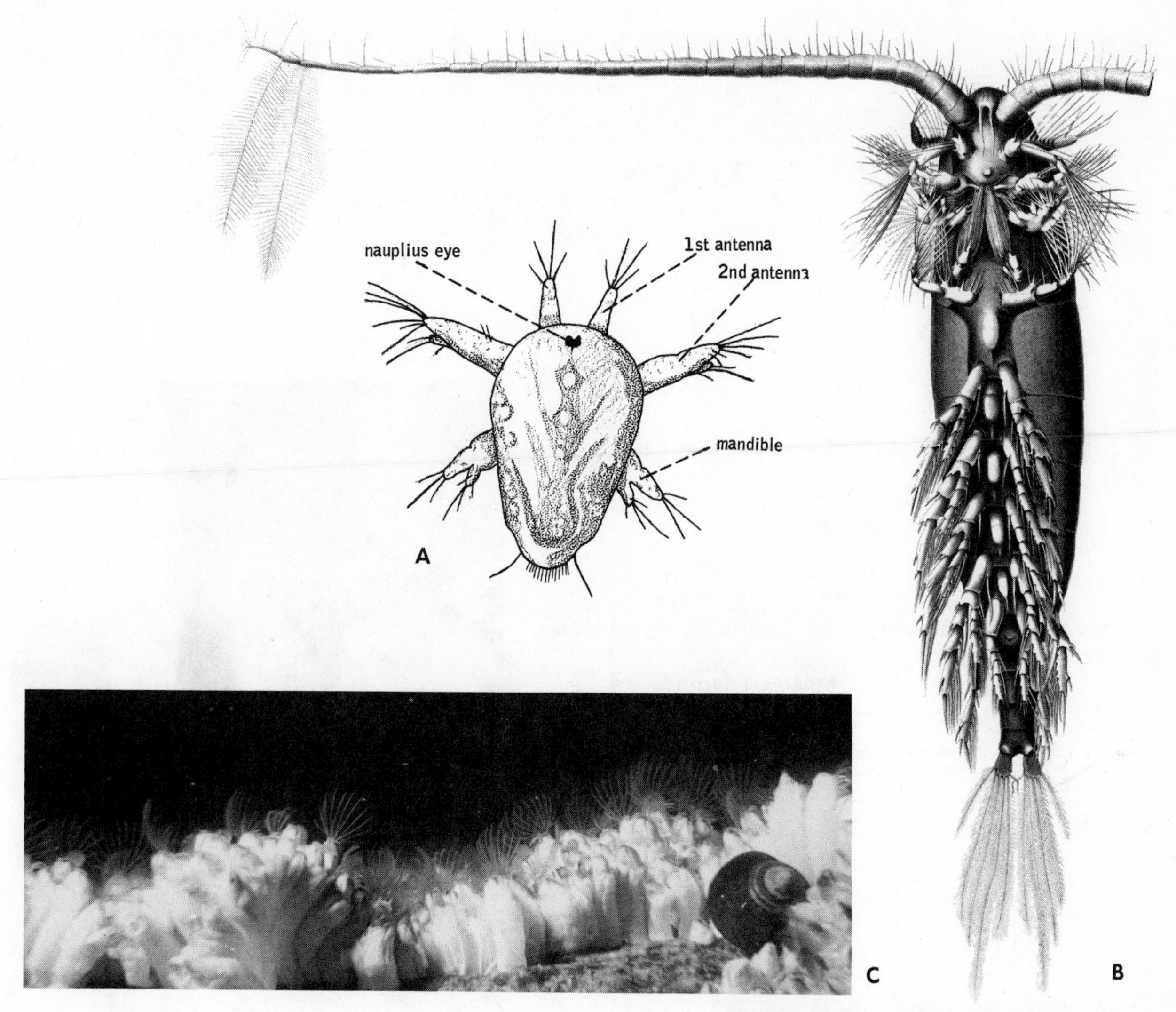

FIGURE 10.30 *A*, Typical copepod larva. *B*, Adult copepod *Calanus*. *C*, Barnacles feeding. (*A* after Green, 1961; *B* after Giesbrecht, 1892; *C* photo by Harold Wes Pratt.)

like a vacuum cleaner. They are frequently found browsing among deep-sea brittle stars (Fig. 10.33).

Sea lilies, or crinoids, may be common in sublittoral and deeper zones. Many are attached to the substrate by a stalk and resemble plants, but others are quite mobile and may even swim. They are apparently all suspension-feeders, eating detritus and plankton that get stuck in mucus on their branched arms. The food is carried toward the mouth in slender grooves that run along the arms. Most species of sea lily are extinct. The general shape and method of feeding is thought to be typical of the earliest echinoderms that lived hundreds of millions of years ago.

Text continues on page 266

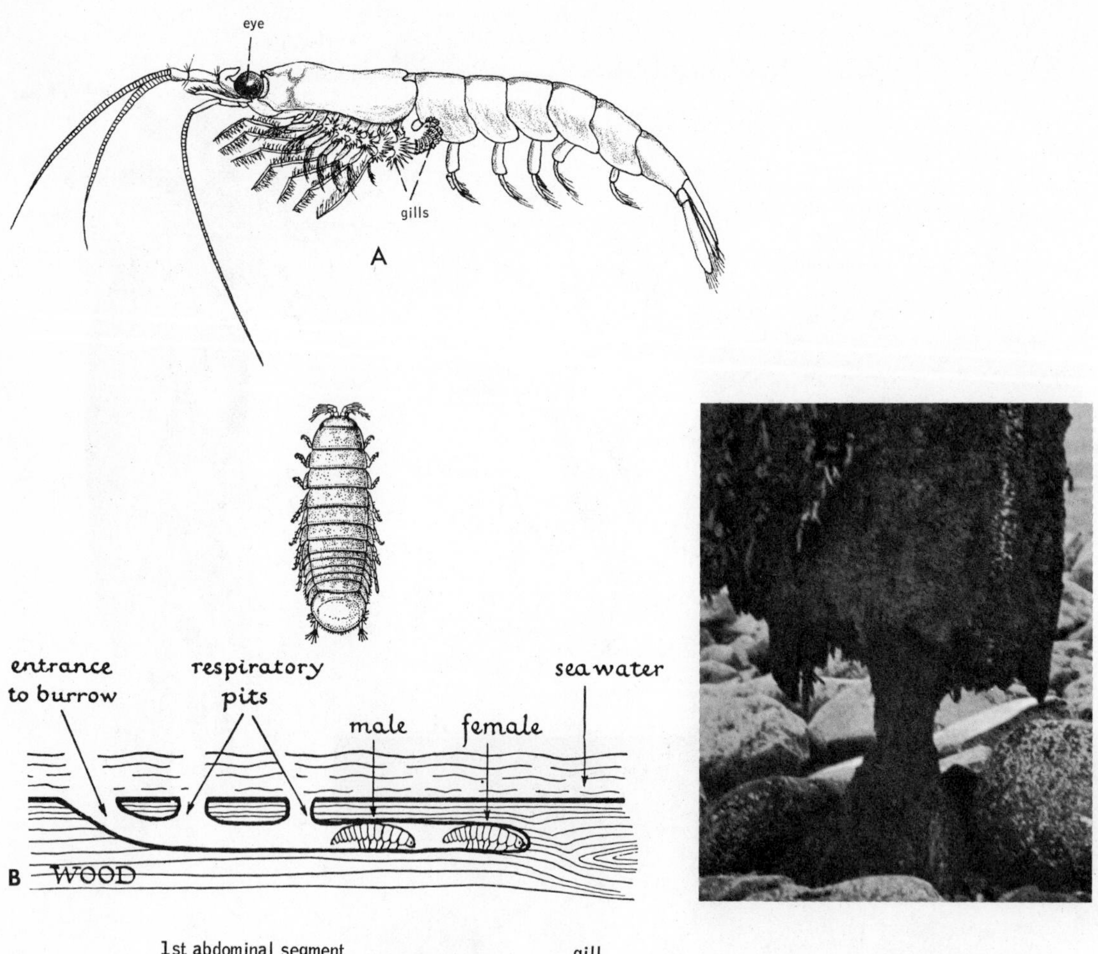

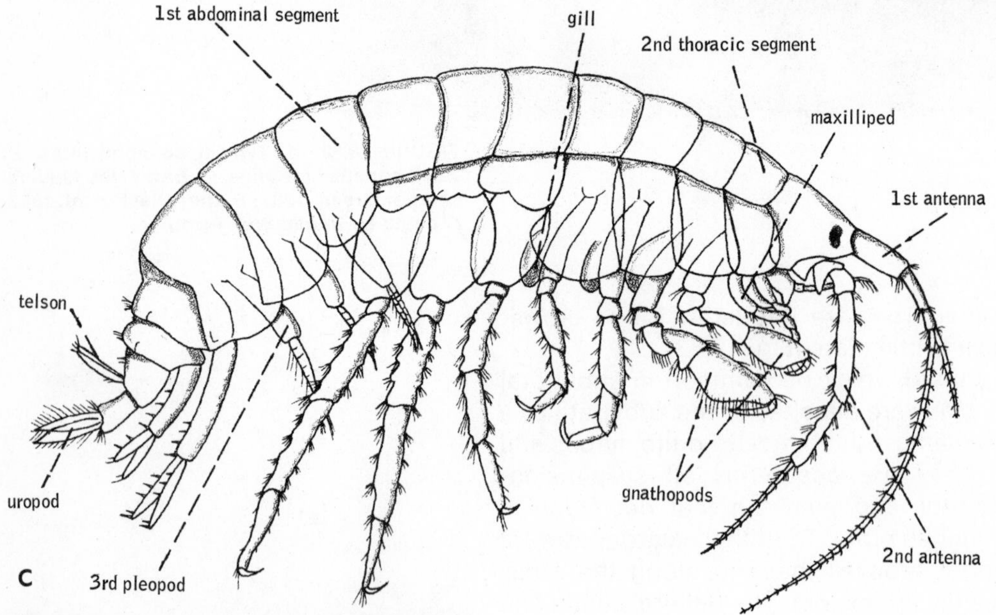

FIGURE 10.31 Crustacean arthropods. ***A*, The shrimplike euphausid krill, *Meganictiphanes*; *B*, The boring isopod, *Limnoria*, or "gribble," and the damage it does to wood. The photograph (by Douglas P. Wilson) shows a piling almost eaten through by gribbles; *C*, A common amphipod, *Gammarus*.**

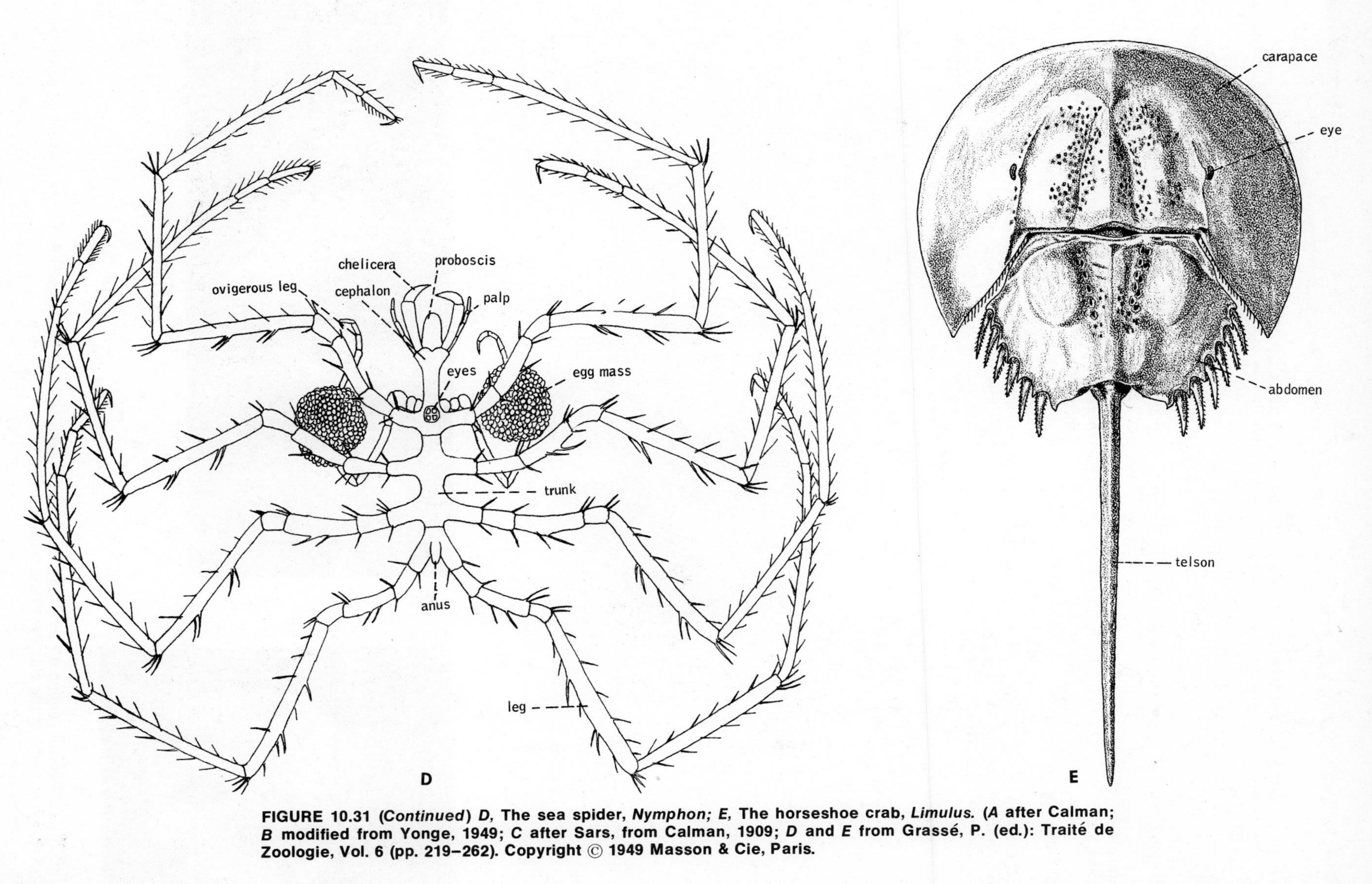

FIGURE 10.31 (*Continued*) *D*, The sea spider, *Nymphon; E*, The horseshoe crab, *Limulus.* (*A* after Calman; *B* modified from Yonge, 1949; *C* after Sars, from Calman, 1909; *D* and *E* from Grassé, P. (ed.): Traité de Zoologie, Vol. 6 (pp. 219–262). Copyright © 1949 Masson & Cie, Paris.

Table 10.2 CHARACTERISTICS OF COMMON CLASSES OF ECHINODERMATA

CLASS	*COMMON NAME*	*CHARACTERISTICS*
Crinoidea "lily-shaped"	sea lily	Branched, jointed arms, many of which are stalked and look somewhat like flowers
Asteroidea "star-shaped"	sea star	Star-shaped; arms with ventral groove; arms taper into central disc of body
Ophiuroidea "shaped like a snake's tail"	brittle star, basket star	Star-shaped; some with branched tentacles; arms without ventral groove; arms joined sharply to central disc
Echinoidea "spiny-form"	sea urchin, sand dollar	Most are biscuit-shaped and have many spines and no arms.
Holothuroidea "sea-cucumber form"	sea cucumber	Most are long and tubular, with tentacles arranged in five longitudinal bands; many resemble cucumbers

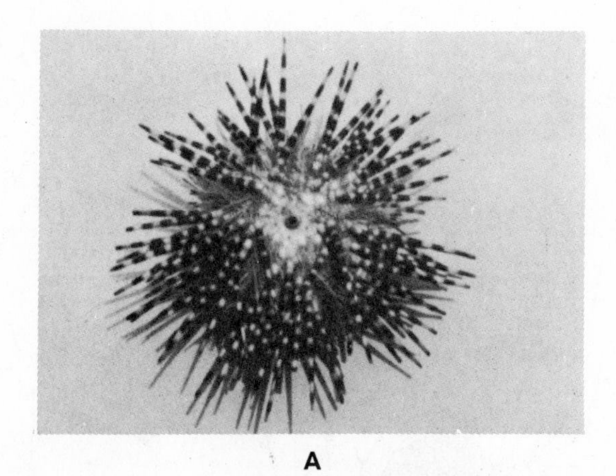

A

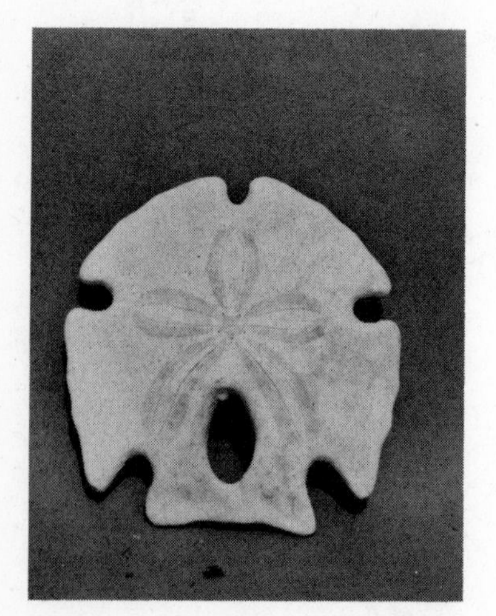

B

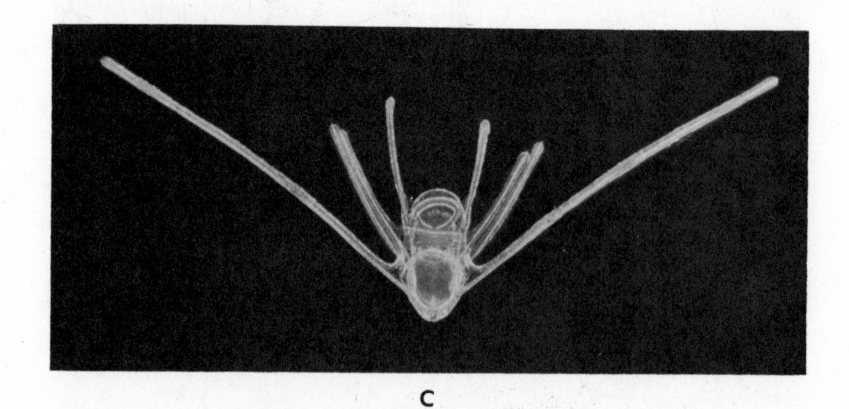

C

FIGURE 10.32 Diversity among the echinoderms. ***A,*** **A Hawaiian sea urchin,** ***Echinothrix; B,*** **The arrowhead sand dollar,** ***Encope; C,*** **Larva of the brittle star** ***Ophiothrix fragilis.***

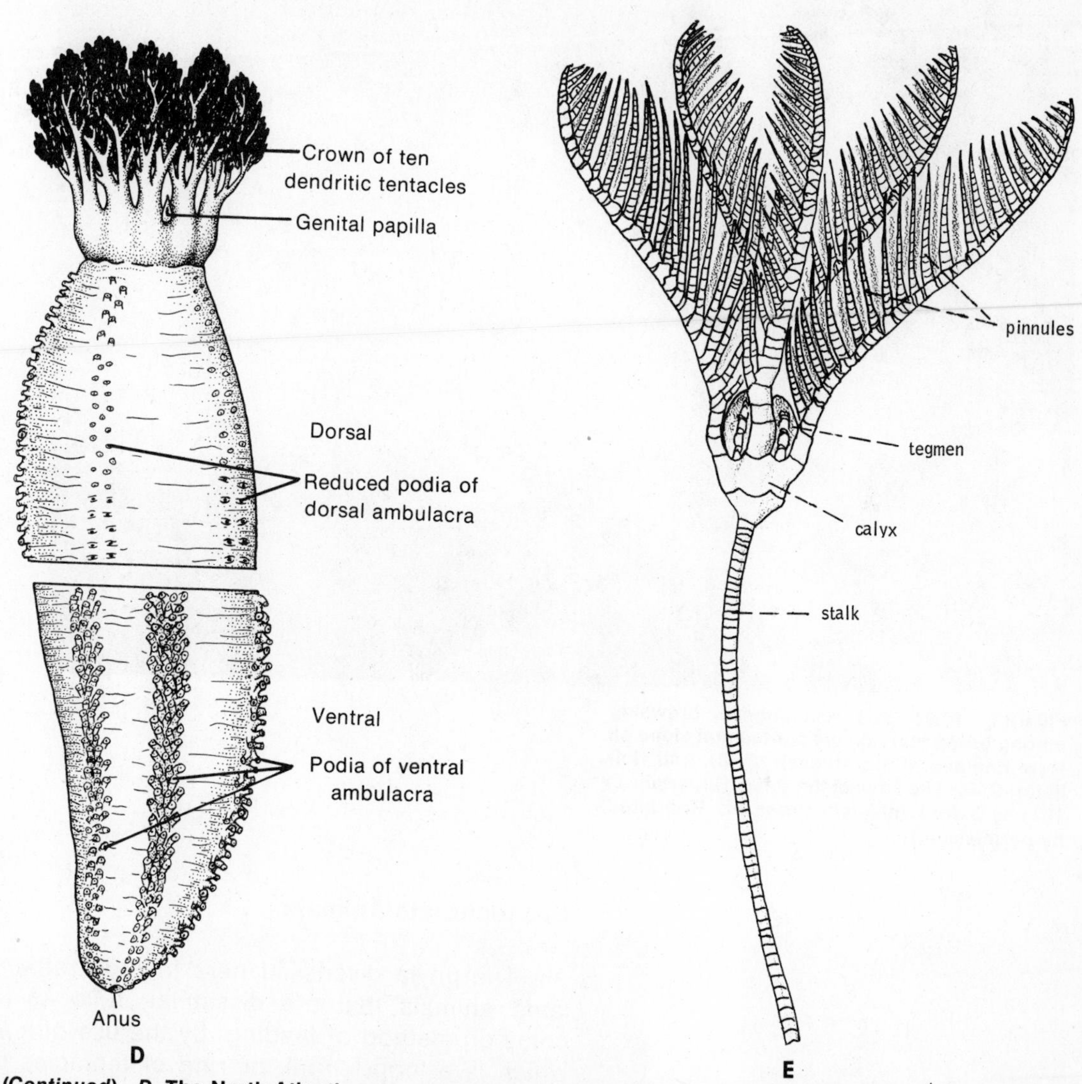

FIGURE 10.32 (*Continued*) *D,* The North Atlantic sea cucumber, *Cucumaria frondosa; E,* The sea lily, *Ptilocrinus.* (*A* and *B* from Barnes, 1974, photo by Betty M. Barnes; *C* photo © Douglas P. Wilson; *D* after Barnes, 1974; *E* after Clark, from Hyman, 1955.) Echinoderms are shown also in Figures 9.8*a* and c, 9.9, 9.10, and 10.33.

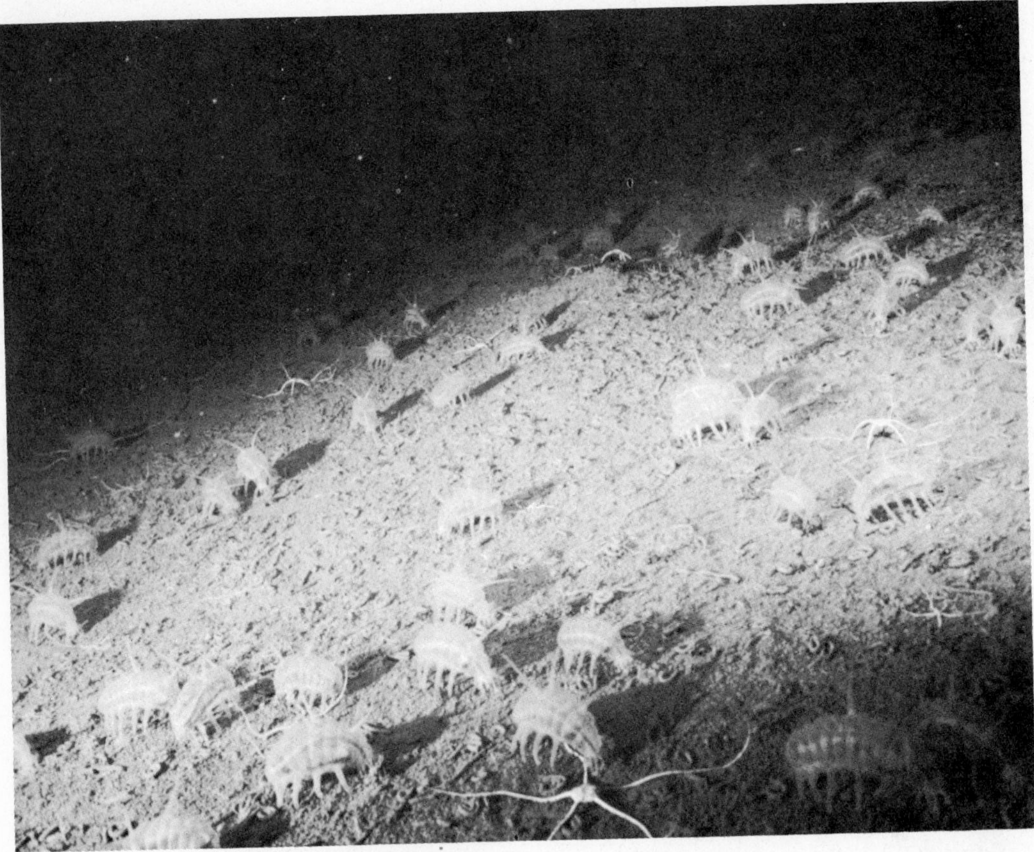

FIGURE 10.33 Sea cucumbers browsing among brittle stars on the continental slope off New England. (From Heezen, B. C., and Hollister, C. D.: The Face of the Deep. Copyright © 1971 by Oxford University Press, Inc. Reprinted by permission.)

Lophophorate Animals

The phyla discussed here include rather loosely related animals that are dissimilar in looks but share a common method of feeding, by the use of a ***lophophore,*** which is a loop, spiral, or ring of tentacles bearing tiny whiplike cilia. The beating of the cilia causes water currents to form, which draw food toward the mouth. They are benthic animals, attached to the bottom, and are unable to go after their food. Consequently they have become adapted to a filter-feeding habit, drawing their food toward them.

Phylum Entoprocta

The small (less than 5 mm) sessile animals belonging to this phylum superficially resemble hydroids because of their tentacles. However, they are filter-feeders and they have no stinging cells (Fig. 10.34*a*). They have a U-shaped gut with a mouth and anus, both opening within the ring of ciliated tentacles. The entoprocts are not generally considered to be truly lophophorate animals because their development differs from that of the lophophores. However, the ring of ciliated tentacles of the entoprocts is similar in structure and function to the true lophophore, borne by the moss animals discussed below. Their food consists of detritus and minute plankton such as diatoms and protozoa. They may be preyed upon by small crustaceans such as sea spiders.

MOSS ANIMALS (PHYLUM BRYOZOA)

The moss animals closely resemble the entoprocts except that the anus opens outside the whorl of tentacles, providing a more efficient and cleaner means of waste disposal. They occupy the same kinds of niches as entoprocts and compete for the same kind of food. Some moss animals have a hard protective shell (Fig. 10.34*b*). Most are colonial, growing on rocks and shells. They may encrust the substrate or they may grow in a branchlike fashion, giving the colony a mosslike appearance (Fig. 10.34*c*).

PHYLUM PHORONIDA

This phylum consists of small wormlike animals that are found mostly in shallow sediments, where they lie partly buried. They secrete a chitinous tube in which they live, filtering the water with their spiral-shaped lophophore. Like the moss animals, they have a U-shaped gut with the mouth within the spiral of ciliated tentacles and the anus just outside the spiral (Fig. 10.34*d*).

LAMP SHELLS (PHYLUM BRACHIOPODA)

Brachiopods are easily mistaken for clams. They are benthic and have two shells (bivalved) and a shape similar to clams. However, they differ from mollusks in that they have double lophophore gills that are also used to filter food from the water. The shells are unequal and frequently give the animal the appearance of an old-fashioned oil lamp (Fig. 10.34*e*) About 99 per cent of the approximately 25,000 known species are extinct and are commonly found as fossils.

Chordates

This phylum includes the most advanced animals. They share the common characteristic of a dorsal strength-

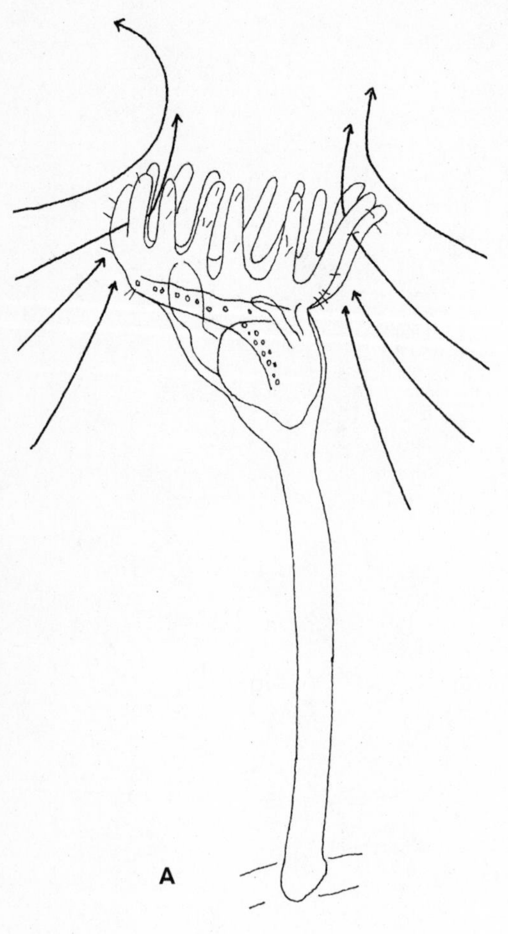

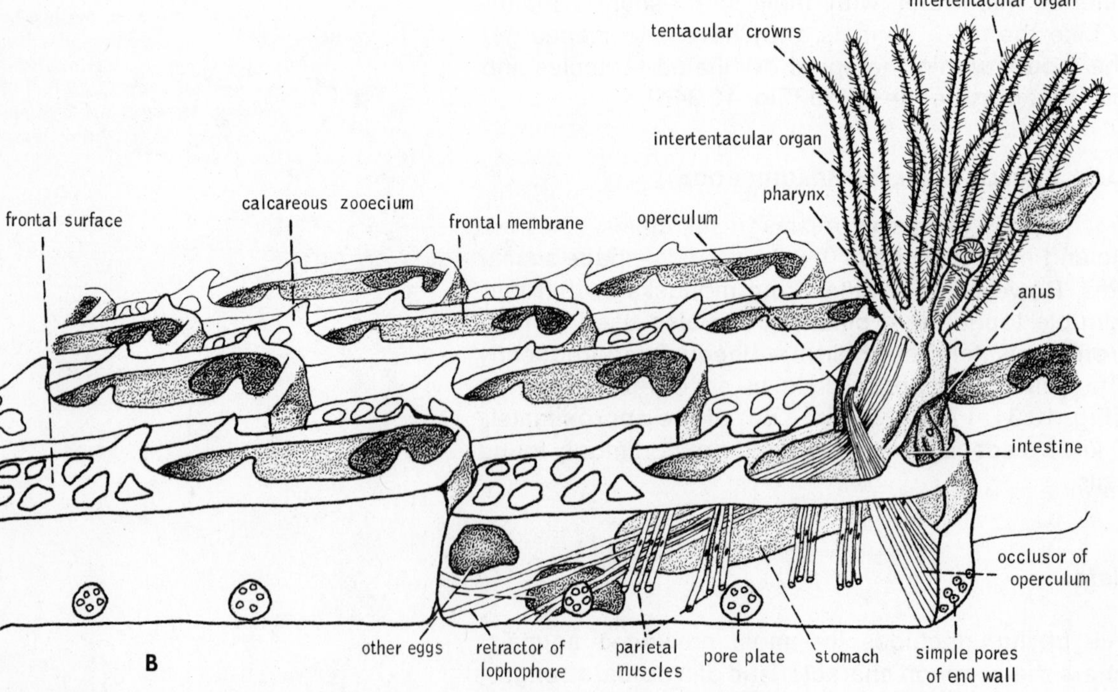

FIGURE 10.34 Lophophorate animals. ***A,*** **An entoproct,** ***Loxosoma.*** **Arrows indicate the flow of water caused by the beating of cilia on its "false lophophore."** ***B,*** **An encrusting bryozoan,** ***Electra.***

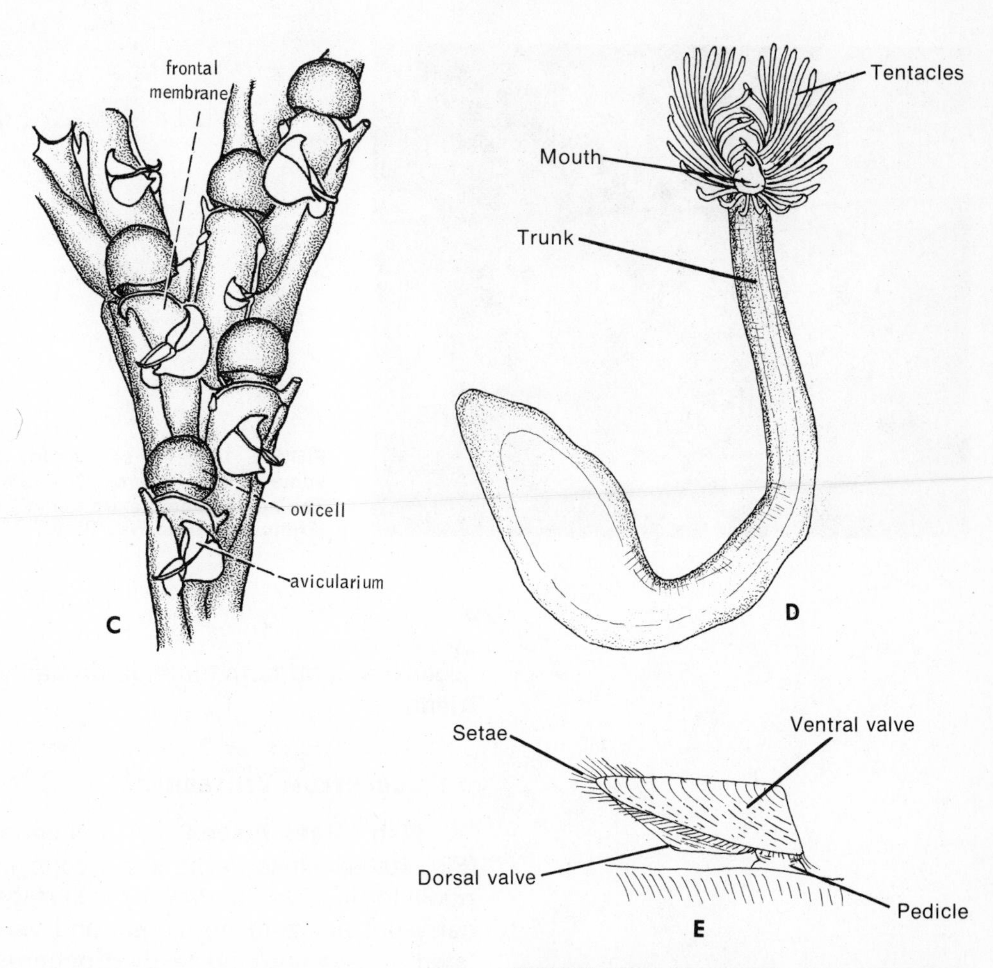

FIGURE 10.34 (*Continued*) *C,* A mossy bryozoan, *Bugula. D,* A tube-dwelling phoronid, *Phoronis,* without its tube. *E,* A brachiopod, *Discinisca.* (*A* after Atkins, from Hyman, 1951; *B* and *C* from Barnes, 1974; *D* after Wilson, from Hyman, 1959; *E* after Morse, from Hyman, 1959.)

ening rod along the back, the ***notochord***. This rod becomes the core of the developing backbone in the vertebrates. These groups are quite diverse, but only one of them includes attached animals, the sea squirts in the Subphylum Urochordata. Some members of the Subphylum Vertebrate (including fish, reptiles, birds, and mammals) can swim. However, some fish may burrow for a time in the sediment.

TUNICATES (SUBPHYLUM UROCHORDATA)

Tunicates are small chordates that have a notochord when they are larvae. Some are attached to rocks, as are the sea squirts (Fig. 10.35); others, such as salps, are free-swimming in the adult stage. They contain a basket-like filter, which strains the water that is drawn through their bodies by ciliary action. The food thus obtained is similar to the plankton food of clams and small crustaceans, and thus tunicates may compete with them for food. Their role in the ecosystem is similar to that of the coelenterates in that neither is important as food for other

FIGURE 10.35 A sea squirt, *Ciona intestinales,* with some sea anemones, *Edwardsia,* in the background. The sea anemones are about 1 centimeter in diameter. (Photo by Harold Wes Pratt.)

organisms, and both function primarily as recyclers of nutrients.

SUBPHYLUM VERTEBRATA

Fish (Class Pisces). A tremendous diversity of fish (Fig. 10.36) exists in the sea, occupying niches from tide pools to deep-sea and oceanic surface waters. Some fish carry out amazing horizontal and vertical migrations related to breeding and feeding patterns. These migrations are discussed in Chapter 9. Many fish are important as food for people; others, even more abundant, play a vital role in the marine food webs.

Some of the most abundant fish in the sea are only a few centimeters in length. These include the anchovy *(Anchoa),* lantern fish (generally known as myctophids), and the bristlemouth (*Cyclothone*). The bristlemouth may be the most abundant fish in the sea, but it is rarely seen because it is usually found in the oceanic realm and is seldom caught by fishing fleets.

Most fish, although nektonic, have tiny planktonic larvae (Fig. 10.37). Epipelagic larvae are frequently transparent. The larvae generally look far different from the adult, and in the case of larger fish that eat other fish the larva's diet is usually also different from that of the adult. The small larvae of these fish often feed on other zooplankton such as copepods.

Many fish that live in the surface waters, such as tuna, mackerel, and many sharks, are voracious predators on other fish and squid. However, many other fish, including the small anchovies and even the large filter-feeding basking and whale sharks, which may reach lengths in

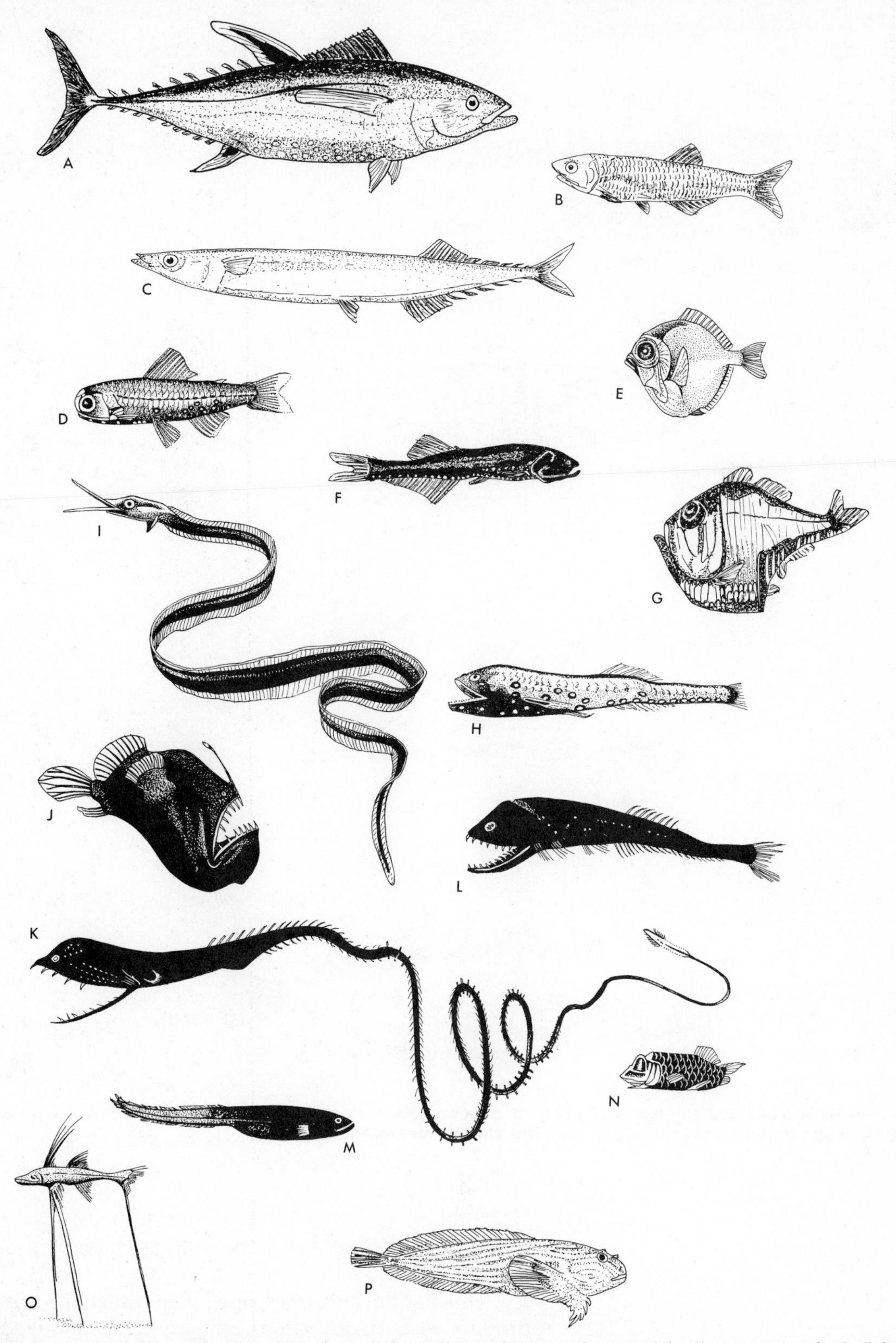

FIGURE 10.36 A variety of marine fish. Epipelagic fish (*A–C*) are shown at the top of the figure, mesopelagic fish (*D–G*) in the middle, and deeper forms below.

A, Yellowfin tuna (*Thunnus*), 1 m. *B*, Anchovy (*Anchoa*), 6 cm. *C*, Pacific saury (*Cololabis*), 40 cm. *D*, Lantern fish (*Diaphus*), 5 cm. *E, Diretmus,* 5 cm. *F*, Bristlemouth (*Cyclothone*), 6 cm. *G*, Hatchet fish (*Argyropelecus*), 2 cm. *H*, Deep-sea lantern fish (*Lampanyctus*), 3 cm. *I*, Snipe-eel (*Nemichthys*), 35 cm. *J*, Angler fish (*Melanocetus*), 5 cm. *K,* Gulper-eel (*Saccopharynx*), 20 cm. *L*, Swallower (*Gonostoma*), 5 cm. *M, Parabrotula,* 4 cm. *N, Opisthoproctus,* 3 cm. *O*, Tripod fish (*Benthosaurus*), 20 cm. *P*, Snail-fish (*Liparid*), 20 cm.

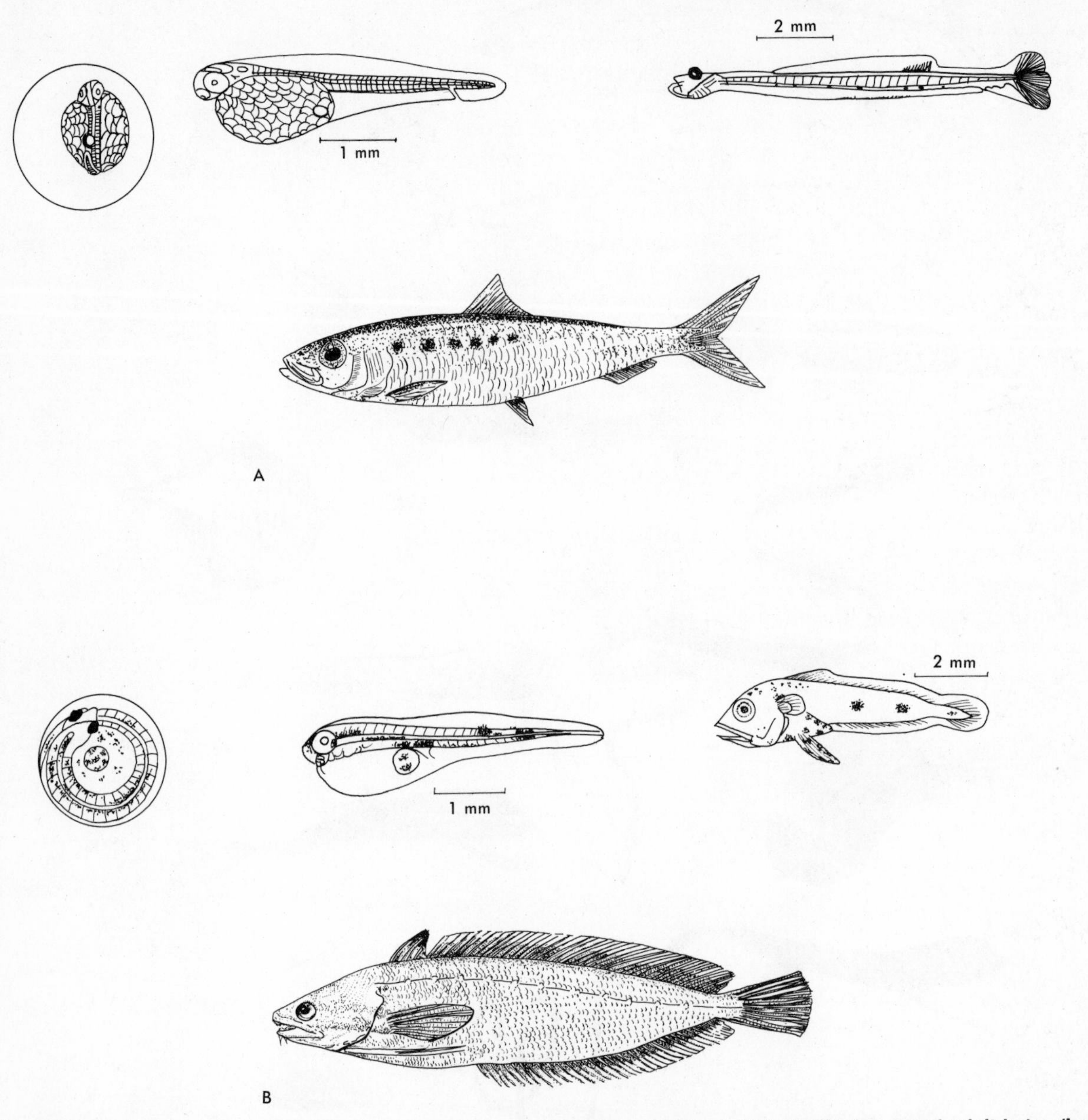

FIGURE 10.37 *A,* **The larva and the adult pilchard, a sardine-like epipelagic fish.** *B,* **The larva and adult hake. (Larvae after Newell and Newell, 1966; adults after Bigelow and Schroeder, 1953.)**

excess of 15 meters, feed on tiny zooplankton such as copepods.

The neritic and epipelagic fish which live in highly productive surface waters and are of economic importance—for example herring, tuna, mackerel, salmon, and sharks—are familiar to most of us that eat them. Actually, about 50 per cent of the fish caught for human consumption are in the herring family (including anchovies and sardines). However, there are many deep-sea fish that we never see in the markets that are well adapted to seemingly inhospitable environments. Many of these bizarre-looking fish, such as the angler fish and the gulper eel *Saccopharynx* (Fig. 10.38), have mouths that are huge relative to the size of the fish. Some can eat prey as large as themselves. This is because their jaws open wide and their guts and body walls can expand to allow room for a large meal. Deep-sea fish do not always have a ready supply of food, therefore they cannot afford to be particular. They must eat whatever they can, whenever they can. Some even have luminous lures that can attract prey in the otherwise total darkness.

Angler fish have evolved an unusual method of mating in the deep, dark sea. The young male bites the female, usually on her underside and becomes a permanently attached parasite on the larger female (Fig. 10.38*c*). As a result, the two sexes are always available to each other for reproduction.

Reptiles (Class Reptilia). Not many reptiles have adapted to life in the sea. Sea turtles, iguanas, and sea snakes are exceptions. However, they still depend on air for breathing, and most still breed on land. Because they may derive much of their food from the sea, they are considered part of the marine food webs.

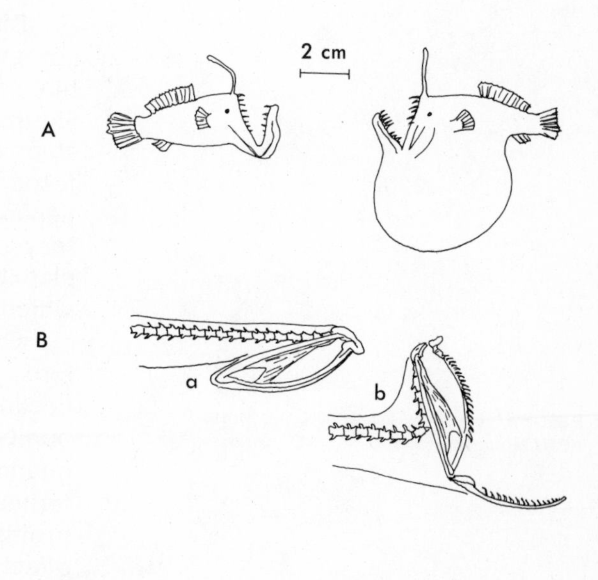

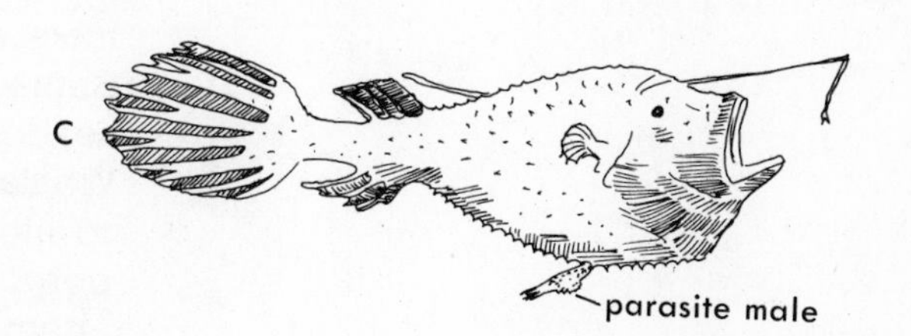

FIGURE 10.38 *A,* The angler fish *Melanocetus* before and after eating a large meal. *B,* The jaws of *Saccopharynx,* the gulper eel, and hinged in a way that allows a large food organism to enter the gut, which expands to hold the food. *C,* The angler fish *Ceratias* carrying a parasitic male on her underside. (After Hardy, 1956.)

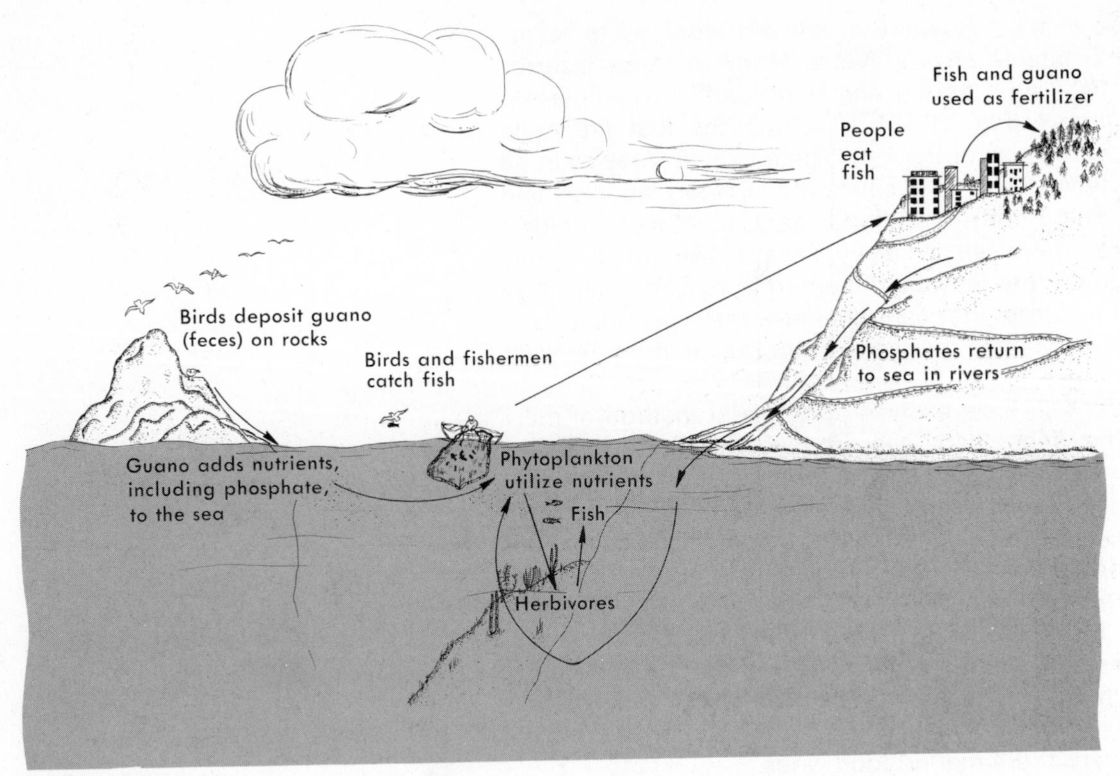

FIGURE 10.39 Birds and the guano (phosphate) cycle.

Birds (Class Aves). As with the reptiles, birds are not strictly speaking aquatic animals. However, many birds feed on marine mollusks, worms, and especially fish. Figure 10.39 illustrates some of the ecological relationships of marine birds. Some birds eat fish and drop their feces (guano) on rocks and on land. The nutrients thus generated may find their way back into the sea, where they enhance the growth of phytoplankton. The phytoplankton are eaten by copepods and other herbivores, which are in turn eaten by fish.

Overfishing may deplete the stock of fish available to birds. The birds then go elsewhere or the bird populations decline. Less guano is deposited and phytoplankton numbers decline. In some areas, such as Peru, much guano is collected and carried away to be used on land as fertilizer, leaving less food for the remaining fish. Thus problems that affect one part of the ecosystem tend to affect the other parts as well. (See Chapter 8.)

Mammals (Class Mammalia). The fur-bearing animals are air-breathing so they are not strictly marine. Some depend on the sea for a supply of food. Sea lions, seals, and walruses move between land and sea (Fig. 10.40). Whales (including porpoises) have become adapted to life in the surface waters of the oceans. Whales and porpoises lack hind legs necessary for movement on land, although their bone structure indicates that a four-legged

FIGURE 10.40 Sea lions along the Oregon coast. (Courtesy of Oregon State Highway Department.)

land mammal was their ancestor. Mammals are the most intelligent of sea creatures, and many have been trained to perform tricks in circus and "seaquarium" shows. Whales and porpoises emit a wide variety of sounds, and efforts are being made to understand their meaning, if any, more fully. Because of their economic importance as a source of food and whale oil, some whales are approaching extinction (see Chapter 11).

DECOMPOSERS

The decomposers are crucial in the sea, as they are involved in breaking down the excess organic matter produced by the rest of the organisms and returning the materials to the sea in the form of inorganic nutrients, which are required for plant growth. If it were not for the decomposers, life in the sea would probably be greatly diminished, because dead organisms and feces would simply accumulate on the sea floor. Some of this material, even in the absence of decomposers, would be consumed by deposit-feeders, but still, much less of the nutrient substances would be recycled. Consequently the primary production of plants would decline, causing less food to be produced for the consumers, which would in turn greatly diminish in abundance. The decomposers can be broadly categorized as bacteria and fungi. These are generally considered to be neither plant nor animal.

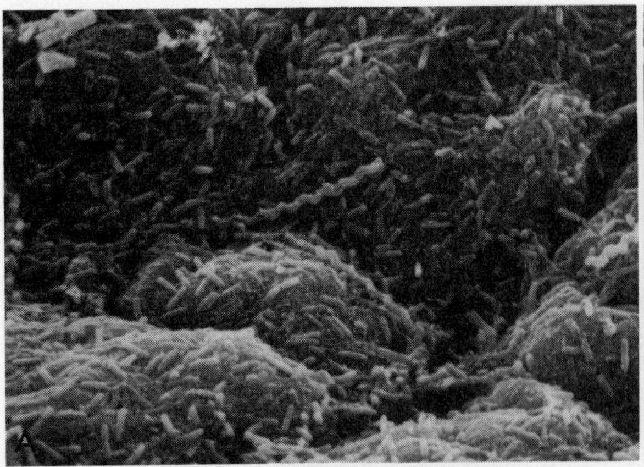

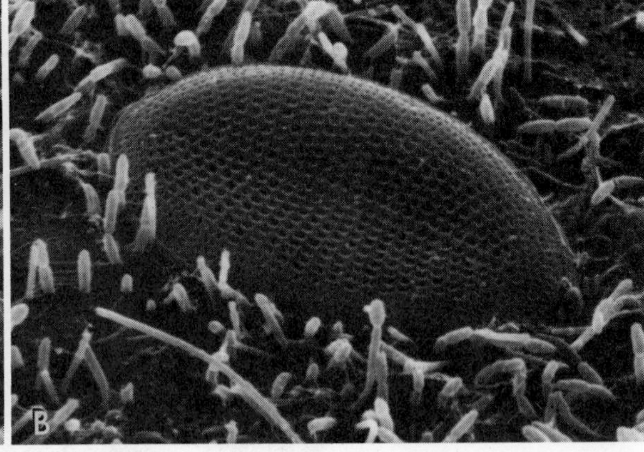

FIGURE 10.41 *A*, Bacteria decomposing the flesh of a young salmon with tail rot. (Scanning electron micrograph, 1500 ×.) *B*, A diatom (*Cocconeis*) and bacteria on turtle grass. (Scanning electron micrograph, 2750 ×.) (Courtesy of P. W. Johnson and J. McN. Sieburth, University of Rhode Island.)

Bacteria

Most bacteria (Fig. 10.41) are found attached to particulate matter, especially detritus. Organic materials tend to adhere to particles suspended in the water or lying in the sediment and provide nourishment for the bacteria associated with them.

Bacteria are found at all depths of the sea and in all kinds of environments, performing their essential role of converting detritus and dissolved organic matter into the inorganic nutrients used as raw materials in photosynthesis. All of the organic remains of marine organisms, including crab shells and oil, are susceptible to decay by bacteria.

Bacteria, in addition to regenerating nutrients, have other functions in the sea. Many bacteria secrete organic compounds, such as vitamin B_{12}, which are essential for the growth of many diatoms and other phytoplankton. In addition, bacteria are important as a source of food for protozoans and sediment-feeding animals such as many worms.

Some bacteria live in pouches on certain fish and produce a cold light or bioluminescence that may aid the fish in attracting prey. This phenomenon is discussed more fully in the previous chapter.

FIGURE 10.42 Fungi colonizing an oak lobster pot. (Scanning electron micrograph 425 ×, courtesy of R. D. Brooks and J. McN. Sieburth, University of Rhode Island.)

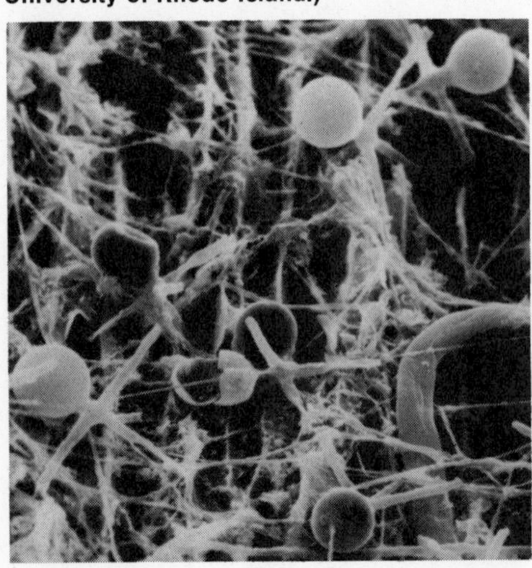

Fungi

Fungi play a role similar to that of bacteria. Many simple fungi (Fig. 10.42) including yeasts and molds, decompose organic matter in the sea. They are either free floating or attached to other marine organisms, especially in the shallow littoral zone, where they may be seen as cottony filaments attached to decaying algae and animals.

SUGGESTED READINGS

Barnes, Robert D. 1973. *Invertebrate Zoology,* 3rd ed. Philadelphia, W. B. Saunders Co.

Caldwell, R. L. and Hugh Dingle. 1976 (January). Stomatopods. *Scientific American.*

Dawson, E. Yale. 1956. *How to Know the Seaweeds.* Dubuque, Iowa, William C. Brown.

Heezen, B. C., and C. D. Hollister. 1971. *The Face of the Deep.* N.Y., Oxford University Press.

Isaacs, J. D. 1969 (September). The Nature of Ocean Life. Scientific American.

Marshall, N. B., and Olga Marshall. 1971. *Ocean Life.* N.Y., Macmillan.

Murphy, Robert C. 1962 (September). The Ocean Life of the Antarctic. *Scientific American.*

Nicol, J. A. Colin. 1967. *The Biology of Marine Animals,* 2nd ed. London, Pitman.

Ricketts, E. F., and J. Calvin. 1968. *Between Pacific Tides,* 4th ed. Revised by Joel W. Hedgpeth. Stanford, California, Stanford University Press.

Shaw, Evelyn. 1962 (June). The Schooling of Fishes. *Scientific American.*

Thorson, Gunnar. 1971. *Life in the Sea.* N.Y., McGraw-Hill.

Biological productivity in the oceans millions of years ago resulted in the formation of oil deposits. Productivity today results in food for human populations. (Photo courtesy of Sun Oil Co.)

CHAPTER 11

Marine Production

Because of growing population on a limited earth, we have long been aware of the need to increase the amount of food we get from the sea. Studies have been conducted to learn where marine food may be collected most easily. Efforts are being made to predict and control the changing populations of food organisms in the sea. Even mariculture (sea farming) is being attempted at this time in semi-enclosed embayments and marine ponds. Pollution, on the other hand, is undermining our attempts to utilize the sea at an optimal level. In addition, poisonous "red tides" pose a dangerous problem in the near-shore waters. In this chapter we will discuss some of the potentials and limitations of the sea as a source of food and recreation.

PRIMARY PRODUCTION

Plants use CO_2, sunlight, and inorganic nutrients to produce organic matter by a process called ***photosynthesis.***[1] This process results in the ***primary production*** of organic matter in the sea. In the sea the primary producers are mostly tiny drifting plants (0.01 to 1 mm) collectively termed ***phytoplankton***. These include coccolithophores (whose skeletons comprise much of the marine sediments), diatoms (sometimes called "the grass of the sea"), and dinoflagellates. On the other hand, some marine plants are quite large. The drifting *Sargassum* weed is a kelplike plant that grows in large masses in the otherwise quite barren Sargasso Sea, an area of warm Atlantic water. The attached kelp, *Macrocystis*, may grow up to 60 meters long (200 feet) off the west coast of the United States.

Life in the sea depends on light from the sun as a source of energy. Most of the sunlight is absorbed by the water, organisms, and other particles in the water before a depth of about 200 meters is reached. Only near the surface is light of value to plants. Plants capture the light and

---

[1]A simplified equation for photosynthesis is as follows: $CO_2 + H_2O$ + light (in the presence of chlorophyll) $\rightarrow$ carbohydrate + O_2. Carbohydrates are converted by the plants into other sugars, starches, proteins, fats, and oils. (Also see Chapter 3.)

convert the radiant energy into the stored energy in food (organic matter) by photosynthesis. This food in the form of plant life is available to herbivores, which are, in turn, food for carnivores.

In order for plants to live, they must have certain raw materials. These include carbon dioxide and water, required in the production of carbohydrates during photosynthesis. In addition, nutrients such as nitrate and phosphate are required if the plant is to produce essential proteins and lipids. Other nutrients essential for certain marine plants include iron, silicon, sulfate, and calcium. Carbon dioxide is present in plentiful supply as a dissolved gas from the air and as a waste product from the respiration[2] of bacteria, plants, and animals. Most nutrients enter the sea as a result of pollution and erosion of land or by being transported into the sea via rivers and rainfall run-off. Nutrients may be recycled by bacterial decomposition of dead organisms and the waste products of living organisms. (Nutrient cycles are discussed in Chapter 3.) Nutrient-rich run-off causes increased production near shore. Since dead plants and animals usually sink toward the sea bottom, many nutrients are recycled in dark, deep water (Fig. 11.1), where they cannot be used by plants until they are brought back to the surface by upwelling and deep-ocean circulation (see Chap. 7).

Biomass refers to the amount of organic matter in pounds or grams per given volume present at any one time; it may be compared to a ***standing crop*** in a field on land. Biomass and productivity are not necessarily related. For example, in a forest on land, the biomass is very

[2]The respiration of both plants and animals may be summarized as follows: carbohydrate + $O_2 \rightarrow CO_2 + H_2O$ + enery (mostly heat).

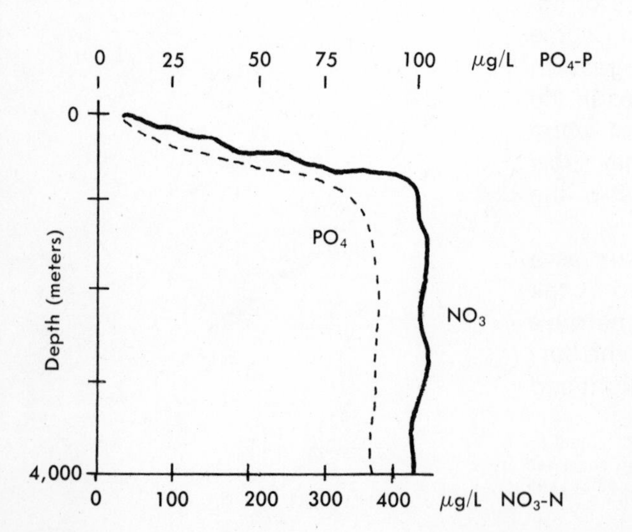

FIGURE 11.1 Distribution of two key nutrients, phosphate (PO_4) and nitrate (NO_3), in relation to depth. Note the depletion of nutrients near the surface and their abundance at greater depths. (After Sverdrup et al., 1942.)

large, but the productivity or growth rate is very low. Similarly, a large standing crop of *Sargassum* weed in the Sargasso Sea has a low productivity. On the other hand, high productivity in a polluted estuary may lead to a high biomass of unwanted organisms.

Productivity is the rate of increase or growth of the standing crop. This can be related to cattle grazing in a field of grass. If the crop of grass remains constant, the productivity of the grass balances the grazing rate of the cattle (and other herbivores). Similarly, many species of shrimplike copepods graze on a crop of diatoms in the sea. Since copepods are generally herbivorous, they are sometimes referred to as the "cattle of the sea" because of their importance as a link between phytoplankton and larger carnivores such as fish.

The primary production due to photosynthesis may be regarded as ***gross production,*** or the total amount of carbon assimilated by plants in a unit of time. Actually, because the plants respire, they use up some of their assimilated energy in their own metabolism, releasing some energy as heat and carbon as CO_2. The amount of production thus available to the herbivores is known as ***net production***. Net production (N) is the gross production (G) minus the respiration (R), or $N = G - R$. Generally N is about one half of G.

Productivity tends to be highest in and around nutrient-rich estuaries, or some enriched areas such as farmlands. Up to 25,000 kcal/m^2/yr of energy may be trapped by plants in these areas.[3] Open ocean waters are much like deserts with respect to productivity at about 500 to 3,000 kcal/m^2/yr (Fig. 11.2). The exploitation of

[3]One kilocalorie (kcal) is the amount of heat energy required to raise the temperature of one kilogram (2.2 lb) of water by 1° Centigrade.

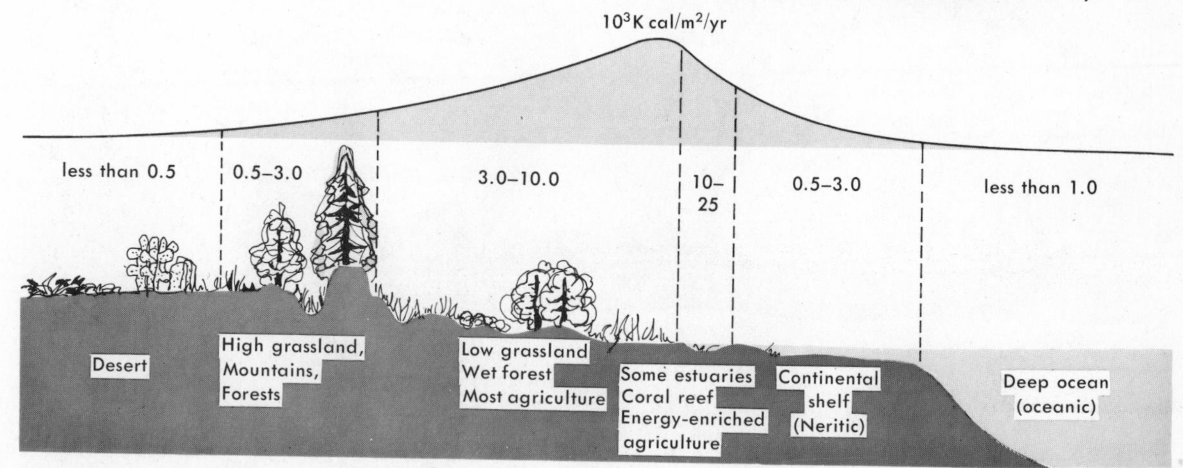

FIGURE 11.2 Distribution of plant production over different types of areas. Values are in millions of calories produced in a square meter of area per year. (After Odum, 1971.)

shelf-fisheries is similar to open-range grazing of cattle, whereas estuarine production and mariculture are comparable to high-intensity agriculture on land.

Figure 11.3 shows the distribution of production over the open sea. We can see that regions of low productivity are in areas with warm surface water. Regions with high productivity include cooler areas such as polar and subpolar regions, where vertical mixing of nutrients occurs, and areas of upwelling. The high production that we find off the west coast of South America may be related to the South Equatorial Current, which flows westward from the land at the equator, and to upwelling along the coast, both of which processes may draw nutrient-rich water upward to replace water moving offshore at the surface.

BIOCHEMICAL OXYGEN DEMAND

The demand placed on the environment for oxygen by bacteria, animals, and plants (as well as its use in other chemical oxidation, such as the rusting of iron) is called ***biochemical oxygen demand*** (B.O.D.). The greater the amount of decaying wastes, the greater the B.O.D. and consequently the greater the amount of oxygen consumed.

As animals and bacteria "feed" and respire, they use up oxygen and release carbon dioxide. Carbon dioxide is

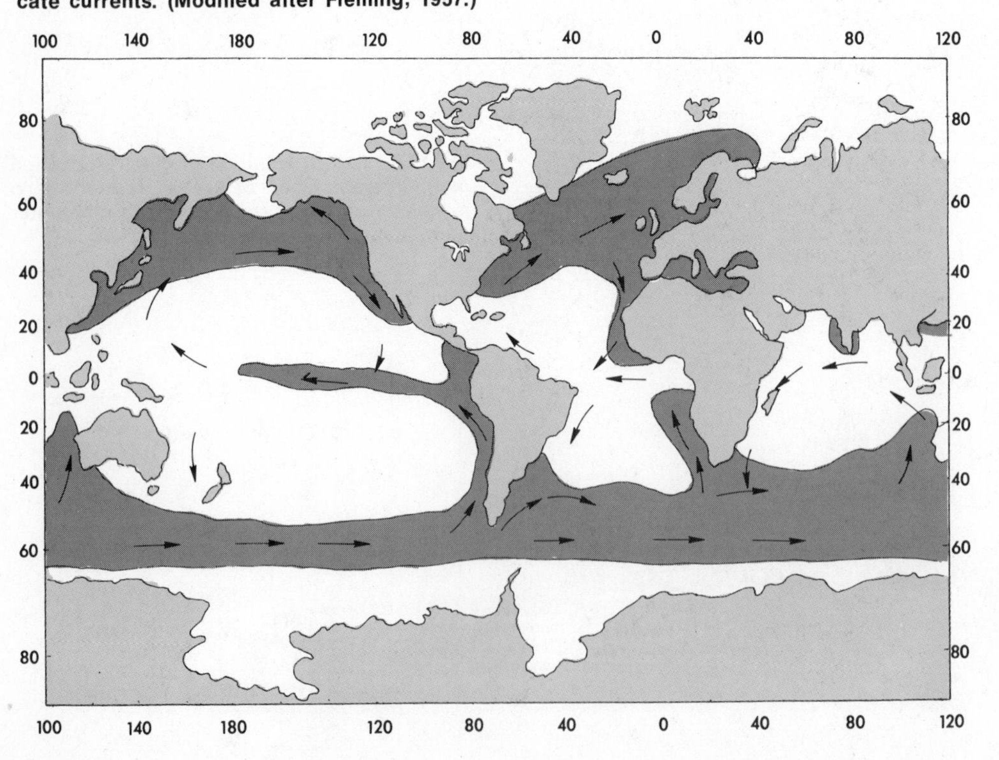

FIGURE 11.3 Areas of the world oceans with greater than (shaded) and less than (unshaded) 100 gm carbon fixed by plants (primary production) per square meter per year. Arrows indicate currents. (Modified after Fleming, 1957.)

taken in by plants, which release oxygen. However, if the plants are unhealthy or receive insufficient light, they may consume more oxygen in respiration than they produce via photosynthesis. If there is an overabundance of dead organic matter (detritus) from plants and animals, including sewage and other biological wastes, animals and bacteria will require more oxygen than is produced by plants through photosynthesis. This is because the detritus is eaten and decomposed by detritus-feeding animals and bacteria, using up oxygen in the process. Consequently, the concentration of oxygen in the water will tend to decrease. If the oxygen level falls below a certain point, some species of fish and other animals will die, further increasing the amount of detritus and the demand for oxygen. Eventually the environment may become ***anaerobic,*** or lacking in oxygen. Anaerobic bacteria thrive in an oxygen-free environment, respiring by a process similar to fermentation. Many of these bacteria produce hydrogen sulfide (H_2S) as a by-product, resulting in a "rotten-egg" odor characteristic of black, stagnant, organic-rich sediments.[4]

NUTRIENTS AND THEIR RELATION TO PRODUCTION AND DIVERSITY

In general, as inorganic nutrients such as nitrates and phosphates are added to a body of water, providing nourishment for the plants, the production of these plants increases, resulting in a higher biomass or standing crop. This may, if the plants are edible, supply food for zooplankton, nekton, or benthic animals. Sometimes, however, the plant species that thrive on the added nutrients are inedible (such as some pond scums) or even harmful (such as some red-tide organisms).

The seasonal temperature structure of the ocean (see Chap. 4) plays a role in the distribution of nutrients. In shallower water in middle or high latitudes, during the winter months the water is about the same temperature at the surface and bottom. Uniform temperature results in little density stratification, and the water tends to be well mixed with regard to nutrients and salinity. During the warm summer, though, the surface water is warmed much more quickly than the deep water, and it becomes lighter. The water becomes stratified with regard to density, and

[4]One example of the sulfate reduction process is as follows:

CH_3COOH +	$SO_4^{=}$ →	$2CO_2$ +	H_2S +	$2OH^-$
acetic acid	sulfate	carbon dioxide	hydrogen sulfide	hydroxyl

little vertical mixing occurs. Nutrients near the surface are depleted, but nutrients are built up near the bottom because the "rain" of dead organisms from above is decomposed in the deeper water. Consequently, there is a reduction in primary production as the surface nutrients are depleted.

Warm, fresher water from an estuary tends to float over the salty ocean water. If the surface water in the estuary is polluted and nutrient-rich, it may trigger blooms of algae. If the nutrient-rich effluent from an ocean sewage outfall is warm and fresh (low salinity), it will tend to rise in the water column and enrich the surface water (Fig. 11.4). If the effluent is colder than the surface water, it will tend to be mixed over a wider area by bottom currents and be more dilute when reaching surface waters. The enrichment of the surface water may be beneficial, increas-

FIGURE 11.4 Flow of sewage effluent from a submerged ocean outfall when density stratification is at a maximum (*A*) and a minimum (*B*).

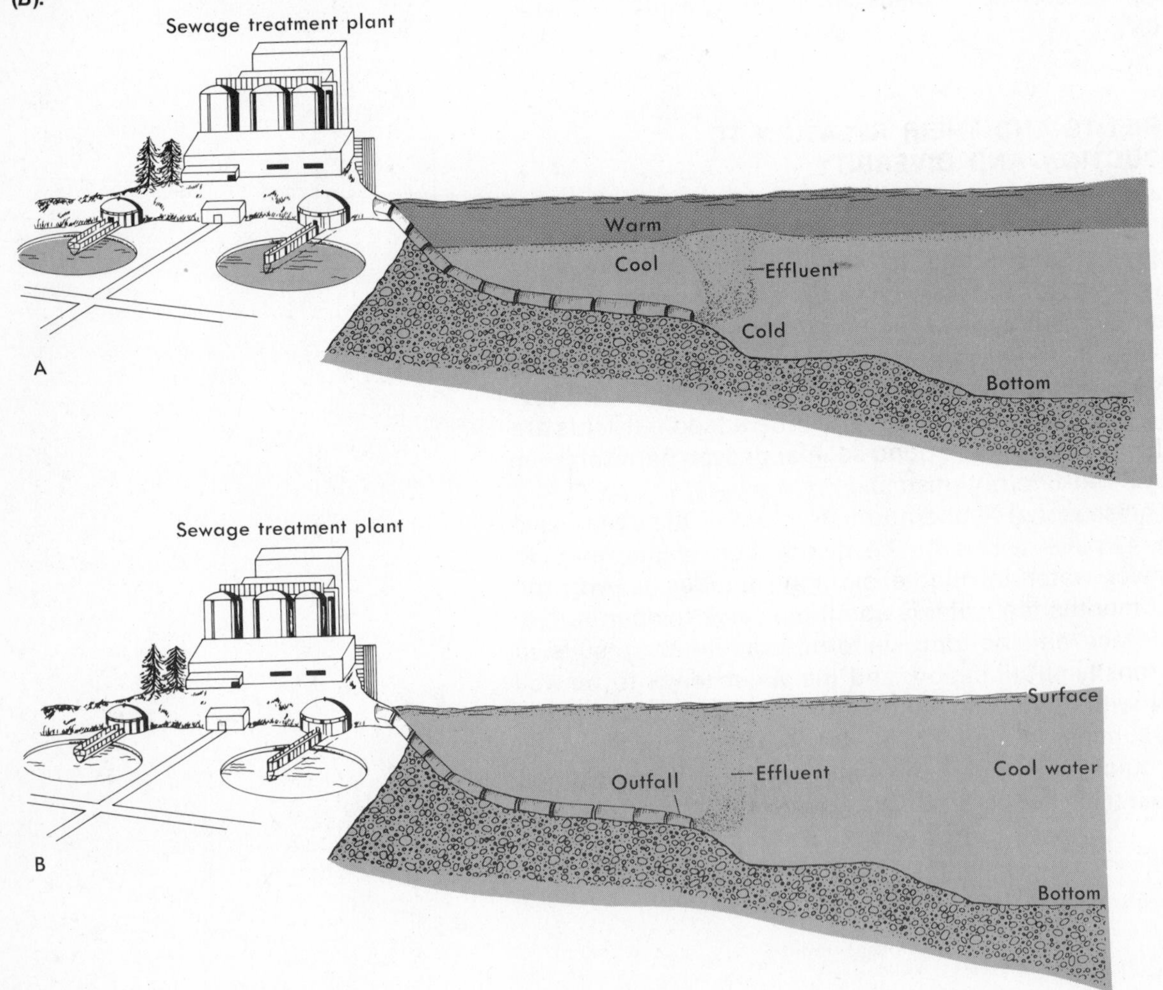

ing the amount of food available for fish. On the other hand, adverse conditions could cause a local red-tide bloom, a condition harmful to people as well as to marine organisms. The red-tide problem is discussed later in this chapter.

Aquatic environments can be characterized as oligotrophic, eutrophic, or dystrophic, depending on the concentrations of nutrients and biomass and the diversity of species.

Oligotrophic (Greek *oligos* = few; Greek *trophe* = food) systems have low nutrient and biomass concentrations but may have many species of organisms. Owing to the rain of detritus into the deep, dark abyss, the surface waters in the oceanic areas contain low concentrations of nutrients. The nutrients are regenerated from the decomposing detritus, mostly in much deeper water. Consequently, production in bodies of water such as the open ocean and mountain lakes is very low.

Eutrophic (Greek *eus* = good) systems are rich in nutrients and biomass but still have many species. However, as the nutrient levels become highly enriched and polluted, there is a reduction in the number of species due to the death of less tolerant organisms. Streams or bays that are enriched by land run-off, especially from agricultural land, or by sewage pollution tend to be eutrophic. Of course, there is a wide range of eutrophism, depending on the level of nutrient addition. Bays enriched by run-off alone would tend to be "somewhat eutrophic." Those that are enriched by both run-off and sewage pollution tend to be more highly eutrophic, as, for example, Long Island Sound. This condition is generally indicated by a high biomass but a lower diversity of organisms than is found in the open ocean.

The next level of enrichment, called ***dystrophic*** (Greek *dus* = bad) or hypertrophic, is characterized by high nutrient levels and biomass but low species diversity. Bays thus affected might be considered highly polluted. Extremely high nutrient levels, high B.O.D., or industrial pollutants may cause the loss of many species that are not tolerant of the high-stress environment implied by these factors. The few species that can survive are present in tremendous numbers because of the large amounts of available nutrients and limited competition. The survivors may have little economic importance or may be unfit for consumption due to contamination by poisons or bacteria. Newark Bay, New Jersey, may be considered an example of a dystrophic bay. As few as three or four species may dominate the plankton; and the common killifish dominates the nekton of Newark Bay.

The distinction between eutrophic and dystrophic water bodies is not easily delineated. It is based mainly on species diversity and nutrient concentration. The point of change also depends on many factors, such as temperature, salinity, tides, and the subjectivity of the observer. The change from a eutrophic to dystrophic condition may not be recognized until great destruction to the environment has occurred.

SUCCESSION

Succession is the change in species composition with time, whether on a substrate such as a rock, piling, or sediment; or suspended in the water, such as plankton or nekton. The change in communities that occurs on a new piling or an automobile body tossed into the water is an example of this process; another example is the seasonal succession of plankton species.

Benthic Succession

As an example of succession, let us look at marine fouling. Even the casual observer can notice that objects that are placed in the sea soon become covered with all sorts of sea life or "fouling" organisms. These organisms attach to lines, anchors, ship hulls, and buoys and then serve as a substrate on which other organisms grow. A ship can be slowed in its movement by adherent organisms because the mass of attached algae and animals increases the resistance to movement through water. Currents pull harder on a fouled buoy and ropes and pilings may be weakened by the burrowing animals that inhabit them.

The problem of marine fouling has been of concern for centuries to people who live near the sea. Efforts are made to reduce fouling of ropes, pilings, and ships' hulls with various chemicals, including creosote and special paints that are toxic to marine organisms. Part of this attempt has involved investigations of the succession process—which organisms settle first, which come later, and which organisms represent the final, stable, well-fouled stage or ***climax community***. The climax community is the one that causes most of the problems, as it is usually the most massive and creates the greatest resistance to currents and destruction of substrates.

The stages of fouling that usually take place may be summarized as follows:

1. *Preparation of the substrate.* The first organisms to settle include bacteria. Many tend to adhere to virtually any solid substance placed in the water, producing a film suitable for certain tiny diatoms to become attached and

grow in mats or chains. This prepares the substrate for the next stage.

2. Random settling. Other plants and animals settle and grow, including larvae of barnacles, mussels, hydroids, and young algae. Some of these survive and grow; others die when the tide goes out and exposes them to air. Some species may be crowded out by other species during the next stage of succession, which we may call a period of selection.

3. Selection. During this period most of the settlers die, are eaten, or are displaced by competition. Only the species best adapted to that particular environment survive. For example, of the settlers from Stage 2, only the mussels may survive. The hydroids and algae may die from exposure. The barnacles may be crowded out. At a higher point on the rock or piling, the mussels may be displaced by the barnacles. However, hydroids and others may arrive and survive at a later stage during a period of development.

4. Community development stage. At this stage the community is slowly developed. For example, new species enter the realm of the mussels and survive in the sheltered spaces or on the mussels themselves. Hydroids and sea squirts may flourish. Bryozoans, foraminiferans, and more diatoms may settle on the hydroids and grow, forming a "jungle" that collects detritus.

5. The climax community. A climax community is ordinarily named for the dominant species that is present. For instance, a fouled substrate in which mussels dominate would be called a mussel community. In such a community, the mussels make up most of the biomass and play a key role in determining the nature of the community. For example, the mussels may provide a substrate for hydroids, protection for amphipods, and food for sea stars. Individuals of smaller species such as amphipods might be present in greater numbers than the mussels, but the success of the mussels would determine the success of the overall community.

On a piling (Fig. 11.5), there actually can be many communities, with the dominant species depending on the position of the piling and the resultant effect of tides, temperature, salinity, light, and—of course—conditions imposed by neighboring organisms. As can be seen in Figure 11.5, bird droppings may be a source of nutrients for the community of blue-green algae. The set of communities on a piling, in fact, is an especially dramatic and condensed example of zonation, which was discussed earlier (Chap. 9).

The climax community may develop within months, as in the case of a rock denuded by ice, or it may take years, as in the case of a clam bed destroyed by dredging.

FIGURE 11.5 Simplified view of a typical zonation on a piling. The bird symbolizes the role of birds as a source of nutrients for the plant life of the sea.

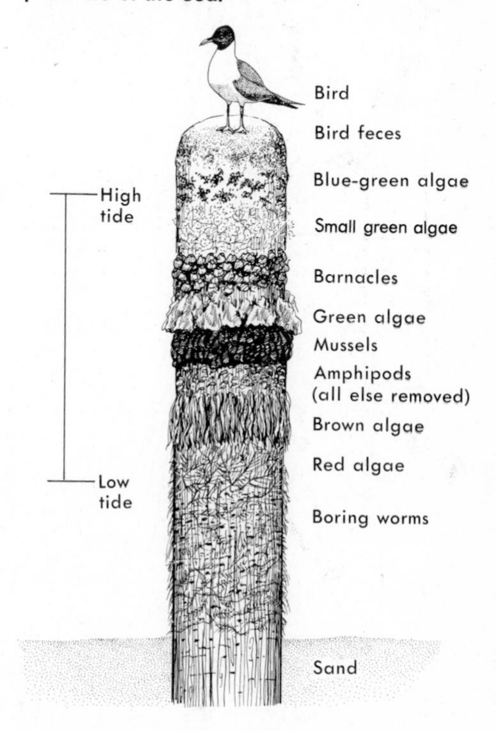

Pelagic Succession

Succession is not limited to growths on pilings and rocks. Succession of plankton communities occurs with seasonal temperature and salinity changes, with one species providing nutrients or food for the next species in succession. In this case, the substrate does not play a great role except that it provides a home for some adults developing from planktonic larvae, such as barnacles, crabs, and clams.

The species composition of the nekton (swimmers), benthic organisms (bottom-dwellers), zooplankton (drifting animals), and phytoplankton (drifting plants) varies with the seasons. Species abundant in the spring may be scarce in summer, fall, or winter (Fig. 11.6*a* and *b*). The changes follow definite patterns that tend to be repeated annually, but extraneous factors such as pollutants or differences in temperature from year to year may alter the

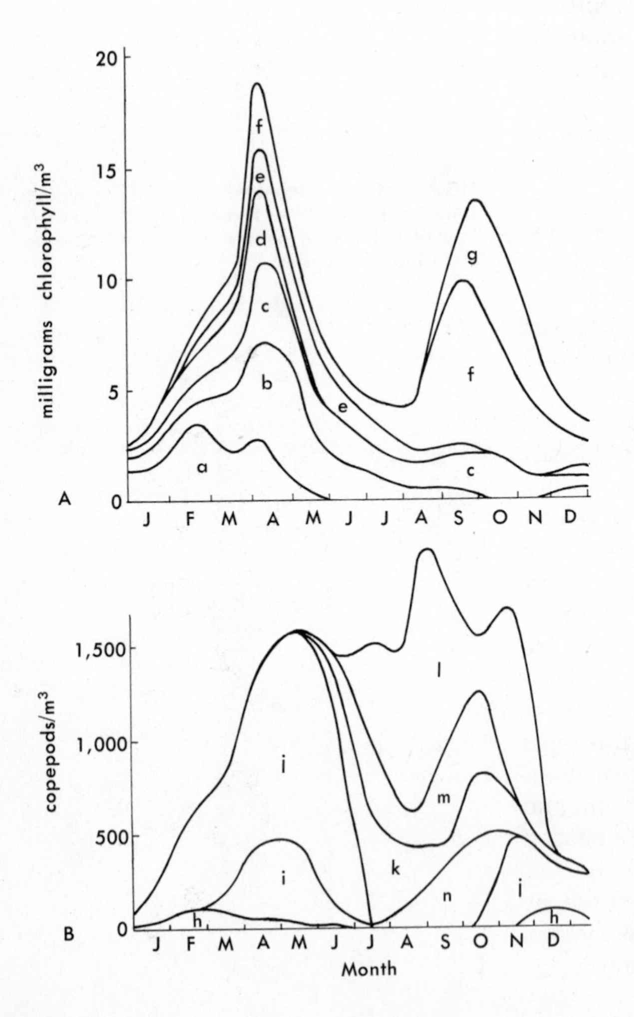

FIGURE 11.6 Hypothetical succession of phytoplankton (*A*) and zooplankton (*B*) species in temperate coastal waters. Top lines represent the total chlorophyll (in *A*) or numbers of copepods (in *B*) present in a cubic meter of water. The differently shaded areas represent the contribution of the individual labeled species present. For example, phytoplankton species "f" contributes approximately half the chlorophyll present in October. (Synthesized from various sources.)

patterns of succession. In fact, it may be difficult to determine what "normal" succession really is in some cases.

A species increases in number when conditions are favorable for its survival and decreases when they are unfavorable. For example, in early spring the surface water in temperate latitudes is rich in nutrients, and the increasing temperature and light cause a sudden growth (bloom) of phytoplankton (Fig. 11.6*a* and 11.7). The phytoplankton soon decline in numbers owing to nutrient depletion in the temperature-stratified surface water and increased grazing by zooplankton. In the fall, the thermocline breaks down and nutrients are mixed into the surface water by turbulence, and another smaller bloom of phytoplankton may occur, sometimes followed by another small burst of zooplankton herbivores. The fall bloom quickly dies out because of cooling of the water and decreasing light. Small blooms may also occur during the summer, when nutrients are brought into the surface water by upwelling.

The phytoplankton blooms really consist of a succession of smaller blooms of several plant species (Fig. 11.6*a*). Each species is adapted to different conditions of temperature, light, and any chemical secreted by the previous bloom. Thus, species that are abundant in the spring are frequently different from those abundant during the other seasons.

The zooplankton most prevalent during the summer in temperate water also exhibit a succession of dominant species (Fig. 11.6*b*). Temperature and salinity requirements and chemicals secreted by previous plant and animal bursts may control the seasonal succession of these species. All of these successions contribute to the ever-changing communities of interacting organisms.

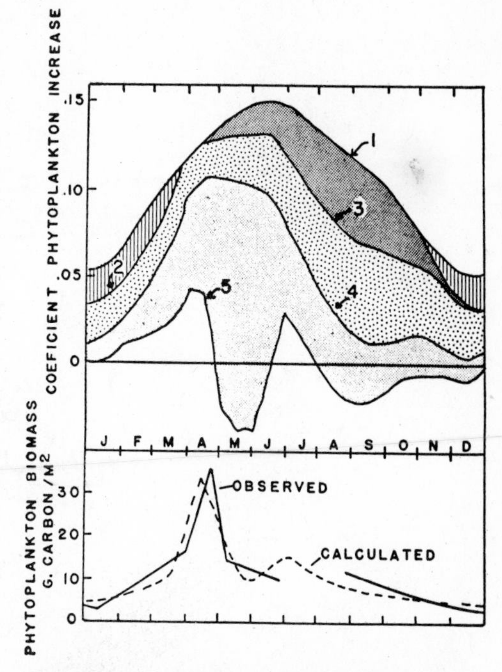

FIGURE 11.7 Effects of six limiting factors on the phytoplankton (upper) and the observed and calculated phytoplankton densities (lower) off the New England coast. The top line is the expected rate of increase in the phytoplankton population considering factor (1), the available light and temperature. Each of the other factors takes away from the expected rate of increase when the previously considered factors have been applied. For example, the dotted area under line (3) indicates the limiting effect of factor (4). Factor (2) = turbulence, which carries cells below the photic zone; (3) = phosphate depletion; (4) = phytoplankton respiration; and (5) = zooplankton grazing, which produces the final coefficient of change in the phytoplankton, allowing for all of these factors. Only during the spring and late summer are conditions favorable for the rapid growth of the population. (After Riley, 1952; from Odum, 1971.)

FOOD WEBS

A food web is simply a pattern of feeding relationships among organisms; an ecologist would call it a diagram of ***trophic relationships***. Each community has a characteristic food web, although communities that are fairly close to each other are actually linked in a much larger web because all communities ultimately depend on the primary production of the local phytoplankton (and other plants), and bacteria decompose detritus from all communities.

Plants capture the sun's energy and store it in the form of organic matter, such as sugar and fat. Seaweed and phytoplankton are eaten by animal herbivores such as some amphipods, copepods, clams, and some fish. The herbivores are, in turn, eaten by primary carnivores such as jellyfish, sea stars, and most fish. Primary carnivores are eaten by larger fish or invertebrates (the secondary carnivores). These, in turn, are eaten by tertiary carnivores, and so on. This sequence may be illustrated by a food chain such as the following:

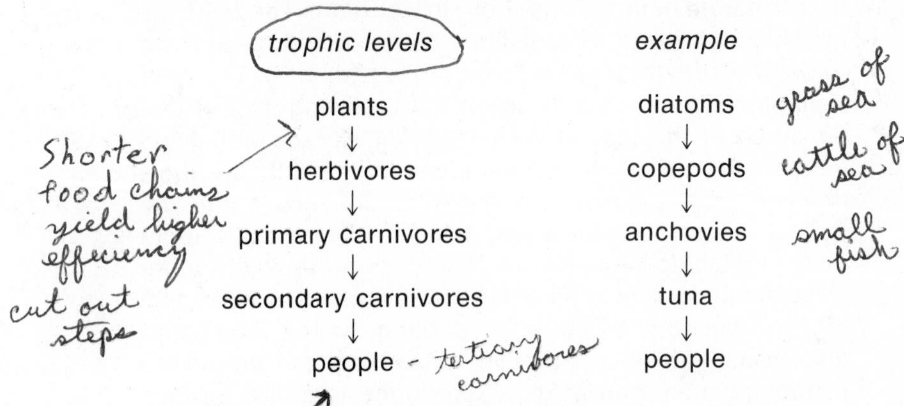

A food chain shows the direction of movement of organic matter or energy as it passes from one organism to another but does not indicate amounts of organic matter (biomass) or energy (calories) required at each level. A ***pyramid of biomass*** or a ***pyramid of energy*** illustrates these factors (see Fig. 11.8).

Each level of the pyramid is called a trophic level (feeding level). If we assume that each transfer of food energy is 10 per cent efficient, it takes about 10 kg of phytoplankton to produce 1 kg of copepods, since copepods may feed directly on phytoplankton. Similarly, it takes about 1,000 kg of phytoplankton, which are transferred through two additional trophic levels, to produce 1 kg of tuna. The copepod or the tuna utilizes only 10 per cent of its food for building its body (production). Efficiencies may vary considerably among species and within a species as well as at different times in the life cycle. Ranges of 10 to 20 per cent efficiency may be normal, but efficiencies as high as 30 per cent or more are possible, especially in areas of nutrient enrichment. Most of the energy consumed is lost as respiratory heat and undigested wastes that are decomposed by bacteria, fungi, and detritus feeders (scavengers). The energy budget of a typical salt-marsh fish is shown in Figure 11.9.

The efficiency of energy transfer from an organism to the organism that eats it (***trophic efficiency***) depends on the nature of the food, or its food value, and its ease of capture. For example, a copepod may be able to eat a large diatom or a colony of diatoms but not the smallest of phytoplankton (***nannoplankton***), which are less than 0.1 millimeter (mm) in diameter. The larger the chains of diatoms, the easier it is for copepods and other larger herbivores to eat them. Less energy is used to capture food.

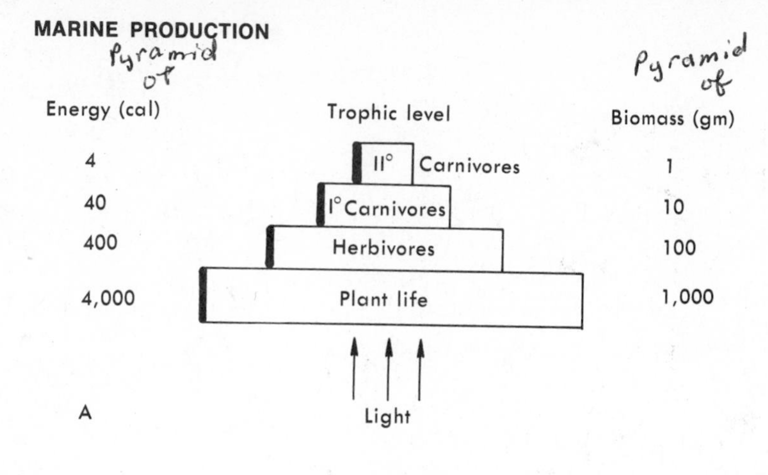

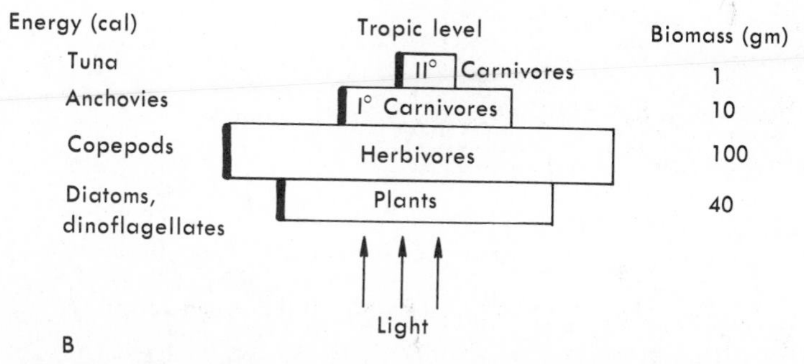

FIGURE 11.8 *A,* Pyramid of energy and biomass for a marine ecosystem, showing quantities required for growth. The quantities given are in approximately correct ratios to the other levels of the pyramid if we assume 10 per cent efficiency throughout.

B, Pyramid of biomass for productive coastal water, showing relative amounts present at one time. The apparent paradox in the greater amounts of herbivores than plant food is due to the fact that the plants are produced very rapidly and quickly replace those eaten.

C, Pictorial representation of a pyramid of biomass of the type shown in *A.*

If the phytoplankton are very small, for example coccolithophores, they may be eaten by protozoans such as foraminiferans and radiolarians, which are then eaten by carnivorous copepods at a higher trophic level. The larger phytoplankton are found mostly in nutrient-rich areas such as coastal or upwelling regions. The minute phytoplankton tend to predominate in the less nutrient-rich offshore oceanic regions.

Energy may pass through as many as six oceanic trophic levels before reaching people. In the rich updwelling areas, only two or three trophic levels are required to produce edible anchovies, and most coastal regions contain at least three or four trophic levels.

Because of the greater efficiency and shorter food chains found in upwelling areas, tremendous amounts of fish are produced in a small area (only about 0.1 per cent of the world's oceans). These relationships are sum-

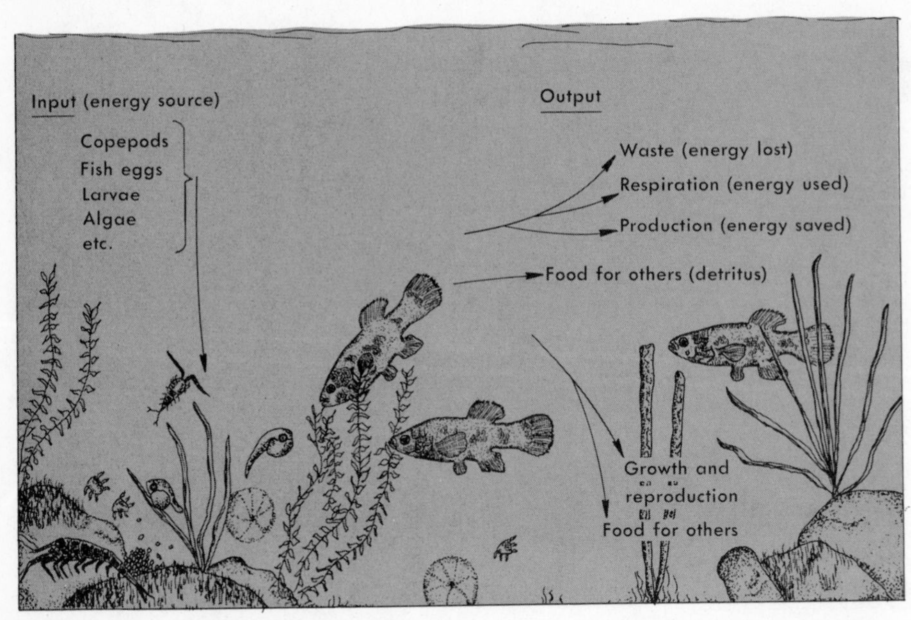

FIGURE 11.9 Energy budget of a salt marsh fish.

marized in Table 11.1. Ten per cent of the ocean (upwelling and other coastal areas) may provide 240 million tons of fish, whereas 90 per cent of the oceans (the oceanic area) may provide only 1.6 million tons of fish per year. The 0.1 per cent of the ocean that lies in upwelling areas equals the fish production of the other 9.9 per cent of the ocean that is regarded as coastal. The high productivity that is possible in upwelling areas is largely due to the nutrient-rich water, high trophic efficiency, and short food chains. These are also areas that are being heavily exploited for food fish because of the small amount of fishing effort needed to catch the abundant fish crops.

The food chains described in the preceding discussion are extremely simplified. Of course, a copepod or a clam eats many different kinds of phytoplankton. Any one species of phytoplankton may be eaten by a number of species of herbivores. Clams may be eaten by starfish, people, fish, or other carnivores. The food chain or pyramid of biomass does not show these trophic details, which are, however, illustrated by the ***food web***. Figure 11.10 shows a greatly simplified marine food web.

All members of the food web eventually die and decompose.[5] The organic compounds (fats, sugars, starches, proteins, and so forth) are converted by bacteria and other detritus-feeders into simple inorganic compounds, which dissolve in seawater and provide the nu-

[5]Diatoms and other phytoplankton that reproduce asexually by division are sometimes considered "immortal." An individual cell may live a long time if it is not eaten or killed under adverse conditions.

Table 11.1 TOTAL ANNUAL POTENTIAL PRODUCTIVITY ESTIMATES FOR THE WORLD'S OCEANS, ASSUMING VARIOUS EFFICIENCIES OF ENERGY TRANSFER*

	OCEANIC AREAS		*COASTAL AREAS*		*UPWELLING AREAS*	
% of Ocean Area	90		9.9		0.1	
Trophic Efficiency (estimated %)	10		15		20	
Mean Production (grams c/m^2/yr)	50		100		300	
Organic Carbon Produced per Year (in millions of tons)						
phytoplankton	16,000	nannoplankton	3,600	diatoms, dinoflagellates	100	colonial phytoplankton
herbivores	1,600	protozoa, copepod larvae	540	copepods, clams	20	copepods, krill, clams, anchovies
I° carnivores	160	copepods	81	anchovies, benthic fish	4	anchovies, larger fish
II° carnivores	16	chaetognaths	12.2	larger fish		
III° carnivores	1.6	small fish				
IV° carnivores	0.16	larger fish				
Average number of Marine Trophic Levels	6		4		$2\frac{1}{2}$	
Total fish Production (in millions of tons *fresh* wt/yr)	1.6		120		120	

*From data of Ryther, 1969.

trient raw materials for plant production. The nutrients are recycled, but the energy that is lost as heat is not recycled. Therefore the propagation of life on earth is dependent on the continuing input of energy from the sun.

Figures 11.11 and 11.12 illustrate the flow of energy through communities in the photic and aphotic zones, respectively. Some of the export of energy as detritus from the photic community may become input of energy to the aphotic community as a "rain" of detritus.

As an ecosystem becomes polluted, some species die out and others replace them; or the remaining "pollution-tolerant" species flourish on the rich pollutants. Figure 11.13 shows a partial food web of a healthy estuarine salt marsh. Some of the changes that occur when the marsh is disturbed by human intervention are illustrated in Figure 11.14, which shows the food web existing now in the Hackensack Meadowlands surrounding Secaucus, New Jersey.

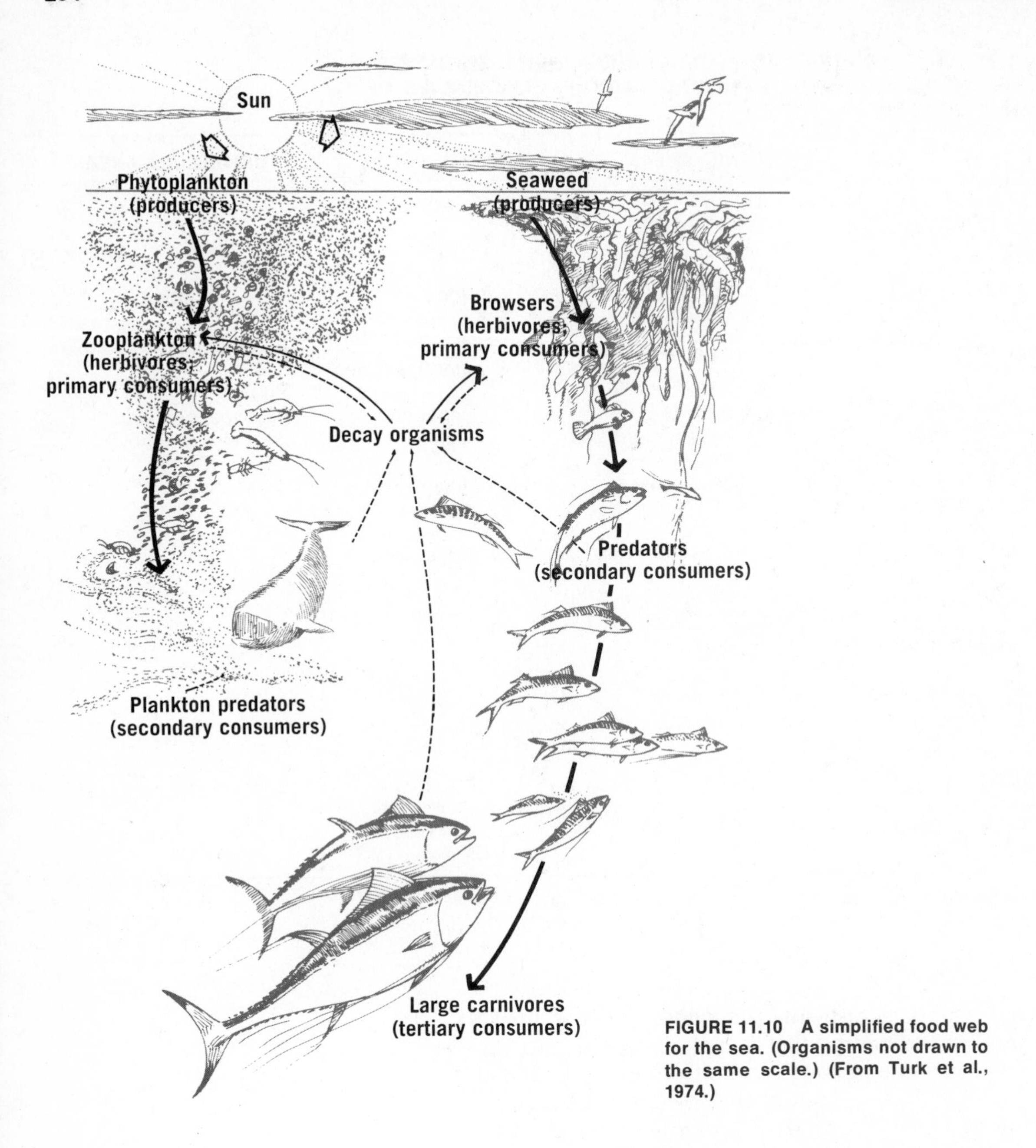

FIGURE 11.10 A simplified food web for the sea. (Organisms not drawn to the same scale.) (From Turk et al., 1974.)

Note that many species have disappeared. However, some rugged species such as killifish have become even more abundant, since they no longer must compete with sticklebacks, silversides, and sheepshead minnows. Important food fish such as bluefish, striped bass, and flounders; and shellfish such as oysters and mussels are lost because they cannot tolerate the highly polluted conditions.

The change from healthy to disturbed marsh is sometimes indicated by the shift in growth from salt-marsh hay

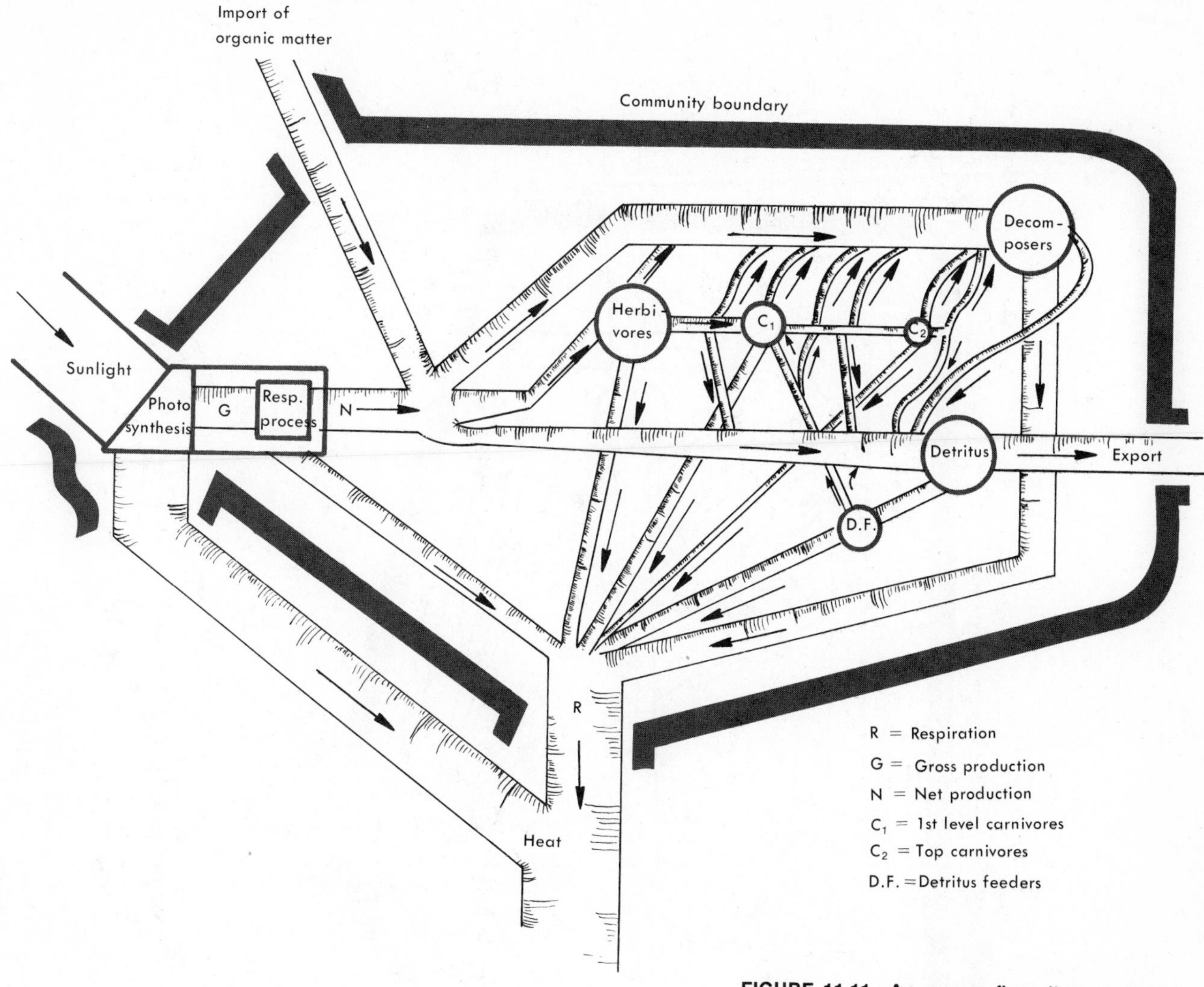

FIGURE 11.11 An energy-flow diagram for a lighted (photic) zone in the sea. (Modified after Odum, 1956.)

(*Spartina*) to plume grass (*Phragmites*) as deposits of detritus accumulate on the marsh. Plume grass is not as desirable a food for marsh animals as is salt-marsh hay, although it is important as a food and/or habitat for muskrats, rabbits, mosquitoes, and killifish, and as a nesting ground for shore birds. The Hackensack Meadowlands of New Jersey are a good example of a disturbed salt marsh where plume grass is a dominant flowering plant and killifish abound.

Disturbed marshes such as the Hackensack Meadowlands may be beyond repair, but they should be protected from needless further destruction because of their present importance as a source of food and habitat for the organisms that live in them and because they help to protect

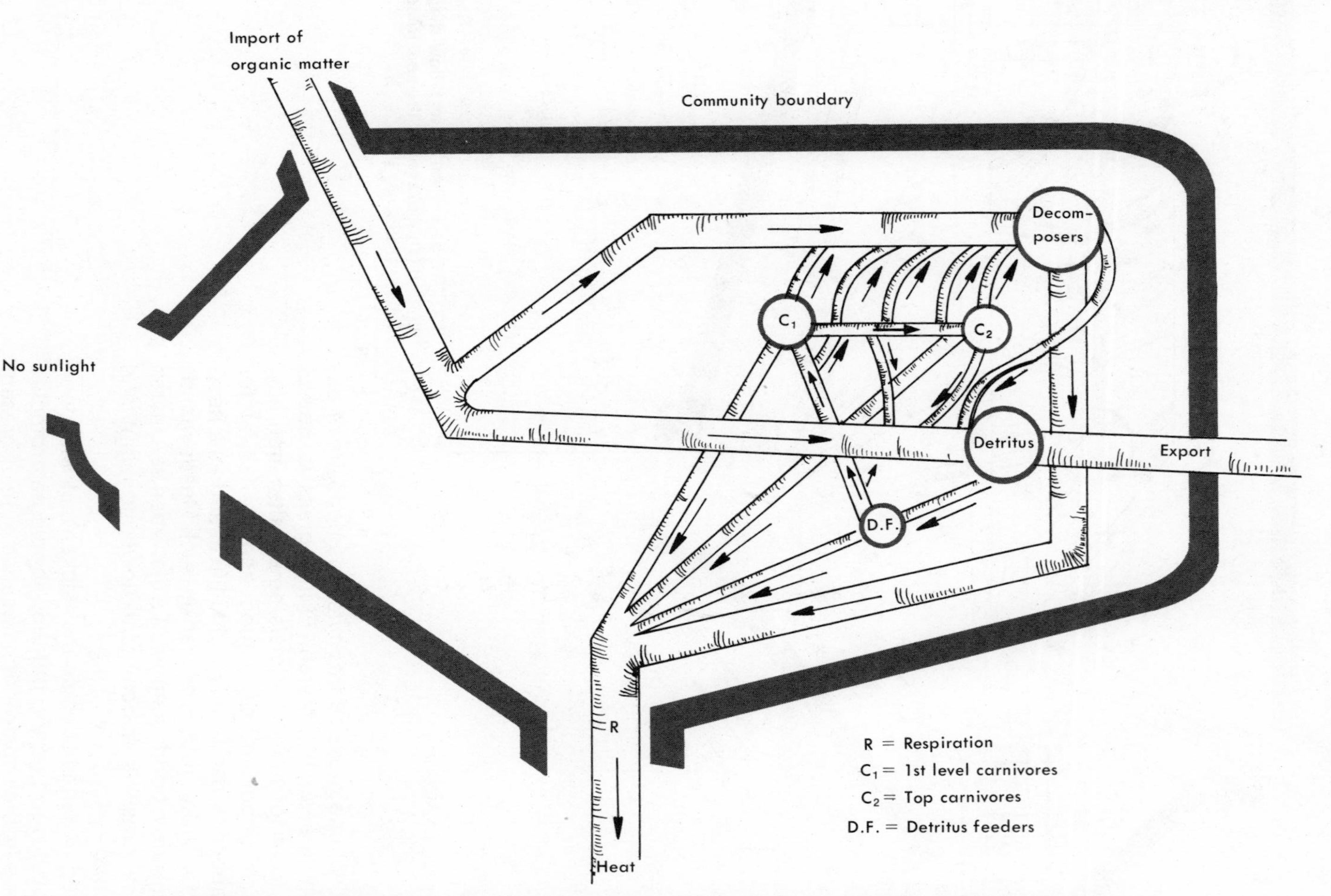

FIGURE 11.12 An energy-flow diagram for the deep sea (aphotic zone). (Modified after Odum, 1956.)

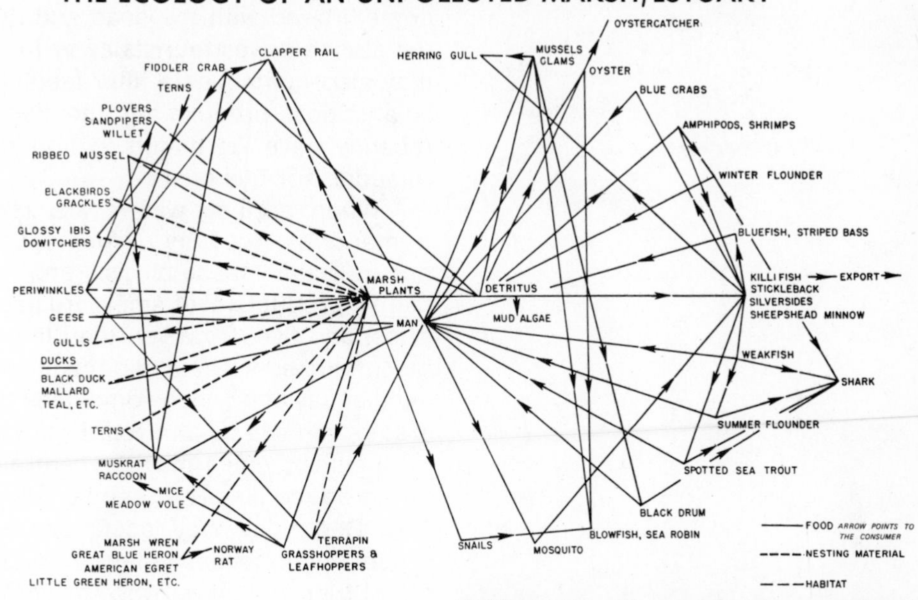

FIGURE 11.13 The ecology of an unpolluted salt marsh estuary. (Food web diagram by E. E. MacNamara, courtesy of Hackensack Meadowlands Development Commission.)

the surrounding waters by absorbing excess nutrients carried by polluted water.

RED TIDES AND OTHER BLOOMS

For centuries people have observed that the surface of the ocean occasionally turns reddish near the shore or in bays and estuaries. Frequently, catastrophic fish kills

FIGURE 11.14 The ecology of a disturbed salt marsh estuary. (Food web diagram by E. E. MacNamara, courtesy of the Hackensack Meadowlands Development Commission.)

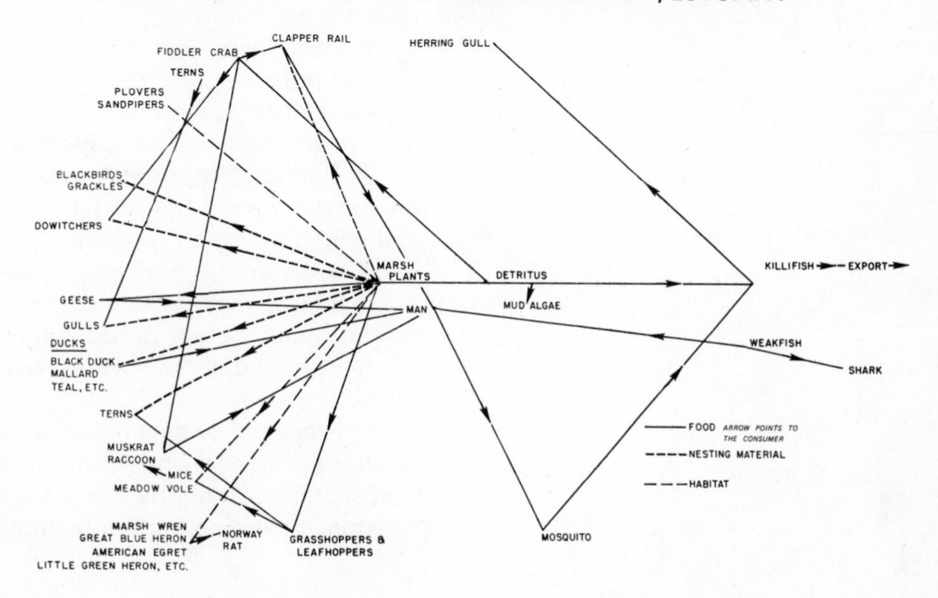

are associated with these red waters. The beaches become littered with the dead and dying bodies of fish and the stench drives tourists away from resort beaches. They may also contaminate filter-feeding shellfish. Clearly this is a critical problem for the fish and for people. Only recently have we begun to find the cause and possible remedies for this crisis.

When surface waters are enriched with nutrients, tremendous growths (***blooms***) of phytoplankton may occur. Usually these blooms are harmless or even beneficial to production of small crustaceans (especially copepods) and fish. However, the bloom occasionally consists of certain species of dinoflagellates that produce toxic substances and which sometimes but not always cause an area of the sea to turn red. Red-tide organisms are eaten by shellfish, which may concentrate the toxic substances and produce illness in people who eat the affected shellfish. Red tides may occur because of natural mixing processes or because of man's activities. Agricultural run-off, nutrient-rich river flow, or sewage effluents may contribute. Factors such as temperature, light, salinity, turbidity, and nutrient content are also involved, so it is not always possible to assign a specific cause to a bloom.

One example of periodically recurring red tides takes place off the west coast of Florida. They have been attributed to increased amounts of phosphate and iron added to the sea by unusually heavy rainfall run-off. Very importantly, tannic acid also enters the sea with the rainfall run-off and increases the solubility of iron in seawater. The extra iron stimulates the growth of *Gymnodinium breve,* a notorious toxic red-tide dinoflagellate.

From many other examples, scientists have pieced together a general model of how most red-tides probably originate. Among the most favorable conditions for red-tides are the following:

1. Upwelling, tidal mixing, and run-off, causing enrichment of nutrients in the surface water. Run-off from agricultural and phosphate-rich land may be especially important, as for example in the Florida red-tides.

2. Diluted seawater (lowered salinity) appears to aid the growth of red-tide organisms. This condition is particularly apt to occur near the mouths of rivers.

3. Vitamin B_{12} from soil, marshes, blue-green algae, and bacteria may aid the growth of dinoflagellates, as many of them cannot produce this vitamin on their own.

4. Iron and tannic acid from industrial and swamp run-off.

5. Presence of small quantities of red-tide organisms which can survive when nutrient concentrations are low.

6. Other conditions favorable for growth of bloom organisms, including optimum light, temperature, organic

and inorganic nutrients. Possible ways to control red-tides include the following:

1. Introduction of vitamin B_{12}–destroying bacteria.
2. Encouragement of natural predators that eat phytoplankton.
3. Copper poisons to control production (copper is toxic to many aquatic plants, but unfortunately also to many animals).

The above controls may have unwanted side effects and may also be rather expensive. A fourth method of control and probably the most scientifically sound one is to limit concentrations of nutrients entering the sea. This might involve the use of certain plants to extract the nutrients from effluents before they enter the sea, or to stringently control the needless use of large quantities of fertilizers.

We should note that not all red waters are toxic. *Noctiluca*, a dinoflagellate, may turn the sea the color of tomato soup, but it is not toxic. The Red Sea and Gulf of California at times get their red color from certain species of "blue-green algae" that actually have a reddish color.

FISHERIES

We have long depended on the sea as a source of food. These resources have been regarded as inexhaustible until at the beginning of the twentieth century it was found that

FIGURE 11.15 A modern stern-trawler towing an otter trawl over the sea floor.

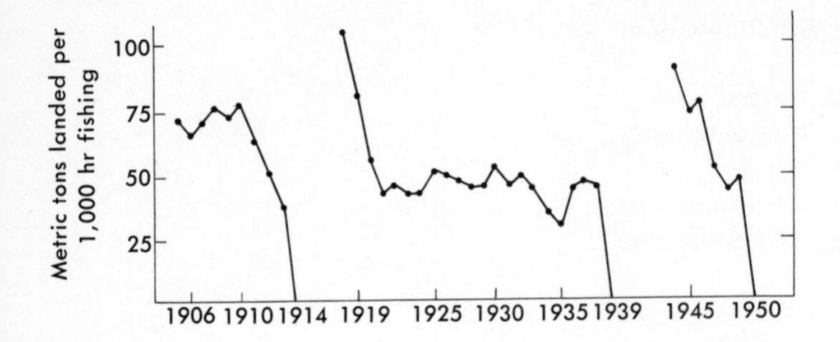

FIGURE 11.16 Catches of haddock by Scottish vessels in the North Sea, showing the effects of the world wars. (After Russell-Hunter, 1970, and Graham, 1956.)

some species of fish were being depleted by "overfishing." Overfishing is generally credited to the development of efficient trawl nets (Fig. 11.15) in the nineteenth century. Toward the middle of the twentieth century, commercial fishermen began to develop "less efficient" nets that would catch only the larger fish and allow the young to remain free to grow and reproduce. However, recent advances in the technology of making food and fertilizer from any size or kind of fish has put tremendous economic pressure on fishermen to sweep up everything in sight.

Fortunately, fish populations apparently can "bounce back" from depletion if the condition is not too severe. During the First and Second World Wars, for instance, the North Sea fishing grounds were closed because of the mines placed in the waters. Much greater catches of fish were made just after the wars as compared to the years just before (Fig. 11.16). Apparently the fish stocks were renewed during the enforced moratorium on fishing.

Declines in fish populations may be due to factors other than fishing. When the ***mortality*** rate increases for any reason, more fish die, and the fish stock declines. Any change in the environment—for instance, the nutrients available, the presence of predators, or increased pollu-

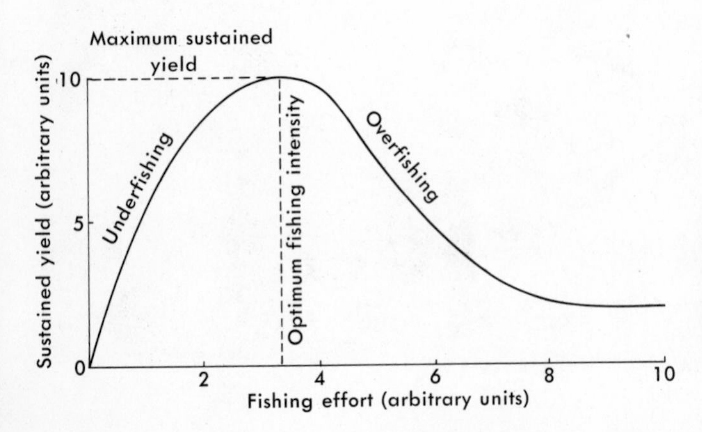

FIGURE 11.17 The relationship between sustained yield and fishing intensity. Notice that as fishing effort is increased, the yield of fish is increased until a maximum is reached beyond which fish catches decline. Beyond the optimum fishing intensity, increased effort results in smaller catches, owing to depletion of the fish population. (After Russell-Hunter, 1970.)

tion—can kill fish and reduce the numbers available for human consumption.

Factors that tend to increase the biomass of fish include ***reproduction*** and ***growth.*** Overfishing occurs if fishing and mortality take place at a rate greater than the natural reproduction and growth rates. On the other hand, if the fishing rate is less than the reproduction and growth rates, we can say that the fishing grounds are "underfished." A "sustained yield" is maintained when fish catch and mortality equal reproduction and growth. These relationships are illustrated in Figure 11.17 and by the following equation developed in 1946 by E. S. Russell, an expert on overfishing:

$$S = S_0 + (A + G) - (M + C)$$

where:

S = biomass of fish at the end of the year
S_0= biomass of fish at the beginning of the year
A = amount reproduced for the year
G = growth for the year
M = mortality (natural and pollution-caused) for the year
C = fish catch for the year

During a period of sustained yield, therefore, $(M + C) = (A + G)$. We might infer from this relationship that a fish or whale with a slow growth rate (and generally a long life) and low reproductive rate would be more apt to be overfished than one that grows quickly and produces many offspring. Whales, which grow slowly and produce only one offspring every other year, have in fact fared badly at the hands of whalers.

Intensive whaling has resulted in the near-extinction of several species, including the northern right whale. In the Antarctic, 8,000 blue whales were caught in the summer of 1948; 1,684 in the summer of 1958, and only 20 in the summer of 1965. Although an international commission was set up in 1946 to control Antarctic whaling, it apparently did not benefit the blue whale! If sustained-yield whaling had been practiced in the beginning, an estimated annual catch of 6,000 blue whales could have been maintained. Meanwhile, protective legislation has enabled the population of California grey whales to climb from a few hundred in 1938 to more than 10,000 in 1967, permitting a certain amount of fishery exploitation.

At one time, fishermen threw away fish that they regarded as "trash fish," such as sea robins and sharks. These were usually discarded dead and were recycled into the ecosystem by scavengers. However, the effort expended in catching them was wasted. Many other fish were too small to bother with, except in the case of delicacies such as sardines and anchovies. Today these "unwanted" fish form the basis of a great industry in fish meal, which is fed to poultry, swine, and pets. Further treatment of the fish meal to extract the protein is being done on a small scale. This product, known as ***fish protein concentrate (FPC),*** may go a long way toward improving the nutrition level of the hungry people of the world, if maximum sustained yield can be achieved. FPC is nutritious, almost tasteless, and simple to store without refrigeration. It can be easily added as a nutritious "extender" to other foods such as bread and soup. Further exploitation of the sea's resources without overfishing will probably involve more technological advances like protein extraction. If this protein can be extracted from the lower trophic levels, the protein yield of the sea could be greatly increased.

New potential sources of food include shrimplike krill (*Euphausia superba*), a staple food of baleen whales; lantern fish (myctophids); and the pelagic "red crab" (*Pleuroncodes*), an important food of tuna in the Pacific.

Another by-product of fish that cannot be ignored is edible oil. We all know of iron-rich cod-liver oil. But how many realize that the sea is a major source of the hydrogenated oil used in the production of some margarine? Anchovies, herring, and whales are important sources of this oil.

Despite the decline in the catch of certain "overfished" species, the total world catch of marine organisms (plant and animal) increased from about 20 million metric tons in 1938 to about 40 million metric tons in 1960 and to almost 60 million metric tons in 1966. This trend parallels the development of large factory ships and more sophisticated fishing techniques, such as the use of special nets and electronic lures.

Estimates made in 1968 indicate that perhaps food from the sea might be increased 2½ times, or to about 150 million metric tons per year. This total seems quite realistic, perhaps even conservative, considering the great potential of fish protein concentrate and the further use of marine algae. About 30 per cent of the world's protein requirement for a population of 6 billion in the year 2000 would be satisfied by 150 million tons of seafood. Of course, this assumes moderate success in population control. Can we or will we be sensible enough to control our future?

FIGURE 11.18 ***A,*** **Raft with submerged rack for growing oysters.** ***B,*** **Raft with submerged ropes and attached oysters. (After Bardach et al., 1972.)**

A

B

MARICULTURE

Whereas fisheries biologists are most concerned with sustained yield and fishing rates, mariculturists are inter-

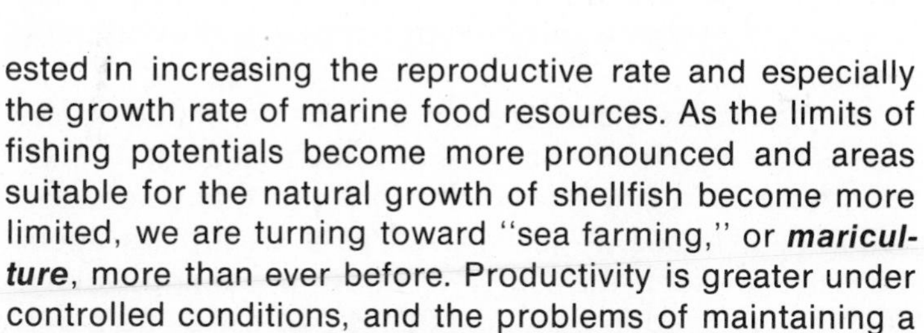

ested in increasing the reproductive rate and especially the growth rate of marine food resources. As the limits of fishing potentials become more pronounced and areas suitable for the natural growth of shellfish become more limited, we are turning toward "sea farming," or ***maricul-ture***, more than ever before. Productivity is greater under controlled conditions, and the problems of maintaining a fleet of large fishing vessels are eliminated.

In Java, about 60 per cent of the fish protein comes from the milk fish (*Chanos chanos*). The fish are collected at sea as juveniles and placed in shallow brackish ponds. Organic fertilizers such as sewage are added. The fertil-izer enables a mat of algae to grow on the bottom of the pond, providing food for the fish, which are grown to ma-turity in captivity.

Oysters usually grow in nature on the bottom, on a hard substrate such as a shell or rock in shallow water, especially in estuaries rich in phytoplankton food. Pro-duction may be limited by pollution, including industrial and domestic sewage, and by the effects of dredging, which may increase sedimentation on the oyster beds themselves. In 1967, 29,000 tons of eastern oysters were harvested in the United States, whereas the production in 1880 was 76,000 tons. Increased pollution, dredging, and covering of large areas of suitable oyster habitats with dredge spoils or other fill caused much of the decline over the years. Pollution abatement may increase production of oysters. The growth of oysters is also limited by the area that is populated and by natural siltation, which may bury the oysters or may clog their filtering apparatus.

Oysters are frequently grown either on racks sus-pended from rafts or attached to ropes suspended from rafts or floats (Fig. 11.18). The racks keep the oysters out of the sediment but in the tidal waters, which are rich in phytoplankton food. Whereas bottom-grown oysters are limited by the ***area*** covered, raft-grown oysters are limited by the ***volume*** of water, since more than one oyster may be grown on a vertically hung rope. Many more oysters may be grown under a given area of rafts than on the bot-tom, assuming a plentiful source of food. By growing oy-sters on ropes or racks, it is possible to increase produc-tion from about 800 pounds per acre for bottom-grown

FIGURE 11.19 The giant kelp *Macrocystis pyrifera*, a major source of algin.

oysters to as much as 64,000 pounds per acre. They are also protected from predation by sea stars.

In the past few years, experimental culture of various crustaceans, including shrimp and lobsters, indicates that (1) plenty of food and (2) water that is warmer than normal for the animals contribute to very rapid growth of these organisms (four times the normal growth rate for lobster). This suggests that commercial cultivation of these crustaceans in tanks or impoundments, possibly utilizing the now wasted heated water from power plants, may soon be possible.

Some marine plants are also of economic importance. Most of the ice cream that we eat is stabilized with a substance called algin, extracted from the giant seaweed (kelp), *Macrocystis*. Algin is used as a smoothing agent and helps to keep the ice cream from becoming watery. Algin is also used in salad dressings and candy bars and in the production of synthetic fibers and plastics. *Macrocystis* is harvested off the coast of California by huge barges that cut the tops of the kelp with blades, much as a lawn mower trims a lawn. This seaweed (Fig. 11.19) grows up to 60 meters long in water from 8 to 30 meters deep, off Southern California. The new shoots and reproductive organs grow from the base, which is attached to rocks on the sea bed. The flexible stem with leaflike fronds is held upright by gas-filled floats which the plant produces. In order to maintain the "forest" in a healthy condition, only the portion within four feet of the surface is harvested. Harvesting reduces the shade on the young shoots, probably increasing their rate of production by exposing them to more sunlight.

Another plant, the red alga *Porphyra*, is grown by the Japanese on sticks placed in the sea. It is periodically harvested, dried, and eaten in dishes such as seaweed soup.

Mariculture may even derive benefit from human pollution. From highly populated and industrial shore cities, tremendous amounts of nutrients are dumped into our estuaries and near-shore ocean as sewage. If uncontrolled, these nutrients and organic matter can be dangerous. Proper treatment, however, can make these nutrients available to grow phytoplankton in tanks as a food for clams and oysters. The attached algae that grow on the sides of the tanks may be grazed upon by snails, which in turn may someday be eaten by people, or at least ground up as animal feed.

OTHER USES OF MARINE ORGANISMS

Marine organisms are tremendously important in biomedical research. The giant nerves of the squid are ideal for the study of the travel of nerve impulses. The basic mechanisms of reproduction, fertilization, and development have been discovered by using echinoderms such as sea urchins and sea stars.

Many chemicals of pharmacological interest have been extracted from marine animals. The simple sponge produces a chemical that appears to be effective in the treatment of some virus infections and of leukemia in mammals. Poisons found in sea stars and sea cucumbers include digitalis, an important drug used in treating heart disorders. Other chemicals from echinoderms have slowed the growth of tumors in laboratory animals. The potential of marine organisms as useful sources of medical information and drugs has only begun to be realized.

Marine organisms will continue to be a vital source of food as well as a source of drugs. Mariculture may ease the pressure on the sea fisheries, but even mariculture is a finite source of food. Productivity is limited by the energy reaching the sea from the sun. Therefore, the production of fish and other economically important marine organisms is also limited.

SUGGESTED READINGS

Idyll, C. P. 1973 (June). The Anchovy Crisis. *Scientific American.*

Holt, S. J. 1969 (September). Food Resources of the Oceans. *Scientific American.*

Hunter, S. H., and John McLaughlin. 1958 (August). Poisonous Tides. *Scientific American.*

Pinchot, Gifford B. 1970 (December). Marine Farming. *Scientific American.*

Raymont, J. E. G. 1963. *Plankton and Productivity in the Oceans.* N.Y., Macmillan.

Russell-Hunter, W. D. 1970. *Aquatic Productivity.* N.Y., Macmillan.

Ryther, John. 1969. Photosynthesis and Fish Production in the Sea. *Science, **178***: 72–76.

The Soviet factory ship *Trudovaya Slava*, one of the mother ships of a large foreign fleet trawling off the Atlantic Coast of the United States in 1969. (Official U.S. Coast Guard photo.)

CHAPTER 12

Political Oceanography

". . . under no circumstance, we believe, must we ever allow the prospects of rich harvests and mineral wealth to create a new form of colonial competition among the maritime nations. We must be careful to avoid a race to grab and to hold the lands under the high seas. We must ensure that the deep seas and the ocean bottoms are, and remain, the legacy of all human beings."

Lyndon B. Johnson, July 13, 1966

Human activites may result in beneficial or detrimental modifications of the oceans, especially near shore. We have already discussed the effects of industrial and domestic pollution and the building of shoreline structures (see Chapter 8), and we will now discuss some of the decision-making processes that affect the use and misuse of the seas. These are largely political in nature because they are controlled by governmental processes at the local, national, and international levels. For example, pollution of the coastal oceans may be prohibited by laws but permitted when the laws are not enforced, or permitted in the absence of appropriate laws.

The nature of the oceans and shores is increasingly dependent on the uses to which they are put. Technology has reached the state at which no part of the ocean is immune from human interference. Conflicts range from the right of a private citizen to own an ocean beach to the right of a nation to mine a given portion of the sea floor.

Historically, ocean waters lying within three miles of shore (the distance that a cannonball could be fired) were claimed by the nearest coastal nation, whereas the high seas have been regarded as belonging to no nation but free to be used by all. In 1967, Ambassador Arvid Pardo of Malta proposed to the United Nations an extension of this concept, allowing that the seabed beyond the limits of national jurisdiction be internationalized and used for peaceful purposes and for the benefit of all mankind. He further proposed that the United Nations be given jurisdiction over the international seabed.

CONFLICT ALONG THE COAST

Beaches: Public or Private?

Local problems may develop when there are conflicts between groups of people who have different views concerning the use of the coastal ocean. For example, conflicts arise between advocates of private and public ownership of beaches. Sunbathing, swimming, surfing, fishing, and ocean-watching are all activities that make use of the beach and are certainly important to the coastal tourist industry.

Laws regarding the ownership of beaches vary from state to state and among nations. Beaches on the coasts of California, Oregon, Washington, Mexico, and Puerto Rico are open to the general public, and free access to virtually all of these beaches is maintained by law. On the other hand, many beaches along the Atlantic coast of the United States are controlled by individuals or municipalities. There is a trend toward making these beaches public, caused by the expense of maintaining the beaches. Natural processes such as storms frequently erode beaches, and states may provide funds for beach restoration. The state of New Jersey, for example, has withheld funds for beach erosion control projects from municipalities that do not open their beaches to the general public. Conversely, in many resorts in Florida, beaches are virtually inaccessible to anyone except paying guests at beachfront hotels. In such cases, the expense of beach maintenance is borne by the hotels themselves.

Coastal Development

Some coastal areas are really unsuitable for homes or industrial sites. Nonetheless, developers frequently fill marshes, build canals, or even build on barrier islands or unstable land above cliffs. Frequently, these areas are subject to coastal landslides or to inundation by storms. The necessary rebuilding may require millions of dollars of public funds. Perhaps what is needed is better land use planning to avoid unnecessary property damage and loss of life in the future.

The coastal regions are subject to a wide variety of uses that may be in conflict. Some bays have been partially filled to create more land for port facilities and industrial development that may pollute productive breeding grounds for clams, oysters, and fish. The filling of San Francisco Bay is a prime example. In 1835, the Bay occupied roughly 1760 square kilometers (680 square miles). By 1975, the area had decreased by some 40 per cent, to only 1040 square kilometers (400 square miles). Today, through local and national coastal management efforts as well as pollution abatement, some of these problems are being alleviated and future problems avoided.

Pollution

The welfare of the coastal regions also depends on activities that take place far inland. For example, pollution of streams far inland, even in land-locked states, may pollute the coastal ocean. Pollution may affect the success of breeding of anadromous salmon and hence the size of the adult population in the sea. Pollution may even be a cause of toxic red-tides along the coast.

Pollution of the coast may result from a wide variety of activities including ocean dumping of wastes, oil spills from ships, and offshore oil drilling. The ships that carry oil are much larger than in years past, and the magnitude of potential oil spills is consequently much greater than before. The transfer of oil to and from these supertankers frequently requires new offshore oil ports (Fig. 12.1) in deeper water than is found in most coastal harbors. Offshore drilling for oil will continue to increase for many years, and pollution associated with it may occur unless proper safeguards are maintained.

In the United States, both coastal states and the Federal government have been under pressure from environmentalists to devise and enforce very strict regulations. This concern was greatly heightened, if not originated, by two particularly severe accidents involving petroleum—the wreck of the tanker *Torrey Canyon* in 1967 and the spill from a Union Oil well in the Santa Barbara Channel in early 1969, both of which were discussed in Chapter 8.

Just how much damage is really done by human-caused pollution is questioned by some scientists. First of all, it is not clear that the damage from a disastrous oil spill is irreversible. At least one report in 1972 claimed that the marine ecosystem in the Santa Barbara Channel had returned to normal, and similar observations have been made regarding the beaches fouled by the *Torrey Canyon.* Also, at least some of the oil on many beaches results from natural seepage through faults in the sea floor. There are a minimum of 60 known natural seeps in the Santa Barbara Channel, for instance, so at least some oil found on the beaches today comes from sources beyond human control. The situation in the Gulf of Mexico is even more dramatic—some 12,000 natural seeps in an area of about 2300 square kilometers (900 square miles)

FIGURE 12.1 *A,* An Arabian American Oil Company off-shore oil port at Ras Tanura, Saudi Arabia, with several tankers at various stages of loading. *B,* The largest tanker ever built in the United States, which is more than 300 meters long and draws over 20 meters of water when loaded, is too deep to enter any United States port. After its maiden voyage it transferred its oil to smaller tankers bound for the United States at Rotterdam, Netherlands. (*A* courtesy of ARAMCO; *B* courtesy of the American Petroleum Institute.)

in the northwestern part of the Gulf. Thus, despite the tremendous amount of activity in oil drilling, shipping, and refining in the Gulf, we must attribute a fair amount of the asphalt blobs on Texas Gulf beaches to natural causes.

Still, some regulation is needed, and local governments must cooperate with the national regulating agencies. National anti-pollution legislation is aimed at protecting the environment for marine food resources. Fish do not abide by political boundaries and may migrate thousands of kilometers and through coastal waters of several states or even nations. Their survival depends on clean water and controlled fishing in all of these waters. In other words, their survival depends on cooperation among states as well as among nations.

TERRITORIAL SEAS

Historically, the leading naval powers have stressed the freedom of the high seas with limited jurisdiction over coastal waters. The seventeenth-century Dutch legal scholar Hugo Grotuis was one of the earliest proponents of this concept. In 1609, he published an essay, *Mare Liberum*, which presented the arguments in favor of freedom of the seas. He is frequently referred to as the "father of international law."

The narrow territorial sea claimed by the early naval powers was only 3 nautical miles (5.55 km), the distance that could be controlled by cannonball fire. Frequently these were nations that wanted free access to widely distributed colonies. Nations with restricted access to the sea or with weaker naval forces have strived to protect their interests with broad territorial waters over which they would retain sovereignty. The concept of a three-mile territorial sea was widely accepted until fairly recently, when some nations extended their territorial control to as much as 200 nautical miles (370 km). Some of the commonly claimed limits have been 3, 6, 10, 12, 18, and 200 miles. The 1975 Draft Treaty of the Law of the Sea Conference in Geneva proposes a 12 nautical mile (22.2 km) territorial sea for all nations that have a coastline. The treaty would, of course, give the right for unimpeded passage through the many international straits whose widths are less than 24 miles.

ECONOMIC ZONES

Some countries, in addition to their territorial seas, have claimed economic zones that extend national jurisdiction, mainly concerning fish stocks and pollution, out to 200 miles from land. This control, which has been exercised for quite a few years now, is essentially as proposed in the 1975 Draft Treaty of the Law of the Sea Conference. The Draft Treaty considers all of the natural resources (living and nonliving) within 200 miles of shore regardless of the width of the continental shelf as belonging to the coastal nations. The economic zones cover about 35 per cent of the ocean area. If a nation has a continental shelf wider than 200 miles, it would also own and control the resources of the seabed out to the edge of the shelf.

USES OF THE SEABED

The seabed is of great economic importance to us because of the wealth of natural resources that it contains. However, its ownership has been in doubt. In 1945, President Truman proclaimed that the United States owned the resources of the continental shelf adjacent to the United States.

The Geneva Convention of the Continental Shelf, in 1958, declared that the resources of the continental shelf belong to the adjacent nation. This is basically in agreement with the Truman Proclamation. When two countries share a continental shelf, they are to divide the shelf between them. Some shelves are as much as 900 miles wide, providing resources far out to sea. Other nations have essentially no shelf at all. Therefore, the realm is further declared to extend "to a depth of 200 meters or, beyond that limit, to where the depth of the superjacent waters admits of the exploration of the natural resources of the said areas . . ." Today, this statement has led to problems because no part of the ocean is technologically beyond the reach of exploitation. Resources such as petroleum and manganese nodules may be found far from shore and are attractive lures for exploitation by advanced technology. Thus, conflicts among nations may arise. Further modification of this convention is certainly warranted.

As stated earlier, the Draft Treaty proposed that coastal nations would own the resources of the seabed out to 200 miles from shore (the Economic Zone) or out to the edge of the continental shelf, whichever is farther. The Draft Treaty also proposed the creation of an International Seabed Authority (ISA) to deal with resources of the sea floor, located beyond the areas of ownership and control of the coastal nations. The ISA may also conduct its own research and mining in the open sea and may issue permits for exploration and mining of the international seabed. The more than 60 per cent of the ocean area that is represented by the international seabed is to be used for the benefit of all.

FISHERIES CONFLICTS

The many ways in which the ocean provides food for the world population frequently result in conflicts among those who fish. For example, local fishing disputes arise between commercial and sports fishermen. The sports fishermen frequently claim that the commercial fishermen (Fig. 12.2) deplete the game fish stocks by their efficient techniques or that they catch too many of the fish that serve as food for the game fish. International disputes may also arise, as when foreign fishing vessels disturb lobster traps along the East Coast of the United States. Today, fishing fleets (Fig. 12.3) from several countries

FIGURE 12.2 A very efficient purse-seine surrounding a large school of menhaden in Long Island Sound. Sports fishermen have protested that this technique removes too many of the fish that serve as food for game fish. Most game fish, however, have varied diets, and the menhaden population appears to have been rather stable in recent years. (Photo by Charles R. Meyer.)

including Russia, Poland, Japan, and Italy fish off the coast of the United States, and ships are occasionally seized for catching lobsters and are escorted to shore. The skipper may even be fined. In addition, American lobstermen have reportedly been "harassed" as foreign fishing ships trawl through their lobster pots in quest of fish. This problem has been partially resolved through a bilateral agreement between the Soviet Union and the United States, in which the Russians have agreed to avoid disturbing or collecting creatures of the continental shelf including lobsters and to permit boarding of their vessels by United States enforcement officials. In addition, a U.S.–U.S.S.R. Fishery Claims Board has been set up to deal with conflicts that arise and to award damages where appropriate.

Similarly, United States fishermen catch tuna in the waters claimed by Peru and Ecuador. Both countries claim a 200 mile fishing limit. Since Ecuador also owns the Galapagos Islands, it claims fishing rights as far as 800 miles to sea. Many boats have been seized and the owners and captains fined as much as $400,000 or more,

FIGURE 12.3 The Polish base ship *Pomorze,* with a side-trawler alongside, off the coast of Virginia. (Official U.S. Coast Guard Photo.)

depending on the capacity of the boat. Some United States tuna fishermen stay outside the 200 mile limit of other countries but require more time to fill their boats with fish. Others consider the risk to be worthwhile and continue to fish in the rich waters off South America. Some even pay high fees for Ecuadorian licenses that permit them to fish as close as 40 miles from shore.

In 1966, the United States extended its exclusive fishing rights from 3 to 12 miles. An extension of this limit to 200 miles was approved by the United States Congress in January, 1976. If the United States should enforce a 200 mile limit to protect the fish stocks off its shores and to ensure an ample catch for its fishermen, this would also imply acceptance of the 200 mile limit claimed by countries like Ecuador and opposed by United States tuna fishermen. However, as mentioned previously, the United Nations Draft Treaty provides for a 200 mile economic zone.

Governmental control of fisheries activities, including net design and catch limits, are necessary in order to maintain the living stock of fish and invertebrates, especially over the continental shelf. This requires international cooperation not only through observance of economic zones but also through international commissions such as the International Pacific Halibut Commission (IPHC), the Inter-American Tropical Tuna Commission (IATTC), and the International Commission for North Atlantic Fisheries (ICNAF). International cooperation is especially essential in controlling the fisheries for migratory fish such as the tuna.

FREEDOM FOR OCEAN RESEARCH

There has been great concern expressed regarding the freedom of ocean research in an ocean restricted by 200 mile economic zones (over 35 per cent of the ocean area). Some coastal nations claiming jurisdiction over a broad economic zone have refused permission to the United States and other developed nations wishing to do ocean research in the shelf waters of those countries. The 1975 Draft Treaty provides freedom of research in the economic zone beyond the territorial sea with the following restrictions:

(1) that the research be fundamental in nature;
(2) that scientists and observers from the coastal nation be provided the opportunity to participate in the research;
(3) that the data and findings from the research be made available to the coastal nation.

If the research is directly related to the resources of the

economic zone, the coastal nation may refuse permission, although it might reach an agreement with the visiting nation to their mutual advantage. These restrictions, of course, may result in considerable red tape and could delay the proposed research. On the other hand, international cooperation, especially with smaller, developing nations, might be further promoted. Such cooperation has proved successful in the past. For example, in 1974, scientists and technicians from 72 countries participated in the Atlantic Tropical Experiment of the Global Atmospheric Research Program (GARP). The work required coordination among workers of varied backgrounds, speaking many languages, and working from 40 ships, 12 aircraft, and numerous land stations. The host country was Senegal, a small nation in westernmost Africa, which has the best port facilities in that part of the Atlantic, the port of Dakar. Participating nations included Brazil, Canada, Finland, France, East and West Germany, Mexico, Netherlands, Portugal, the Soviet Union, United Kingdom, and the United States. A great deal of information was gathered regarding the origin of tropical rain clouds, surface ocean characteristics, and the equatorial undercurrent. The success of this venture is an encouraging note for further international research.

Many other international projects have been launched in the 1970s, especially through the International Decade of Ocean Exploration (IDOE). These projects have involved scientists and ships (Fig. 12.4) from many nations and have included research on marine resources (biology and geology), chemistry, pollution, ocean circulation, and weather. Several developing coastal nations, such as Senegal, Gabon, Brazil, Peru, and Ecuador, and some landlocked nations such as Switzerland and Bolivia have participated in these projects. Some of these projects have obvious economic overtones and have involved research in the territorial waters claimed by coastal states such as Peru.

Research and exploitation, even in the open ocean, will be restricted in the future. The Draft Treaty suggests that the International Seabed Authority be notified of any proposed research in the open ocean. However, permits would be required for much of the research involving the international seabed. If, indeed, there is a net benefit for all mankind, the inconveniences experienced in the years to come will be worthwhile.

WHAT ABOUT THE FUTURE?

The message of this book is that the oceans are not only interesting but also valuable. They have a role to play

FIGURE 12.4 The *R. V. Chain* from the Woods Hole Oceanographic Institution was one of the many research vessels that have participated in programs of the International Decade of Ocean Exploration. (Courtesy of Woods Hole Oceanographic Institution.)

in supplying energy (such as petroleum and tidal power), food (fisheries activities and perhaps mariculture), avenues of navigation and trade, minerals, and even recreation. It is a sad but well established fact that people constantly squabble and occasionally go to war over virtually anything that has value, and throughout history this has certainly been true of the oceans. The threat we all face today is that the oceans are becoming more valuable—petroleum is more scarce and more expensive; we need and can obtain more food from the sea; and our ability to extract minerals from the oceans and seabeds continues to improve. Thus, the threat of conflict or simply of selfish misuse of marine resources is constantly increasing.

Particularly through the United Nations, there is an increasing awareness of the dangers, and signs of international cooperation during the mid-1970s have been heartening. Similarly, the governments of most coastal nations have begun to take steps to assure the continued integrity of the marine environment under their jurisdiction.

The question now is whether these hopeful trends will continue. The answer is beyond the scope of this or any other textbook and really lies with informed individuals and their governments.

SUGGESTED READINGS

Anderson, A. 1973 (November). The Rape of the Seabed. *Saturday Review/World.*

Cadwalader, George. 1973. Freedom for Science in the Oceans. *Science.* **182**:15–20.

Gullion, Edmund A. (ed.). 1968. *Uses of the Seas.* Englewood Cliffs, N. J., Prentice-Hall.

Hammond, Allen L. 1975. Probing the Tropical Firebox: International Atmospheric Science. *Science.* **188**:1195–1198.

———. 1976. Deep Sea Drillers: Entering a New Phase. *Science,* **191**:168–169.

Meyer, Charles. 1975 (May). The Menhaden War. *National Fisherman.*

Ross, Carol. 1976 (February). Passage of 200-mile Limit Seen Adding New Problems. *National Fisherman.*

Schiller, Ronald. 1975 (November, December). The Grab for the Oceans, Part I and Part II. *Reader's Digest.*

Shapley. Deborah. 1975. Now a Draft Sea Law Treaty—but What Comes After? *Science,* **188**:918.

Wenk, Edward, Jr. 1972. *The Politics of the Ocean.* Seattle. University of Washington Press.

Glossary

abyssal hills. Small extinct volcanoes (usually less than a kilometer in height) distributed over large areas of the ocean floor.
abyssal plain. Almost featureless or flat region of the ocean floor, produced by the burial of abyssal hills by sediments.
abyssopelagic. Pertaining to the deep oceanic pelagic environment between 3,000 and 6,000 meters of depth.
aerobic. Requiring free oxygen for organic growth.
alga (pl., **algae**). The simplest plants, having no stems, roots, or leaves. They may be single-celled (such as diatoms) or quite large (such as the seaweeds). The term *algae* is most commonly applied to the seaweeds.
amphidromic point. A point at which there is no tidal fluctuation but around which tides (cotidal lines) rotate.
anaerobic. Lacking in free oxygen. Organisms not requiring free oxygen are also termed *anaerobic.*
Annelida. Phylum comprising segmented worms, including clamworms and earthworms.
aphotic. Devoid of light.
atoll. A ringlike island made up of coral reefs.
azoic zone. A zone in the sea originally postulated to be lifeless. We now know that life exists at all depths.

basalt. A fine-grained, dark-colored, igneous rock, commonly produced by volcanic eruptions. The oceanic crust and the lower continental crust are believed to be basaltic in nature.
bathyal. Pertaining to the environments of the continental slope. Organisms living at this depth are called *bathyal* organisms.
bathypelagic. Pertaining to the oceanic pelagic environment between 1,000 and 3,000 meters.
bathythermograph. A device that measures and records the change in temperature with depth.
beach. The region of unconsolidated materials extending from the low tide line to the uppermost limit of wave action, usually represented by cliffs, sand dunes, or permanent vegetation.
benthic. Pertaining to the environments of the sea floor.
berm. A nearly horizontal surface located above the high tide line and produced by the deposition of beach materials by wave action.
biochemical oxygen demand (BOD). A drain on the oxygen supply of an environment produced by chemical oxidation and biological respiration.
bioluminescence. The production of a cool light by certain plants, animals, and bacteria.
biomass. The quantity (weight) of living organisms present in a given volume or area.
bloom. A sudden burst or growth of phytoplankton, leading to massive populations of the bloom organism and sometimes resulting in discolored or toxic water (red-tides).
brackish. Pertaining to water, generally estuarine in which the salinity ranges from about 0.50 to 17 parts per thousand (‰) by weight.

chlorinity. The chloride content of seawater expressed in parts per thousand (‰) by weight. It is now defined as the weight in grams of silver required to precipitate the chloride, bromide, fluoride, and iodide in 0.3285233 kg of seawater.
ciliate. A single-celled animal that moves by beating many tiny hairlike threads called cilia.
climax community. The temporally stable assemblage of interrelated organisms sharing a common habitat.
coccolith. Calcareous ($CaCO_3$) buttonlike plates covering the single-celled plants known as coccolithophores.
continental rise. A broad sedimentary wedge that may be present at the foot of the continental slope.
continuous seismic profiling (sparker profiling). Method of obtaining a continuous profile record of geological structures below the sea floor by the use of a strong energy (sound) source.
core. The central region of the earth, extending from a depth of 2,900 to 6,370 km (the center of the earth). The outer core (2,900 to 5,100 km) resembles a liquid, and the inner core (below 5,100 km) resembles a solid.
cotidal line. A line connecting points of simultaneous high tide in the ocean.
crust. The thin cover of the earth above the mantle. In oceanic regions, the crust has an average thickness of about 10 km, and in continental areas it has an average thickness of 35 km.

decomposer. An organism that causes the breakdown of organic matter and the production of inorganic nutrients.
deep scattering layer (DSL). A subsurface stratified population of animals that scatter sound. They may cause the appearance of a "false bottom" on echograms.
detritus. Dead organic matter.
diatom. A single-celled plant with two overlapping, siliceous (SiO_2), porous shells. Diatoms are sometimes called the "grass of the sea" because of their importance as food for marine herbivores.
diatomaceous earth. A sedimentary rock, largely composed of diatom shells.
dinoflagellate. A single-celled, flagellated plant. Some are bioluminescent; others may cause toxic red-tides.
disphotic. Pertaining to regions where some light exists but not enough for plant growth.
diurnal. Daily, as in *diurnal* tides, which comprise one high tide and one low tide each day.
dust cloud. A diffuse cloud of dust and gas, composed mostly of hydrogen. According to the *dust cloud hypothesis,* all celestial bodies were formed from dust clouds.

echinoderm. A spiny-skinned, radially symmetrical animal. This group includes sea stars, sea urchins, and sea cucumbers.
epi-. Prefix meaning: *on, upper,* or *attached to.* From the Greek *epi* = on.
epifauna. Animals that live upon the sea floor.
epipelagic. Pertaining to the well-lighted surface waters of the sea.
estuary. A bay or river mouth where seawater and fresh water meet and mix and where seawater is measurably diluted.
euphotic. Pertaining to the well-lighted regions where plants may live and grow.
eutrophic. Pertaining to nutrient-enriched conditions, with large biomass and moderately high diversity of species.

feldspar. A mineral composed of silicates of potassium, sodium, or calcium.
fetch. The distance over which the wind blows in the wave-generating area.
flagellate. A single-celled organism that moves by beating a whiplike thread, or flagellum.
food web. The feeding (trophic) relationships among members of a community.
foraminiferan. A single-celled animal with pseudopodia and a calcareous ($CaCO_3$), porous shell.

granite. A coarse-grained, light-colored igneous rock, formed within the earth by the cooling of molten materials. The upper continental crust is believed to be granitic in nature.
guyot. A flat-topped seamount. The flat surface is believed to have been produced by wave erosion.
gyre. A large, closed system of surface current circulation.

hadal. Pertaining to the environment of the trenches.
hadopelagic. The pelagic environments of the trenches (generally at depths greater than 6,000 meters).
halocline. A sharp change in salinity with depth.
heat capacity. The ratio of the heat in calories per gram absorbed (or released) by a substance to the corresponding temperature (°C) rise (or fall). For example, one calorie is required to raise the temperature of one gram of liquid water by 1°C.

infauna. Animals that live within the sediment.
in situ. A Latin term meaning *in place;* in the orginal or natural position.
ion. An atom or group of atoms bearing an electrical charge, either positive (cation) or negative (anion).
iso–. Prefix meaning *alike, equal* or *same.* From the Greek *isos* = equal.
isohaline. A line or plane of equal salinity.
isotherm. A line or plane of equal temperature.

larva. The young, generally planktonic stages in an animal's life cycle.
lime. Calcium carbonate ($CaCO_3$); for example, that which is found in the shell of a clam.
lithosphere. The entire region of the earth located above the low velocity layer.
littoral. Intertidal; the benthic zone between the high and low tide levels.
littoral transport. The movement of suspended material parallel to the shore by longshore currents.
longshore current. A current flowing parallel to the shore, formed when waves strike the coast obliquely.
lophophore. A ciliated looped, spiraled, or ringlike structure, used in feeding and respiration in the lophophorate animals.
low velocity layer. A region in the upper mantle of the earth (between about 100 and 300 kilometers below the earth's surface) where seismic wave speeds are lower than in regions above and below it.
luciferase. An enzyme that promotes the oxidation of luciferin in the process of bioluminescence.
luciferin. A chemical substance that may be oxidized to produce a cool blue-white light during bioluminescence.

mantle. 1. The region of the earth between the crust and the core. The mantle extends downward to a depth of about 2,900 kilometers.

2. The fleshy sheetlike structure that surrounds the internal organs of a mollusk; it also secretes any shell that these organisms may have.

mariculture. Sea farming, or the cultivation of marine organisms in the ocean, estuaries, or saltwater tanks, frequently with artificial nutrient enrichment.

meroplankton. Plankton that become nektonic or benthic at a certain stage in their life cycle.

mesopelagic. Pertaining to the "twilight zone" of the sea, between 200 and 1,000 meters.

Moho (Mohorovičić discontinuity). The boundary between the crust and the mantle.

mollusk. A soft-bodied animal with gills and a mantle, which in most mollusks, secretes a shell. This group includes clams, snails, and squid.

nannoplankton. Minute plankton that are less than 0.1 mm (100 microns) in size and which thus can pass through the finest plankton nets.

neap tide. The tide that occurs when the tide-producing forces of the sun and the moon are at right angles to each other. Neap tides occur every two weeks (about a week after full and new moons), and they have the smallest tidal ranges.

nekton. The swimmers, such as fish and squid, that can swim against ocean currents.

neritic realm. The water near shore, or over the continental shelf.

niche. The set of functional relationships of an organism to the environment that it occupies.

nodal line. A line (across a basin) along which no tidal fluctuations occur.

notochord. A flexible strengthening rod located along the back in the members of the Phylum Chordata, present at least in the larval or embryonic stages.

nutrient. A chemical, such as nitrate or phosphate, required for the growth of plants.

oceanic realm. The waters of the open ocean, beyond the continental shelf.

oligotrophic. Pertaining to nutrient-poor conditions, with low biomass and a high diversity of species.

ooze. Fine-grained sediments composed of at least 30 per cent skeletal remains of organisms.

opal. Hydrated silicon dioxide, similar to glass. Opal is a major constituent of the siliceous shells or skeleta of some marine organisms (e.g., diatoms).

osmosis. Diffusion across a selectively permeable membrane.

osmotic pressure. The pressure exerted on a membrane by osmosis.

P-Wave. See *Primary wave.*

paleomagnetism. The ancient magnetism of rocks that was produced when they were initially formed.

paleontology. The study of ancient life, through the use of fossil evidence.

pelagic. Pertaining to the water between the surface and the sea floor. The pelagic environment is distinct and separate from the sea floor itself. Pelagic sediments are formed primarily in the pelagic environment.

photophore. A light-producing organ.

photosynthesis. The utilization of light energy in the synthesis of organic matter by a plant.

plankton. Drifting organisms that cannot swim against ocean currents, including phytoplankton (plant plankton) and zooplankton (animal plankton).

plate. A broken piece of the lithosphere. The lithosphere of the earth is believed to consist of about seven major plates and several smaller ones.

primary producer. A plant that produces organic matter by photosynthesis.

primary wave (P–wave). A seismic wave in which the particles vibrate parallel to the direction of wave advancement. See also *Secondary wave.*

productivity. The rate of production of organic matter by the organisms in an area.

pseudopodia. The threadlike or fingerlike projections of the semifluid body of certain single-celled animals in the Phylum Protozoa. They are used in locomotion and in the entrapment of food.

pycnocline. A sharp change in density with depth.

quartz. A crystalline mineral consisting of silicon dioxide (SiO_2).

radiolarian. A single-celled animal with pseudopodia and a siliceous (SiO_2) skeleton.

respiration. The consumption of oxygen and the release of energy from stored organic matter by a plant or animal.

rift valley. A valley that sometimes characterizes the axis of a mid-ocean ridge.

S-wave. See *Secondary wave.*

salinity. The dissolved salt content of water, usually expressed in parts per thousand (‰) by weight.

Sargasso Sea. A region of warm, sluggish, nutrient-poor water in the west central Atlantic, north of the equator.

sea. 1. Ocean or a subdivision of an ocean, such as a semi-enclosed body of water connected to an ocean. 2. Waves in the generating (storm) area.

seamount. Undersea volcanic peak having a height of over one kilometer.

secondary producer. Herbivore, or plant-eater.

secondary wave (S-wave). A seismic wave in which the particles vibrate perpendicular to the direction of wave advancement. They can propagate through solids only. See also *Primary wave.*

seismic waves. Waves (vibrations) produced by earthquakes, explosions, or other energy sources and transmitted through the earth; a type of sound wave.

seismic sea wave. See *Tsunami.*

semidiurnal. Occurring twice daily, as in semidiurnal tides, which comprise two high and two low tides each day.

shadow zone. A region into which little sound energy penetrates when the sound source is at or near the surface. It is produced when a sound velocity maximum occurs near the surface (usually at a depth of about 100 meters).

significant wave height. The average wave height of the highest 1/3 of the waves resulting from a particular storm.

sill. A shallow ridge across the mouth of a basin or a deep estuary (such as a fiord).

siphon. A fleshy tube, found in some molluscs, through which water flows. It is frequently used to carry water and food into (incurrent) and out of (excurrent) the animal.

SOFAR. See *Sound channel.*

sound channel. A region in the ocean where sound velocity is at a minimum (usually at a depth of about 1,000 meters). Sound originating in this region can be propagated thousands of kilometers without appreciable loss of energy. Also known as the SOFAR (Sound Fixing and Ranging) channel.

sparker profiling. See *Continuous seismic profiling.*

splash zone. The sea spray zone above high tide levels.

spring tide. The tide that occurs when the earth, sun, and moon are on a straight line. It occurs every two weeks (during full and new moons) and has the greatest tidal ranges.

standing crop. The total amount (biomass) of organisms in a volume of water at one time.

storm surge (storm wave, storm tide). An abnormal rise in water level caused by a hurricane or other severe storm.

sublittoral. Pertaining to the environment of the continental shelf below low tide.

submarine canyon. A V-shaped, rocky canyon located across the continental shelf and slope.

surface tension. The attraction of molecules near the surface of a liquid toward the interior of the liquid, tending to result in decreased surface area and a measurable tension at the surface.

swell. Ocean waves outside the generating area, generally having longer periods and flatter crests than those within the generating area (sea).

swim bladder. A gas-filled sac found in certain fish. It is used in the regulation of the fish's buoyancy. Owing to the difference in density between water and the gas, swim bladders reflect sound well and may cause deep scattering layers.

thermocline. A sharp change in temperature with depth.

terrigenous sediment. Land-derived sediment.

tidal flow. 1. The total volume of seawater flowing into or out of an estuary with the incoming or outgoing tide. 2. The difference between the mean high water volume and the mean low water volume of an estuary, also known as the *tidal prism.*

trophic. Pertaining to feeding or food.

tsunami (seismic sea wave). A surface ocean wave produced by an abrupt vertical displacement of the sea floor caused by an earthquake, submarine landslide, or volcanic explosion.

turbidity. Reduced water clarity owing to the presence of suspended material.

turbidity current. A dense fast-flowing body of sediment-laden water.

viscosity (internal friction). The tendency of a liquid to resist flowing.

Appendix A

MAJOR OCEANOGRAPHIC INSTITUTIONS IN THE UNITED STATES*

INSTITUTION	ACADEMIC PROGRAMS OFFERED
University of Alaska Institute of Marine Science Fairbanks, AK 99701	Biological, chemical, geological, and physical oceanography; ocean engineering; marine science.
University of California, San Diego Scripps Institution of Oceanography La Jolla, CA 92037	Oceanography, marine biology, and earth sciences.
Columbia University Lamont-Doherty Geological Observatory Palisades, NY 10964	Geology and biology.
University of Connecticut Marine Science Institute Groton, CT 06340	Oceanography.
Florida State University Department of Oceanography Tallahassee, FL 32306	Marine biology, oceanography, and geophysical fluid dynamics.
University of Hawaii Marine Programs Honolulu, HA 96822	Oceanography, ocean engineering, marine biology, and geophysics.
Johns Hopkins University Department of Earth and Planetary Sciences Baltimore, MD 21218	Oceanography. (Johns Hopkins University operates the Chesapeake Bay Institute.)
Louisiana State University Department of Marine Sciences Baton Rouge, LA 70803	Marine sciences, fisheries and marine resources law. (Louisiana State University operates the Coastal Studies Institute.)

*This is only a partial list of institutions offering academic programs in oceanography and related fields. All institutions listed offer doctoral degrees in marine science areas. For further information consult the following publications.

1. *University Curricula in Marine Sciences 1975* (revised periodically). National Sea Grant Program, NOAA, US Department of Commerce, Rockville, MD.
2. *Directory of Geoscience Departments,* published annually by the American Geological Institute, Falls Church, VA.

University of Maine Orono, ME 04473	Oceanography.
Massachusetts Institute of Technology Joint Program in Oceanography 77 Massachusetts Ave. Cambridge, MA 02139	Oceanography (with Woods Hole Oceanographic Institution); shipping and shipbuilding; marine engineering, naval architecture, naval engineering, and ocean engineering.
University of Miami School of Marine and Atmospheric Science Miami, FL 33149	Marine biological science, marine geology, chemical oceanography, atmospheric sciences, ocean engineering, ocean law, and fisheries.
University of Michigan Department of Meteorology and Oceanography Ann Arbor, MI 48104	Oceanography.
State University of New York at Stony Brook Marine Sciences Research Center Stony Brook, NY 11790	Marine environmental studies, marine biology.
New York University Department of Meteorology and Oceanography University Heights Bronx, NY 10453	Meteorology and oceanography.
Nova University Oceanographic Laboratory Fort Lauderdale, FL 33316	Marine biology, physical oceanography, and chemical oceanography.
Oregon State University School of Oceanography, School of Engineering, and Department of Fisheries and Wildlife Corvallis, OR 97331	Oceanography (biological, chemical, geological, and physical), geophysics, ocean engineering, and fisheries. (Oregon State University operates the Marine Science Center at Newport, OR.)
Princeton University Princeton, NJ 08540	Marine geology and physical oceanography.
University of Rhode Island Graduate School of Oceanography Narragansett, RI 02882	Oceanography, marine affairs, and ocean engineering.
University of Southern California Allan Hancock Foundation Los Angeles, CA 90007	Oceanography, ocean engineering, and biological or geological sciences with marine specialization. (University of Southern California operates the Marine Sciences Center on Santa Catalina Island.)
Stevens Institute of Technology Department of Ocean Engineering Castle Point Station Hoboken, NJ 07030	Ocean engineering.

Texas A&M University Department of Oceanography or Department of Civil Engineering College Station, TX 77843	Oceanography, marine resources management, coastal engineering, and ocean engineering.
Virginia Institute of Marine Science Gloucester Point, VA 23062	Marine science (with College of William and Mary and the University of Virginia.)
University of Washington Division of Marine Resources Seattle, WA 98105	Oceanography, ocean engineering, and fisheries. (The University of Washington operates the Friday Harbor Marine Biological Station on San Juan Island.)
University of Wisconsin Marine Studies Center Madison, WI 53706	Oceanography and ocean engineering.
Woods Hole Oceanographic Institution Woods Hole, MA 02543	Oceanography. The Woods Hole Oceanographic Institution also has cooperative programs with other institutions, including Yale, Harvard, and Massachusetts Institute of Technology.

Appendix B

REFRACTION OF WAVES

To a diver in the water, the sun appears to be closer to the vertical than it really is. This is because the light waves are ***refracted***, or bent, as they travel from a medium in which they have a higher speed (air) into one in which they have a slower speed (water). In fact, all waves, whether they are light waves, water waves, or sound waves, undergo refraction under certain conditions. The principle governing the refraction of waves (Fig. B.1) is known as ***Snell's Law*** and is stated as follows:

$$\frac{\text{Sine } i}{C_a} = \frac{\text{Sine } r}{C_b}$$

where

i is the angle of incidence, and equals the angle of reflection (R),
r is the angle of refraction,
C_a is the speed of the wave in the first medium, and
C_b is the speed of the wave in the second medium.

Thus waves tend to be refracted toward the perpendicular when they slow down and are refracted away from the perpendicular when they speed up. However, no refraction occurs as the waves enter the second medium perpendicular to the interface: that is, when the angle of incidence (i) is zero, the angle of refraction (r) is zero. Note that only a portion of the wave energy enters the lower medium. Part of it is reflected by the surface. The "media"

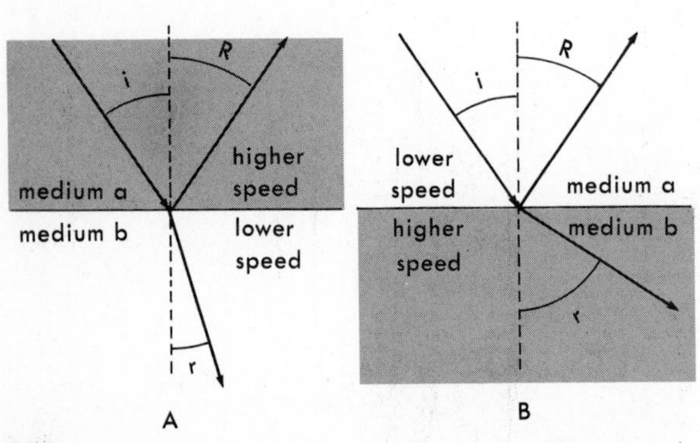

FIGURE B.1 *A,* **The refraction of waves as they enter a medium in which they have a lower wave speed.** *B,* **Refraction as they enter a medium in which they have a higher wave speed. The angle of incidence = i; r is the angle of refraction; and R is the angle of reflection and equals i.**

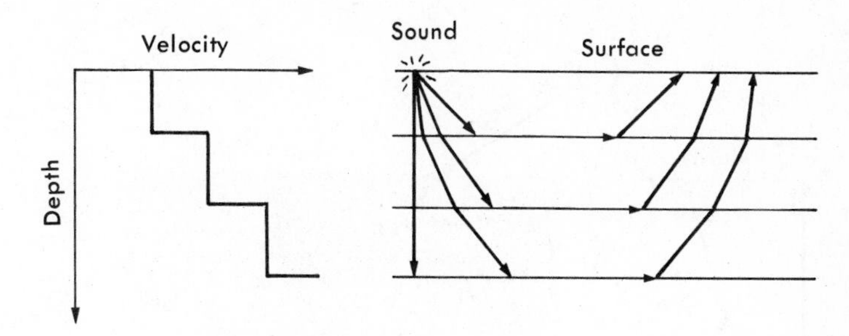

FIGURE B.2 **The paths of some sound waves through layers (rock or water) of increasing density and consequently of increasing wave speeds.**

that we are discussing here need not be as distinctly different as air and water; they may be waters or rocks of slightly different densities. Light waves decrease in speed, whereas sound waves increase in speed, as they enter media of greater densities. In the case of water waves, "the media" refers to regions of different water depths. These waves slow down as they enter shallower water. Water waves, however, are refracted *only* when they "feel the bottom"; that is, when the water depth is less than half the wave length.

Figure B.2 illustrates the refracted paths of some seismic (sound) waves through different rock layers of increasing density with depth. In general, increasing rock density results in an increasing seismic wave speed. Note that each seismic ray, after refraction, penetrates to a maximum depth at which it may travel along the interface with the velocity that it would have in the lower layer. As it travels along the interface it sends seismic waves back toward the surface, as shown in the figure.

Not all media are arranged in discrete layers. In most cases, density changes continuously with depth. Figure B.3 illustrates the paths of some seismic waves as they travel through rocks in which the density (and, consequently, the velocity) increases gradually with depth. The resulting

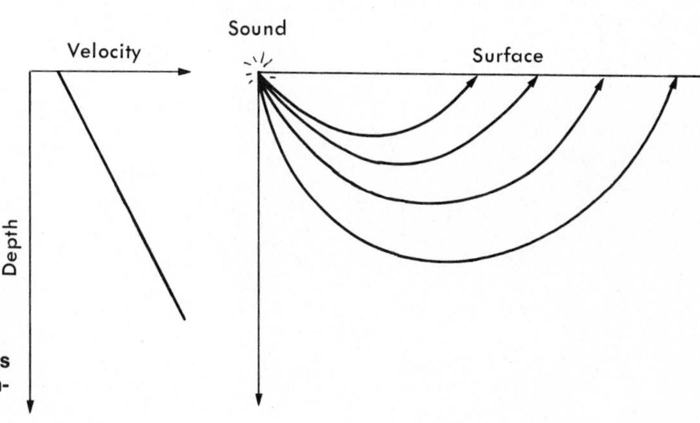

FIGURE B.3 **The paths of some sound waves through rock (or water) in which density increases gradually with depth.**

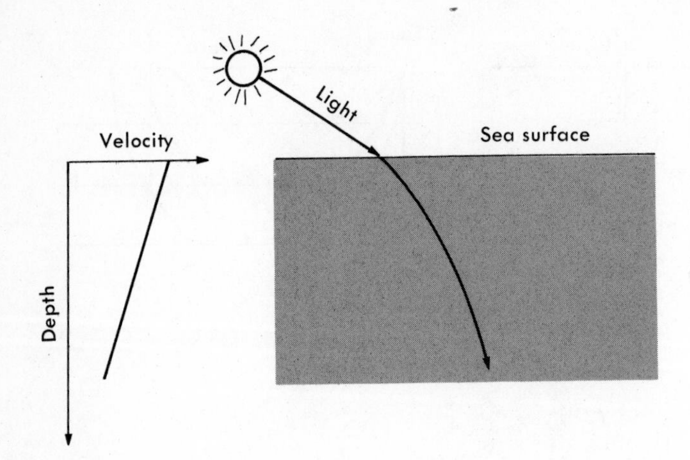

FIGURE B.4 Path of a light ray in water in which density increases gradually with depth and the velocity of the light decreases with depth. As water waves enter shallower water (lower speed) near the shore, they are refracted in a similar manner. After refraction these waves are more nearly perpendicular to the shore than before refraction.

wave paths are curved, as shown in the figure. Sound waves in the ocean behave in a similar way.

The refraction of light in the sea is shown in Figure B.4. As light penetrates deeper into the water its speed decreases gradually, owing to the increasing density with depth, and hence the light rays are curved downward. Water waves are refracted in a manner similar to that of light waves. As they enter shallower water near the shore, they slow down and are refracted as shown in the figure. After refraction, water waves tend to arrive perpendicular to the shore.

Appendix C

CORIOLIS FORCE

Moving objects, such as air, water, airplanes, and satellites, are affected by the earth's rotation. The result is an apparent deflection in the paths of the objects, which is due to the fact that the moving objects are observed from a frame of reference fixed to a rotating earth. For example, a missile fired initially in a straight direction will appear to an observer on the earth to be deflected to the right in the Northern Hemisphere and to the left in the Southern Hemisphere (Fig. C.1). However, to an observer in space the missle would appear to be moving in a straight path over a rotating earth. Of course, there would be no deflection at all if the earth were not rotating. Water particles associated with ocean currents also are deflected to the right in the Northern Hemisphere and to the left in the Southern Hemisphere on the rotating earth.

The force responsible for this phenomenon is known as the ***Coriolis force***, named after Gaspard Gustav de Coriolis, the French mathematician who described it in 1835. The Coriolis force is defined by the following equation:

$$\text{Coriolis force per unit mass} = 2\Omega V \text{ sine } \phi$$

where Ω is the angular velocity of the earth (360°, or 2π radians, per 24 hours), V is the velocity of the object relative to the earth, and ϕ is the latitude. The Coriolis force is zero for an object moving directly over the equator ($\phi = 0$) and increases with increasing latitude. Coriolis force is directed perpendicular to the direction of the object's movement and is proportional to its velocity. However, the amount of deflection becomes less as the velocity of the object increases (at a given latitude). This is because an object moving rapidly (such as a missile) over a given distance allows less rotation of the earth than would an object moving much more slowly (such as an ocean current) over the same distance. Let us now take a closer look at Coriolis deflection.

An object fixed or resting on the earth's surface will be subjected to a gravitational force toward the earth's

FIGURE C.1 The path of a missile as it travels around the earth. Note that it is deflected toward the right in the Northern Hemisphere and toward the left in the Southern Hemisphere.

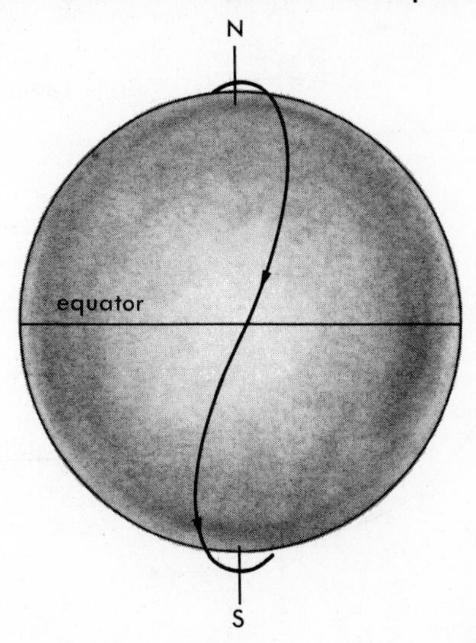

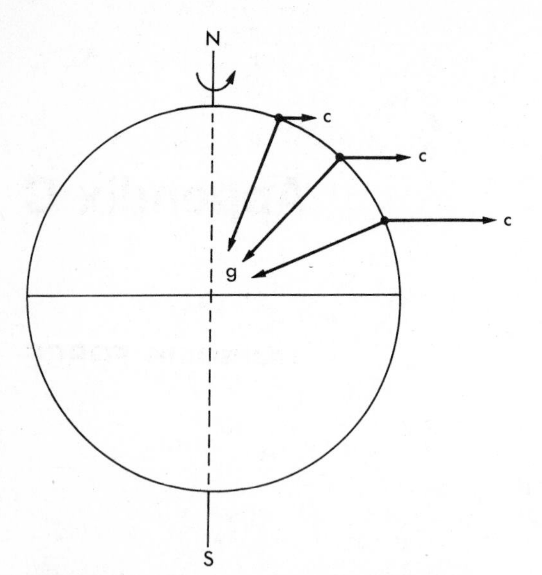

FIGURE C.2 The centrifugal (c) and gravitational (g) forces acting on particles at rest on the earth's surface. The centrifugal forces shown are highly exaggerated, as they are actually much smaller in magnitude than the gravitational force.

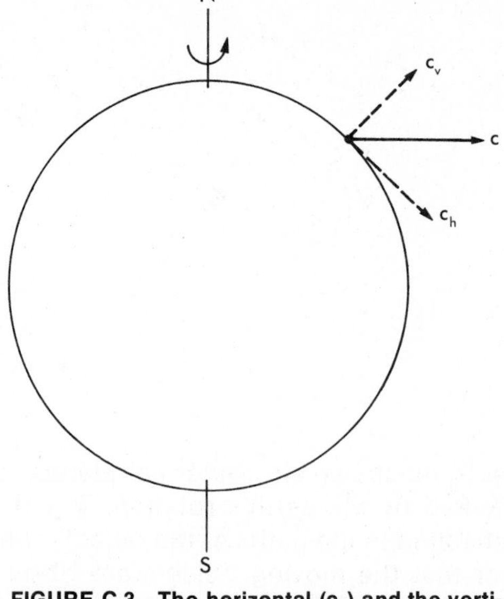

FIGURE C.3 The horizontal (c_h) and the vertical (c_v) components of the centrifugal force (c) acting on a particle at the earth's surface.

center and to a centrifugal force due to the earth's rotation directed away from the earth's axis of rotation, as shown in Figure C.2. The centrifugal force increases from zero at the poles to a maximum at the equator. It is the horizontal component of the centrifugal force (Fig. C.3) that has resulted in the flattening of the earth at the poles (the polar radius of the earth is about 22 km shorter than the equatorial radius) and a bulging at the equator. This probably occurred very early in the earth's history, before even the oceans were formed. Every particle of the earth, including water particles, is in equilibrium, provided that it is not in motion.

Consider a water particle moving from west to east in the Northern Hemisphere. It will be moving eastward at a speed equal to that of a fixed point on the earth at that latitude (Fig. C.4) plus the particle's own speed with respect to the earth. Thus, the eastward-moving water particle will be subjected to an increased centrifugal force (and hence a greater horizontal component of this force) than that of a fixed point at the same latitude. On the other hand, a particle moving due west will experience a decreased centrifugal force as compared to a particle at rest.

FIGURE C.4 The linear (tangential) velocities of fixed particles at different latitudes as the earth rotates on its axis.

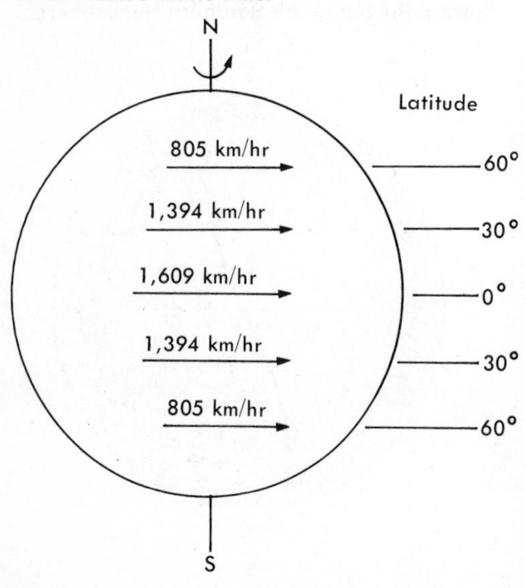

Figure C.5 shows the differences in centrifugal force between particles that are moving eastward or westward and those that are at rest at the same latitude in the Northern Hemisphere. Their horizontal components, directed 90° toward the right of the particle motion, represent the Coriolis force.

The Coriolis deflection of objects moving northward

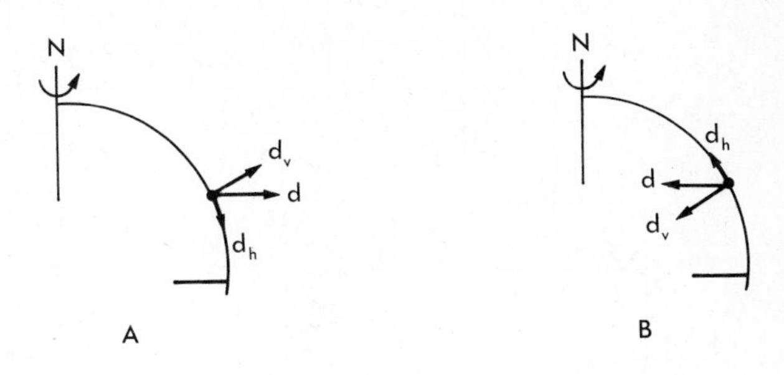

FIGURE C.5 *A,* The difference (d) in centrifugal force of a particle that is moving eastward and one that is resting at the same latitude. *B,* Difference in centrifugal force of a westward-moving particle and a resting particle. d_h and d_v represent the horizontal and vertical components of these differences, respectively. d_h also represents the Coriolis force.

or southward over the earth's surface can be explained in the following manner. Consider an object moving southward in the Northern Hemisphere. In addition to its southward speed, it also maintains an eastward motion equivalent to that of a fixed particle on the earth at the starting point. The object will be deflected to the right of its initial path, since the earth's surface under its path has a greater eastward speed than does the moving object (Fig. C.6). On the other hand, if the object moves northward, its eastward component of velocity is retained as it travels toward a point where the eastward motion of the earth's surface is less. The object tends to drift eastward, or to the right.

It is important to note that Coriolis force acts on all moving objects, regardless of the direction in which they are moving, and that it causes them to be deflected to the right in the Northern Hemisphere and to the left in the Southern Hemisphere.

FIGURE C.6 Polar view of the Coriolis forces (CF) acting upon particles initially moving southward (V_s) and northward (V_n) in the Northern Hemisphere. The heavy arrows indicate the paths of the particles' movements.

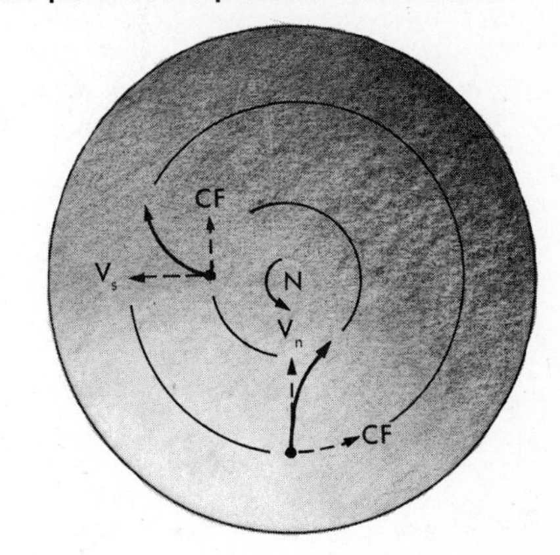

Bibliography

BAKER, B. B., Jr., W. R. DEEBEL and R. D. GEISENDERFER (eds.). 1966.
Glossary of Oceanographic Terms, SP 35. Washington, D.C., U.S. Naval Oceanographic Office.

BARAZANGI, M. and J. DORMAN. 1969.
World Seismicity Maps Compiled from ESSA, Coast and Geodetic Survey, Epicenter Data 1961–1967. *Seismological Society of America Bulletin,* **59**:369–380.

BARDACH, J. E., J. H. RYTHER and W. O. McLARNEY. 1972.
Aquaculture. New York, Wiley.

BARNES, R. D. 1974.
Invertebrate Zoology, 3rd ed. Philadelphia, W. B. Saunders Co.

BASCOM, W. 1960. (August).
Beaches, *Scientific American.*

BASCOM, W. 1964.
Waves and Beaches. Garden City, N.Y., Doubleday.

BIGELOW, H. B. and W. C. SCHROEDER. 1953.
Fishes of the Gulf of Maine, Fisheries Bulletin 74. Washington, D.C., U.S. Fish and Wildlife Service.

BLAXTER, J. H. S. 1970.
Light: Animals; Fishes. In: *Marine Ecology,* Vol. I, Part 1. Otto Kinne (ed.). New York, Wiley-Interscience.

BOWDITCH, N. 1965.
American Practical Navigator. Washington, D.C., Government Printing Office.

CALMAN. W. T. 1909.
Crustacea. In: *Treatise on Zoology,* Vol. 8. E. R. Lankester (ed.). London, A & C Black.

CARMODY, D. J., J. B. PEARSE and W. E. YASSO. 1973.
Trace Metals in Sediments of New York Bight. *Marine Pollution Bulletin.* **4**:132–135.

CARR, A. 1955 (May).
The Navigation of the Green Turtle. *Scientific American.*

CARRIKER, M. R. 1969.
Excavation of Boreholes by the Gastropod, *Urosalpinx. American Zoologist.* **9**:917–933.

CARTWRIGHT, D. E. 1969.
Deep Sea Tides. *Science Journal.* **5**:60–67.

CLARKE, G. L. and E. J. DENTON. 1962.
Light and Animal Life. In: *The Sea,* Vol. 1. M. N. Hill (ed.). New York, Wiley-Interscience.

COUNCIL ON ENVIRONMENTAL QUALITY. 1970.
Ocean Dumping—A National Policy. Washington, D.C., Government Printing Office.

CULKIN, F. 1965.
The Major Constituents of Seawater. In: *Chemical Oceanography,* Vol. 1. J. P. Riley and G. Skirrow (eds.). New York, Academic Press.

DEFANT, A. 1961.
Physical Oceanography. New York, Pergamon Press.

DEWEY, J. F. and J. M. BIRD. 1970.
Mountain Belts and the New Global Tectonics. *Journal of Geophysical Research.* **75**:2625–2645.

DIETRICH, G. 1963.
General Oceanography. New York, Wiley.

DIETZ, R. S. 1961.
Continent and Ocean Basin Evolution by Spreading of the Sea Floor. *Nature.* **190**:854–857.

DOTY, M. S. 1957.
Rocky Intertidal Surfaces. In: *Treatise on Marine Ecology and Paleoecology,* Vol. 1, Ecology. J. W. Hedgpeth (ed.). New York, Geological Society of America.

EUKEN, A. 1948.
Zür Struktur des Flüssigen Wassers. *Angew. Chemie A.* **60**:166.

FITZGERALD, R. A., D. C. GORDON and R. E. CRANSTON. 1974.
Total Mercury in Sea Water. *Deep-Sea Research.* **21**:139–144.

FLEMING, N. C. 1969.
Archaeological Evidence for Eustatic Change of Sea Level and Earth Movements in the Western Mediterranean During the Last 2000 Years. Special Paper 109. Boulder, Colorado, Geological Society of America.

FLEMING, R. H. 1957.
Features of the Oceans. In: *Treatise on Marine Ecology and Paleoecology,* Vol. 1, Ecology. J. W. Hedgpeth (ed.). New York, Geological Society of America.

GARLAND, G. D. 1971.
Introduction to Geophysics. Philadelphia, W. B. Saunders Co.

GIESBRECHT, W. 1892.
Systematik und Faunistik der Pelagischen Copepoden. *Fauna und Flora Golfes Neapel.* **19**:1–831.

GOLDBERG, E. D. 1965.
Minor Constituents in Seawater. In: *Chemical Oceanography,* Vol. 2. J. P. Riley and G. Skirrow (eds.). New York, Academic Press.

GRAHAM, M. 1956.
Sea Fisheries. London, Edward Arnold, Ltd.

GREEN, J. 1961.
A Biology of Crustacea. Chicago, Quadrangle Books.

GROEN, P. 1969.
The Waters of the Sea. London, Van Nostrand Reinhold.

HARDY, A. C. 1956.
The Open Sea. London, Collins.

HAYES, D. E. and A. C. PIMM. 1972.
Bathymetric, Magnetics, and Seismic Reflection data. *Initial Reports of the Deep Sea Drilling Project.* **14**:341–376.

HEDGPETH, J. W. 1953.
An Introduction to the Zoogeography of the Northwestern Gulf of Mexico, with Reference to the Invertebrate Fauna. *Publications of the Institute of Marine Science of the University of Texas.* **3**:109–224.

HEDGPETH, J. W. (ed.). 1957.
Treatise on Marine Ecology and Paleoecology, Vol. 1, Ecology. New York, Geological Society of America.
HEEZEN, B. C. and M. EWING. 1952.
Turbidity Currents and Submarine Slumps, and the 1929 Grand Banks Earthquake. *American Journal of Science.* ***250***:849–873.
HEEZEN, B. C. and C. D. HOLLISTER. 1971.
The Face of the Deep. New York, Oxford University Press.
HEIRTZLER, J. R., G. O. DICKSON, E. M. HERRON, W. C. PITMAN, III, and X. LE PICHON. 1968.
Marine Magnetic Anomalies, Geomagnetic Field Reversals, and Motions of the Ocean Floor and Continents. *Journal of Geophysical Research.* ***73***:2119–2136.
HEIRTZLER, J. R., X. LE PICHON and J. G. BARON. 1966.
Magnetic Anomalies over the Reykjanes Ridge. *Deep Sea Research.* ***13***:427–443.
HESS, H. H. 1962.
History of Ocean Basins. In: *Petrologic Studies.* A. E. J. Engel et al. (eds.). New York, Geological Society of America.
HYMAN, L. H. 1940–1959.
Invertebrates, Vols. 1–5. New York, McGraw-Hill.
ISACKS, B., J. OLIVER and L. R. SYKES. 1968.
Seismology and the New Global Tectonics. *Journal of Geophysical Research.* ***73***:5855–5899.
KING, C. A. M. 1960.
Beaches and Coasts. London, Edward Arnold, Ltd.
KINNE, O. (ed.). 1970–1972.
Marine Ecology, Vol. 1, in 3 parts. New York, Wiley-Interscience.
LE PICHON, X. 1968.
Sea-floor Spreading and Continental Drift. *Journal of Geophysical Research.* ***73***:3661–3697.
LOWRIE, A. and E. ESCOWITZ. 1969.
Global Ocean Floor Analysis and Research Data Series, Kane-9. Washington, D.C., U.S. Naval Oceanographic Office.
MACGINITIE, G. E. and N. MACGINITIE. 1968.
Natural History of Marine Animals, 2nd ed. New York, McGraw-Hill.
MARSHALL, N. B. and O. MARSHALL. 1971.
Ocean's Life. New York, Macmillan.
MORGAN, W. J. 1968.
Rises, Trenches, Great Faults, and Crustal Blocks. *Journal of Geophysical Research.* ***73***:1959–1982.
NATIONAL ACADEMY OF SCIENCES. 1972.
Understanding the Mid-Atlantic Ridge. Washington, D.C., National Academy of Sciences.
NEUMANN, G. 1968.
Ocean Currents. Amsterdam, Elsevier.
NEUMANN, G. and W. J. PIERSON, Jr. 1966.
Principles of Physical Oceanography. Englewood Cliffs, N.J., Prentice-Hall.
NEWELL, G. E. and R. C. NEWELL. 1966.
Marine Plankton; A Practical Guide. London, Hutchinson.
ODUM, E. P. 1971.
Fundamentals of Ecology, 3rd ed. Philadelphia, W. B. Saunders Co.

ODUM, H. T. 1956.
Efficiencies, Size of Organisms, and Community Structures. *Ecology,* ***37***:592–597.

ODUM, H. T. and H. P. ODUM. 1955.
Trophic Structure and Productivity of a Windward Coral Reef Community on Eniwetok Atoll. *Ecological Monographs.* ***25***:291–320.

OTHMER, D. and O. ROELS. 1973.
Power, Freshwater and Food from Cold Deep Seawater. *Science.* ***182***:121–125.

PEARSE, A. S., H. J. HUMM and G. W. WHARTON. 1942.
Ecology of Sand Beaches at Beaufort, North Carolina. *Ecological Monographs.* ***12***:136–190.

PICKARD, G. L. 1963.
Descriptive Physical Oceanography. New York, Pergamon Press.

PIERSON, W. S., G. NEUMANN and R. W. JONES. 1955.
Practical Methods for Observing and Forecasting Ocean Waves by Means of Wave Spectra and Statistics. H.O. Publication 603. Washington, D.C., U.S. Naval Oceanographic Office.

PRITCHARD, D. W. 1952.
The Physical Structure, Circulation, and Mixing in a Coastal Plain Estuary. *Chesapeake Bay Institute, The Johns Hopkins University, Technical Report 3.*

RAFF, A. D. and R. G. MASON. 1961.
Magnetic Survey of the West Coast of North America, 40°N.–52°N. Latitude. *Bulletin of the Geological Society of America.* ***72***:1267–1270.

RAYMONT, J. E. G. 1963.
Plankton and Productivity in the Oceans. New York, Macmillan.

RILEY, G. A. 1952.
Biological Oceanography. In: *Survey of Biological Progress.* **2**:79–104. New York, Academic Press.

RILEY, J. P. and G. SKIRROW (eds.). 1965.
Chemical Oceanography. New York, Academic Press.

RUSSEL, F. S. 1927.
The Vertical Distribution of Marine Macroplankton. V. The Distribution of Animals Caught in the Ring-trawl in the Daytime in the Plymouth Area. *Journal of the Marine Biological Association of the United Kingdom.* ***14***:557–608.

RUSSEL-HUNTER, W. D. 1970.
Aquatic Productivity. New York, Macmillan.

RYTHER, J. 1969.
Photosynthesis and Fish Production in the Sea. *Science.* ***178***: 72–76.

SEGERSTRÅLE, S. G. 1957.
Baltic Sea. In: *Treatise on Marine Ecology and Paleoecology,* Vol. 1, J. W. Hedgpeth (ed.). New York, Geological Society of America.

SHEPARD, F. P., G. A. MacDONALD and D. C. COX. 1950.
The Tsunami of April 1, 1946. *Bulletin of the Scripps Institution of Oceanography.* **5**:391–455.

SHEPARD, F. P. 1973.
Submarine Geology, 3rd ed. New York, Harper and Row.

SMITH, F. WALTON. 1973.
The Seas in Motion. New York, Thomas Y. Crowell Co.

SMITH, R. L. 1966
Ecology and Field Biology. New York, Harper and Row.

STOMMEL, H., E. D. STROUP, J. L. REID and B. A. WARREN. 1973.
Transpacific Hydrographic Sections at Latitudes 43°S. and 28°S.: the SCORPIO Expedition-I. *Deep Sea Research.* **20**:1–7.
SVERDRUP, H. O., M. W. JOHNSON and R. H. FLEMING. 1942
The Oceans. Englewood Cliffs, N. J., Prentice-Hall.
SWALLOW, J.C. 1955.
A Neutral-buoyancy Float for Measuring Deep Currents. *Deep Sea Research.* **3**:74–81.
TAIT, R. S. and R. V. DE SANTO. 1972.
Elements of Marine Ecology. New York, Springer-Verlag.
TÉTRY, A. 1959.
Classe des Sipunculiens. In: *Traité de Zoologie,* Vol. 5. P. Grassé (ed.). Paris, Maison et cie.
THORSON, G. 1971.
Life in the Sea. New York, McGraw-Hill.
TREWARTHA, G. T. 1954.
An Introduction to Climate, 3rd ed. New York, McGraw-Hill.
TURK, A., J. TURK, J. T. WITTES and R. WITTES. 1974.
Environmental Science. Philadelphia, W. B. SaundersCo.
U.S. ARMY COASTAL ENGINEERING RESEARCH CENTER. 1966.
Shore Protection, Planning and Design, TR-4, 3rd ed. Washington, D.C., Government Printing Office.
U.S. ARMY COASTAL ENGINEERING RESEARCH CENTER. 1973.
Shore Protection Manual. Washington, D.C., Government Printing Office.
U.S. NAVAL OCEANOGRAPHIC OFFICE. 1944.
Breakers and Surf: Principles in Forecasting, H.O. Publication 234. Washington, D.C., Government Printing Office.
U.S. NAVAL OCEANOGRAPHIC OFFICE. 1944.
Wind Waves at Sea, Breakers and Surf, H.O. Publication 602. Washington, D.C., Government Printing Office.
U.S. NAVAL OCEANOGRAPHIC OFFICE. 1950.
Sea and Swell Observations, H.O. Publication 606-e. Washington, D.C., Government Printing Office.
U.S. NAVAL OCEANOGRAPHIC OFFICE. 1965.
Oceanographic Atlas of the North Atlantic Ocean, H.O. Publication 700. Washington, D.C., Government Printing Office.
U.S. NAVAL OCEANOGRAPHIC OFFICE. 1966.
Handbook of Oceanographic Tables, S.P. 68. Washington, D.C., Government Printing Office.
U.S. NAVAL OCEANOGRAPHIC OFFICE. 1968.
Instruction Manual for Oceanographic Observations, 3rd ed., H.O. Publication 607. Washington, D.C., Government Printing Office.
VETTER, R. C. (ed.). 1973.
Oceanography. New York, Basic Books.
VILLEE, C. A., W. F. WALKER and R. D. BARNES. 1973.
General Zoology, 4th ed. Philadelphia, W. B. Saunders Co.
VINE, F. J. 1966.
Spreading of the Ocean Floor: New Evidence. *Science.* **154**: 1405–1415.
VINE, F. J. 1968.
Magnetic Anomalies Associated with Mid Ocean Ridges. In: *The History of the Earth's Crust, A Symposium.* R. A. Phinney (ed.). Princeton, Princeton University Press.

VINE, F. J. 1969.
Sea-floor Spreading—New Evidence. *Journal of Geological Education.* **17**:6–16.
VINE, F. J. and D. H. MATTHEWS. 1963.
Magnetic Anomalies Over Oceanic Ridges. *Nature.* **199**:947–949.
VON ARX, W. S. 1962.
Introduction to Physical Oceanography. Reading, Mass., Addison-Wesley.
WARREN, B. A. 1966.
Oceanic Circulation. In: *Encyclopedia of Oceanography.* R. W. Fairbridge (ed.). New York, Reinhold.
WILLIAMS, J., J. J. HIGGINSON and J. D. Rohrbough. 1968.
Oceanic Surface Currents. In: *Sea and Air, the Naval Environment.* Annapolis, Naval Institute Press.
WILSON, B. W. and A. TØRUM. 1968
The Tsunami of the Alaskan Earthquake, 1964: Engineering Evaluation, Technical Memorandum No. 25. Washington, D.C., U.S. Coastal Engineering Research Center.
WATTENBERG, H. 1933.
Über die Titrations-alkalinität und der Kalziumkarbonatgehalt des Meerwassers. *Deutsche Atlantische Exped. Meteor 1925–1927. Wiss. Erg.* **8**:122–231.
WEGENER, A. 1966.
Origin of Continents and Oceans. New York, Dover Publications. (A translation of the original German edition of 1929.)
WÜST, G. 1936.
Die Stratosphere des Atlantischen Ozeans. *Deutsche Atantische Exped. Meteor 1925–1927.* **6**:109–251.
WÜST, G. 1954.
Gesetzmässige Wechselbziehungen Zwichen Ozean und Atmosphäre in der Zonalen Verteilung von Oberflächenzalzgehalt, Verdunstung und Niederschlag. *Archiv Für Meteorologie Geophysik und Bioklimatologie.* **7A**:305–328.
YONGE, C. M. 1949.
The Seashores. London, Collins.

Index

Note: Page numbers in *italics* refer to illustrations.
Page numbers followed by the letter "t" refer to tables.